Soil Bioremediation

Soil Bioremediation

An Approach Towards Sustainable Technology

Edited by

Dr. Javid A. Parray
Department of Environmental Sciences
Govt Degree College Eidgah, Srinagar
Jammu and Kashmir, India

Dr. Abeer Hashem Abd Elkhalek Mahmoud
Botany and Microbiology Department, College of Science
King Saud University
Riyadh, Saudi Arabia
Mycology and Plant Disease Survey Department
Plant Pathology Research Institute
Agriculture Research Center
Giza, Egypt

Prof. Riyaz Sayyed
Department of Microbiology
PSGVPM'S ASC College
Shahada, India

WILEY Blackwell

This edition first published 2021

Registered Office(s)
John Wiley & Sons, Inc., 111 River Street, Hoboken, NJ 07030, USA
John Wiley & Sons Ltd, The Atrium, Southern Gate, Chichester, West Sussex, PO19 8SQ, UK

Editorial Office
The Atrium, Southern Gate, Chichester, West Sussex, PO19 8SQ, UK

For details of our global editorial offices, customer services, and more information about Wiley products visit us at www.wiley.com.

Wiley also publishes its books in a variety of electronic formats and by print-on-demand. Some content that appears in standard print versions of this book may not be available in other formats.

Library of Congress Cataloging-in-Publication Data

Names: Parray, Javid Ahmad, editor. | Abd Elkhalek Mahmoud, Abeer Hashem, editor. | Sayyed, Riyaz, editor.
Title: Soil bioremediation : an approach towards sustainable technology / edited by Dr. Javid A. Parray, Dr. Abeer Hashem Abd Elkhalek Mahmoud, Prof.Riyaz Sayyed.
Description: Hoboken, NJ : Wiley-Blackwell, 2021. | Includes bibliographical references and index.
Identifiers: LCCN 2021005025 (print) | LCCN 2021005026 (ebook) | ISBN 9781119547952 (cloth) | ISBN 9781119547969 (adobe pdf) | ISBN 9781119547938 (epub)
Subjects: MESH: Soil Microbiology | Biodegradation, Environmental | Environmental Pollution–prevention & control | Soil Pollutants | Conservation of Natural Resources
Classification: LCC QR111 (print) | LCC QR111 (ebook) | NLM QW 60 | DDC 579/.1757–dc23
LC record available at https://lccn.loc.gov/2021005025
LC ebook record available at https://lccn.loc.gov/2021005026

Cover Design: Wiley
Cover Image: KATERYNA KON/SCIENCE PHOTO LIBRARY/Getty Images

Set in 9.5/12.5pt STIXTwoText by SPi Global, Pondicherry, India

10 9 8 7 6 5 4 3 2 1

Contents

List of Contributors

Samrat Adhikary
Department of Agricultural Chemistry and Soil Science
Mohanpur
West Bengal
India

Saroj Belbase
Rajiv Gandhi South Campus, IAS
Banaras Hindu University
Mirzapur
Uttar Pradesh
India

Suhail Ahmad Bhat
Department of Biochemistry
Pondicherry University
Puducherry
India

Asma Absar Bhatti
Division of Environmental Sciences
Sher-e-Kashmir University of Agricultural Science and Technology
Srinagar
Jammu and Kashmir
India

Liliana Cepoi
Institute of Microbiology and Biotechnology
Chisinau
Republic of Moldova

Puneet Singh Chauhan
CSIR- National Botanical Research Institute
Rana Pratap Marg
Lucknow
Uttar Pradesh
India

Devendra Kumar Choudhary
Amity Institute of Microbial Technology (AIMT)
Amity University Campus
Noida
Uttar Pradesh
India

Gousia Jeelani
Centre of Research for Development
University of Kashmir
Srinagar
Jammu and Kashmir
India

Vijay Kant Dixit
CSIR- National Botanical Research Institute
Rana Pratap Marg
Lucknow
Uttar Pradesh
India

Department of Environmental Science
Kanya Gurukul Campus, Gurukul Kangri University
Haridwar
Uttarakhand
India

Zia Ur Rahman Farooqi
Institute of Soil and Environmental Sciences
University of Agriculture Faisalabad
Faisalabad
Pakistan

Hera Fatima
Centre of Excellence, DST-FIST Supported Department of Microbiology
Dr. Rammanohar Lohia Avadh University
Ayodhya
Uttar Pradesh
India

B.A. Ganai
Centre of Research for Development
University of Kashmir
Srinagar
Jammu and Kashmir
India

A.K.Gangawane
Faculty of Applied Sciences
Parul University
Vadodara Gujarat
India

Saima Hamid
Centre of Research for Development
Department of Environmental Sciences
University of Kashmir
Srinagar
Jammu and Kashmir
India

Ansar ul Haq
Department of Chemistry
University of Kashmir
Srinagar
Jammu and Kashmir
India

Shamsul Haq
Division of Environmental Sciences
Sher-e-Kashmir University of Agricultural Science and Technology
Srinagar
Jammu and Kashmir
India

Muhammad Mahroz Hussain
Institute of Soil and Environmental Sciences
University of Agriculture Faisalabad
Faisalabad
Pakistan

Shekhar Jain
Faculty of Life Sciences
Mandsaur University
Mandsaur
Madhya Pradesh
India

Magdalena Jastrzębska
University of Warmia and Mazury
Olsztyn
Poland

Namita Joshi
Department of Environmental Science
Kanya Gurukul Campus
Gurukul Kangri University
Haridwar
Uttarakhand
India

Azra N. Kamili
Centre of Research for Development
Department of Environmental Sciences
University of Kashmir
Srinagar
Jammu and Kashmir
India

Rangasamy Kirubakaran
Department of Biotechnology
Vysya College
Salem
Tamil Nadu
India

Ravindra Kumar
Indian Agricultural Research Institute
Regional Station
Karnal
Haryana
India

Shailendra Kumar
Centre of Excellence, DST-FIST
Supported Department of Microbiology
Dr. Rammanohar Lohia Avadh University
Ayodhya
Uttar Pradesh
India

Shrvan Kumar
Rajiv Gandhi South Campus, IAS
Banaras Hindu University
Mirzapur
Uttar Pradesh
India

Ritesh Kundu
Department of Agricultural
Chemistry and Soil Science
Mohanpur
West Bengal
India

Junaid Latif
College of Natural Resources and
Environment
Northwest Agriculture and
Forestry University
Yangling
China

Uttara Mahapatra
Department of Chemical
Engineering
National Institute of Technology
Agartala
Tripura
India

Ajay Kumar Manna
Department of Chemical
Engineering
National Institute of Technology
Agartala
Tripura
India

Mohammad Yaseen Mir
Centre of Research for Development
Department of Environmental Sciences
University of Kashmir
Srinagar
Jammu and Kashmir
India

Shafat Ahmad Mir
Division of Environmental Sciences
Sher-e-Kashmir University
of Agricultural Science and
Technology
Srinagar
Jammu and Kashmir
India

Brajesh Kumar Mishra
Rajiv Gandhi South Campus, IAS
Banaras Hindu University
Mirzapur
Uttar Pradesh
India

Shashank Kumar Mishra
CSIR- National Botanical Research
Institute
Rana Pratap Marg
Lucknow
Uttar Pradesh
India

Sankalp Misra
CSIR- National Botanical Research
Institute
Rana Pratap Marg
Lucknow
Uttar Pradesh
India

Muhammad Umair Mubarak
Institute of Soil and Environmental
Sciences
University of Agriculture Faisalabad
Faisalabad
Pakistan

Athiappan Murugan
Department of Microbiology
Periyar University
Salem
Tamil Nadu
India

Ruqeya Nazir
Centre of Research for Development
University of Kashmir
Srinagar
Jammu and Kashmir
India

Philiswa N. Nomngongo
Department of Chemical Sciences
University of Johannesburg
Doornfontein Campus
Johannesburg
South Africa

Dhaneshwar Padhan
Department of Agricultural Chemistry
and Soil Science
Mohanpur
West Bengal
India

Scientist-B
Central Silk Board
Bangalore
Karnataka
India

Purbasha Priyadarshini Padhi
Department of Soil Science and Agricultural Chemistry
Indira Gandhi Krishi Vishwavidyalaya
Raipur
Chhattisgarh
India

Ayantika Pal
Department of Zoology
Women's College
Tripura University
Agartala
Tripura
India

Rajeev Pandey
Department of Environmental Sciences
Dr. Rammanohar Lohia Avadh University
Ayodhya
Uttar Pradesh
India

Javid A. Parray
Department of Environmental Sciences
Govt Degree College
Eidgah, Srinagar
Jammu and Kashmir
India

Neelam Pathak
Department of Biochemistry
Dr. Rammanohar Lohia Avadh University
Ayodhya
Uttar Pradesh
India

Sabeehah Rehman
Centre of Reserach for Development
University of Kashmir
Srinagar
Jammu and Kashmir
India

Gulab Khan Rohela
Biotechnology Section, Moriculture Division
Central Sericultural Research & Training Institute
Central Silk Board
Ministry of Textiles
Government of India
Pulwama
Jammu and Kashmir
India

Pragyan Paramita Rout
Institute of Agricultural Sciences
Siksha O Anusandhan University
Odisha
India

Dijendra Nath Roy
Department of Bio Engineering
National Institute of Technology
Agartala
Tripura
India

Agnieszka Saeid
Wroclaw University of Science and Technology
Wroclaw
Poland

Durgesh Narain Singh
Department of Zoology
University of Delhi
Delhi
India

Mukesh Kumar Singh
Rajiv Gandhi South Campus, IAS
Banaras Hindu University
Mirzapur
Uttar Pradesh
India

Asha Sinha
Mycology and Plant Pathology, IAS
Banaras Hindu University
Varanasi
Uttar Pradesh
India

Deepa Thomas
Faculty of Applied Sciences
Parul University
Vadodara
Gujarat
India

Manikant Tripathi
Centre of Excellence, DST-FIST Supported Department of Microbiology
Dr. Rammanohar Lohia Avadh University
Ayodhya
Uttar Pradesh
India

Baba Uqab
Department of Environmental Science
University of Kashmir
Srinagar
Jammu and Kashmir
India

Ajit Varma
Amity Institute of Microbial Technology (AIMT)
Amity University Campus
Noida
Uttar Pradesh
India

Fazila Younas
Institute of Soil and Environmental Sciences
University of Agriculture Faisalabad
Faisalabad
Pakistan

Preface

The economic growth and industrialization aimed at better living has resulted in environmental deterioration. The major contributing factors being industries, agriculture practices, and urbanizations, for example, that are a threat to the natural environment in all parts of the world. The fragile water and land ecosystems are under continuous threat due to the accumulation of toxic substances or contaminants thereby limiting the scope of sustainable development. Heavy metal pollution in soil is of particular concern, as it is accumulating further at higher trophic levels. The rigorous use of antibiotics, pesticides, herbicides, drugs, and other chemical forms with their persistence and disposal in a concentrated environment is also of great concern. The environmental self-perpetuating process for transformation and degradation of xenobiotic substances is one of the key processes in natural systems. However, the rate of degradation may be reduced with the rise in pollutant concentrations. A major challenge in the current scenario is to improve the quality of the environment through effective utilization and conservation of natural resources, waste minimization, and adoption of the 3Rs, that is, reduce, recycle, and reuse. Besides, an increased demand for the remediation of wastes and contaminated areas has produced a new demand for improved and newer remediation methods that are appropriate at low cost and offer a broader application of waste management. Therefore, research into biological approaches for waste degradation has received great attention with bioremediation being one of the most important new innovations for the treatment of contaminated sites. Bioremediation mainly concerns the use of microbes and microbial agents to improve the quality of pollutant discharge in the environment vis a vis remediating contaminated sites. This book discusses effective and sustainable technological approaches for remediation of contaminates via ecofriendly usage of microbes and their products in all physical components of the environment. The main focus is the role of microbes, particularly bacteria and fungi, for degradation and removal of various xenobiotic substances from the environment. The book also emphasizes molecular approaches and the biosynthetic pathways of microbes and their gene and protein expression studies for biodeterioration techniques. As previously noted, increased urbanization and industrialization has a detrimental impact on a fragile environment and with this in mind some chapters focus on the role of biotechnological advances in cleaning the environment, through waste minimization and control, using new innovative and sophisticated green technologies. The book offers environmentalists and researchers new insights and directions in addition to motivations for waste management and control leading to effective conservation of natural resources. The research and their mechanisms presented here will be a major contributing factor to sustainable development. Presently, there is keen interest in environmental research particularly in pollution remediation, adoption of ecofriendly technologies, and the better use of agricultural products/residues as economical substrates for cleaning the environment. The untouched wealth of microorganisms in harsh environments are considered potential candidates for utilization in a better and sustainable future. A major part of this book highlights the potential aspects of microbes for various techniques of bioremediation, for example, biosorption, bioaugmentation, and biostimulation in cleaning a contaminated environment. This book covers established and up-to-date research on emerging trends in bioremediation and there are contributions from experimental and numerical researchers; reports on field trials are of special interest.

Javid A. Parray

1

In-situ Bioremediation

An Eco-sustainable Approach for the Decontamination of Polluted Sites

Shamsul Haq[1], Asma Absar Bhatti[1], Suhail Ahmad Bhat[2], Shafat Ahmad Mir[1], and Ansar ul Haq[3]

[1] *Division of Environmental Sciences, Sher-e-Kashmir University of Agricultural Science and Technology, Srinagar, Jammu and Kashmir, India*
[2] *Department of Biochemistry, Pondicherry University, Puducherry, India*
[3] *Department of Chemistry, University of Kashmir, Srinagar, Jammu and Kashmir, India*

1.1 Introduction

The environment has been severely polluted with chemicals that are poisonous both to the environment and to human beings [1–3]. A polyphasic approach has been adapted to overcome the effect of these toxic pollutants that includes (i) stringent regulations for the production and usage of complex chemicals, (ii) the treatment and safe disposal of harmful chemicals, and (iii) the reclamation of polluted sites [4, 5]. The first two are defensive in nature and minimize damage to environment, while the latter is a restorative mechanism [6, 7]. The methods of bioremediation are used for (i) the conversion of highly toxic to less-toxic substances, (ii) the mineralization of contaminants, and (iii) pollutant immobilization [8–12]. Microorganisms in general, and bacteria in particular, harbor enormous metabolic diversity, allowing them to utilize the complex chemicals as energy sources [11, 13]. Further, due to genetic evolution they attain a new metabolic potential to degrade newly added xenobiotic substances [13–15]. The other major focus area of bioremediation studies has been the characterization of metabolic pathways and their respective molecular regulations [16–18]. The advent of whole genome sequencing and related genomics methods has also given rise to new avenues for genome-wide screening of degradative genetic elements and regulatory sequences among the pollutant-degrading strains [19–22]. The main concerns for using isolated microorganisms are: (i) the portion of microorganisms may have substantial

Soil Bioremediation: An Approach Towards Sustainable Technology, First Edition.
Edited by Javid A. Parray, Abeer Hashem Abd Elkhalek Mahmoud, and Riyaz Sayyed.

potential of degrading pollutants and (ii) true degradation of pollutants is not often a true reflection of the *in-situ* bioremediation [23, 24]. The perfect bioremediation techniques need to be executed in such a manner that microorganisms counter a variety of biotic and abiotic factors [23–26]. These factors greatly affect the efficiency of bioremediation process through various mechanisms [27]. Numerous studies also suggest that these are not only the environmental factors but also the technological advances, which affect the process of bioremediation. From the above studies, a major area of environmental research has emerged that assess the eco-sustainability of *in-situ* bioremediation process. Furthermore, various programs are required to monitor and address the following uncertainties: the expected remediation of hazardous substances, the potential of microorganisms, and the adverse impact of remediation processes on various environmental factors [28]. Previously, only the kinetics of degradation was determined but, with the advancement of ecological techniques, community behavior has also been made mandatory for *in-situ* bioremediation studies [29, 30].

The key target of a bioremediation technique is to improve the effectiveness of the restoration of contaminated sites in a cost-effective and environmental-friendly manner. There is no single technique for restoring contaminated sites but research on the basis of nature and type of pollutants has led to the development of new techniques. Autochthonous microorganisms present in polluted environments hold the key to solving most of the challenges associated with biodegradation and bioremediation of polluting substances [31] provided that environmental conditions are suitable for their growth and metabolism. Bioremediation is ecofriendly and cost effective, which offers major advantages of this process over conventional physical and chemical methods. The bioremediation process mainly depends upon the nature of the pollutant, which include: agrochemicals, chlorinated compounds, dyes, greenhouse gases, heavy metals, hydrocarbons, nuclear waste, sewage, and plastics. The nature, depth, and degree of contamination, location, and cost are considered in any bioremediation technique [32, 33]. Furthermore, O_2 concentration, nutrient content, temperature, pH, and other factors also are very important in considering the bioremediation technique.

1.2 *Ex-situ* Versus *In-situ* Bioremediation

In-situ bioremediation is the removal of pollutants under natural conditions by using microbial potential without excavating for polluted samples [34–36], whereas *ex-situ* bioremediation is the degradation of pollutants in excavated samples [37, 38]. There is a noteworthy difference between the two methods of bioremediation, both in terms of experimental control and the end result. While considering the performance of these two methods of bioremediation, it was found that the degradation process *in situ* is more variable than *ex situ* [37]. The other significant advantage of the application of an *ex-situ* bioremediation method is its independence from environmental factors that could adversely affect the efficacy of the process. Further, "because ex situ bioremediation is carried out in nonnatural environments, the process can be manipulated easily by physico-chemical treatments of the target pollutant before and/or during the degradation" [39]. In spite of the selective advantages of *ex-situ* bioremediation techniques, the *in-situ* bioremediation technique constitutes the most widely used technological treatment for the restoration of polluted environments [36, 37, 40, 41]. One-fourth of all remediation projects make use of *in-situ* bioremediation strategies [36]. *In-situ* bioremediation technology is less expensive because it does not need evacuation and it also releases fewer pollutants. The other important aspect of *in-situ* bioremediation process is its applicability to diverse environmental niches for example, industrial sites, aquifers [42], soil subsurface [43], and groundwater [35, 44]. The significance of *in-situ* bioremediation is increased by abundant presence and activity of microorganisms, thereby enhancing the efficiency of the decontamination process even in non-accessible environments. *Ex-situ* bioremediation is carried out by several methods, which are non-related, e.g., slurry phase bioremediation and solid-phase bioremediation, which are driven by the physico-chemical properties of contaminants [45, 46]. *In-situ* bioremediation techniques can be categorized as (i) biostimulation or (ii) bioaugmentation [36, 40, 41, 47] and focus mainly on speeding up the removal of toxic pollutants. The choice between the two techniques is determined by: the physico-chemical characteristics of the polluted area, the presence of co-contaminants, and the type and concentration of the pollutant, for example. It is suggested that *ex-situ* bioremediation methods are useful for the remediation of (i) soils polluted with recalcitrant pollutants in higher concentrations, (ii) soils rich in clay where the permeability of pollutants is low, (iii) sites where conditions are not favorable for biological activities, and (iv) where microorganisms are not released for a range of reasons [48]. The selection of a bioremediation technique based on the expected outcome is very important. The enhanced degradation by *in-situ* bioremediation can result in increased contamination of lesser hydrophobic metabolites in the water sources in the vicinity of the source contamination [49, 50].

1.3 *In-situ* Bioremediation Techniques

1.3.1 Bioaugmentation

Successful bioremediation processes require the use of various strategies for the specific environmental conditions of polluted sites. The most frequently used bioaugmentation strategy is the addition of a pure bacterial strain that is preadapted or the addition of a preadapted consortium that has been genetically engineered through adding biodegradation relevant genes to them [51]. Feasibility studies are a prerequisite for any planned intervention. They usually revolve around screening, followed by tailoring of a competent microbial formula for a particular site. The screening must be based on the ability of the microbes and also on the factors that enable the cells to be activated and persistent under the particular environmental conditions. In order to select the competent microbe it is crucial to have prior knowledge of microbes [23, 52]. Once the contamination is of both high metal concentrations and organic pollutants, the degradation of organic compounds may be constrained by co-contaminants [53]. The use of a multicomponent system such as a microbial consortium gives a better performance in the environment than single component systems [54]. It is more advantageous to use a microbial consortium instead of a pure culture, because it provides metabolic diversity to the remediation technology and obtains complete degradation of diesel oil and phenanthrene: a reduction of 60% of isoprenoids and an overall reduction of about 75% of the total hydrocarbons in 42 days, using a microbial formula made with selected native strains [55–57]. The indigenous microbes degrade PAH in polluted soil after adding microbial consortia (five fungi: *Phanerochaete chrysosporium*, *Cuuning hamella sp.*, *Alternaria alternate (Fr.) Keissler*, *Penicillium chrysogenum*, and *Aspergillus niger*; and three bacteria: *Bacillus* sp., *Zoogloea* sp., and *Flavobacterium*) that enhanced the degradation rate significantly (41.3%) [57, 58]. Microbial inoculants produced under optimum conditions are homogeneous cell suspensions, which usually suffer stress when introduced in a variety of natural habitats. After these populations are introduced to a diversity of biotic and abiotic stresses they start declining at a rapid rate. These stresses include temperature changes, moisture content, pH, nutrient decline, and also harmful contaminants in polluted soil [59].

The microorganisms have a great ability to decontaminate organic pollutants in cultures but they fail to degrade the pollutants in natural environments. The various possible reasons for the failure of the bioaugmentation process is either: inoculated microorganisms facing several problems in adapting to the natural conditions or competition between inoculated and native organisms; or the use of other biological substances and predation by grazing protozoa [60]. Consequently, seeding alone is not enough but it should be supplemented with some physical and environmental modifications [61]. The various factors that influence the efficiency of bioaugmentation and biostimulation are given in Table 1.1. The use of carrier materials often provides a physical support for biomass, along with improved access to nutrients, moisture, and aeration, which increases the survival rate of the microbes [62]. Encapsulation of microbial cells offer good survival rates under harsh environments and thus makes the rate of degradation very

Table 1.1 Various aspects affecting *in-situ* bioremediation.

Aspects	Depiction	References
Microbial viability loss during inoculation	Sudden changes in environmental conditions	[67]
Death of microbes after inoculation	Presence of toxic pollutants and nutrient depletion	[58, 60]
Competition	Competition for nutrients by autochthonous microbes	[23, 51]
Predation	Overgrowth of protozoa	[68]
pH	pH change inhibits growth of microbes	[69]
Temperature	Affects the nature of oil vis a vis microorganisms	[70]
Moisture	Growth and metabolism are lowered by low moisture content and higher moisture content reduces aeration	[61, 69]

rapid and efficient when compared to free-living microbes [63, 64]. Encapsulation controls the flow of nutrients, lowers the concentration of toxic mixtures in the microenvironment of the cells, minimizes cell membrane damage as it reduces the exposure to the toxic compounds, and protects from predation and competition thereby impersonating a miniature bioreactor in the environment [65]. Various materials like agar, agarose, alginate, gelatin, gellan gum, kappa-carrageenan, acrylate copolymers, and polyvinyl alcohol gel have been studied and verified to immobilize cells [66]. Immobilized cells showed a smaller lag phase and thus a higher degradation of gasoline when compared to free cell counterparts at equal microbial applications [63]. The biodegradation of phenol by free and immobilized cells of *Acinetobacter* sp. XA05 and *Sphingomonas* sp. FG03 strains collected from activated sludge and phenol contaminated soil were compared with pure cultures and it was found that the mixture of *Acinetobacter* sp. XA05 and *Sphingomonas* sp. FG03 strains performed better than pure cultures and encapsulated cells achieved better phenol degradation [58].

1.3.2 Biostimulation

Petroleum hydrocarbons are degraded by native microorganisms, which are preferred by the nutrient presence in the polluted site [71]. The most significant source of native microbes is spilled hydrocarbons, whereas, N and P are limiting factors in almost all environments. Henceforth, biostimulation increases the rate of cleansing, as the potential of microbes is amended by the addition of one or more limiting factors or nutrients [72]. Environmental conditions, e.g., pH, and moisture content can also be improved to achieve optimum microbial degradation conditions [70]. Several researchers have studied the addition of N and P to enhance nutrient level. The biodegradation of petroleum hydrocarbons was enhanced by up to 96% after the addition of biosolids and N and P rich inorganic fertilizers to diesel contaminated soils [73]. Likewise, commercial fertilizers were used to remediate diesel oil in the Antarctic seas. Furthermore, higher concentrations of N and P sources can cause eutrophication, thereby enhancing algal growth and ultimately reducing the dissolved O_2 concentration in the water [72]. Separate from nutrient content numerous other factors can greatly affect the degradation rate of polyaromatic hydrocarbons under natural environmental conditions, for example, physical mixing, use of biostimulation agents, mechanical tilling, manual removal, and C sources. It was observed that factors including the intensity of physical mixing, the pretreatments, and the accessibility of alternative carbon sources effected the degradation potential of microbes after Exxon Valdez oil spill [74]. Temperature has also a considerable effect on degradation potential of microbes, because it affects the viscosity, water solubility, and composition of oil. Furthermore, it also affects the metabolism of hydrocarbons and composition of microorganisms [70]. Subsequent to the spillage at Chedabucto Bay the effect of temperature on the degradation of bunker C fuel oil was studied, with temperature ranging from 5 to 28 °C, using mixed microbial cultures and it was found that 41–85% benzene soluble components disappeared after incubation of 7 days at 15 °C, however, 21–52% degradation was obtained after 14 days of incubation at 5 °C [75]. It was found that nutrient stock is essential for degradation by microorganisms under all environmental conditions. The degradation of the contaminant after 17 weeks was almost three times higher at 20 °C and eight times higher at 6 °C when compared to nutrient-deficient sands [76]. On the other hand, temperature exhibited limited influence on petroleum degradation in Antarctic seawater samples in a laboratory microcosm study, where commercial fertilizer improved bioremediation [71]. Biostimulation aided with biosurfactants enhanced the rate of biodegradation [55, 77, 78]. Biostimulation using N and P fertilizer together with biosurfactants facilitated naturally occurring microbes to adapt better and faster to the oil spill contamination, confirming a relatively shorter lag phase and faster degradation rates [72]. The most promising strategy to enhance the rate of degradation is the use of a combination of biostimulation, bioaugmentation, and biosurfactants [79]. However, any such planned intervention must be followed by ecotoxicity and quality studies of the contaminated site to ascertain that it has regained its natural biological activity and integrity [58, 80]. Thus, toxicity tests and measuring microbial activity must be carried out for monitoring purposes during and after bioremediation of contaminated soils [81].

1.3.3 Bioaugmentation Versus Biostimulation

Application of bioaugmentation or biostimulation techniques for bioremediation processes significantly depends upon the prevailing environmental conditions. Hamdi et al. [80] found that the efficiency of a remediation process depends on the added microorganisms, rather than the nutrient content [80]. Bento et al. [82] compared bioremediation of diesel oil by natural attenuation, biostimulation, and bioaugmentation. Of the three bioremediation techniques, i.e. natural attenuation, biostimulation, and bioaugmentation used to degrade diesel oil, the best results were revealed by bioaugmentation after inoculation of microbes selected from the polluted site. Evidently, native microorganisms have more possibility to endure and procreate when they are reintroduced into the site, as compared to foreign strains [23, 82]. However, several reports suggest that the use of native cultures were not particularly effective in the removal rates of hydrocarbons however, stimulation was very effective in such a case [83]. For native and foreign microorganisms, biostimulation provides suitable nutrients and encouraging conditions. Thus, biostimulation becomes a feasible method in those cases where microorganisms adapt due to exposure to hydrocarbons at polluted sites. Eventually, the population, which has adapted to the conditions, exhibits high bioremediation rates and, consequently, biostimulation is more appropriate in such cases [61, 84]. However, natural acclimatization by the indigenous microbial population often requires a longer time period due to an extended lag phase leading to prolonged bioremediation processes [85]. Bioaugmentation and biostimulation techniques are now developing as complementary techniques due to the various limitations when they are applied separately. Hamdi et al. [80] amended PAH contaminated soil using both bioaugmentation and biostimulation and achieved higher PAH dissipation rates, remarkably for anthracene and pyrene, than those observed in unamended PAH-spiked soils.

1.4 Conclusion

Bioremediation is a more ecofriendly and economical technique as compared to chemical or physical removal of toxic pollutants from the contaminated soil or water. However, certain contradictory results for bioaugmentation and biostimulation have been obtained, these two techniques of bioremediation hold the potential of exemplifying *in-situ* bioremediation. These techniques are very distinct from each other but are used as complementary techniques for the decontamination of oil spills and other severely contaminated sites. The necessary requirements for bioremediation processes like the presence of competent microbes, nutrients, and suitable environmental conditions must be determined by laboratory and field trials. It has been clearly indicated that bioaugmentation and biostimulation are extremely efficient *in-situ* remediation techniques. However, data prediction depends mainly upon the environmental conditions and thus finding appropriate microorganisms and suitable environmental conditions for each polluted site is perhaps the best solution.

References

1 Travis, A.S. (2002). Contaminated earth and water: a legacy of the synthetic dyestuffs industry. *Ambix* 49: 21–50.
2 Ostroumov, S.A. (2003). Anthropogenic effects on the biota: towards a new system of principles and criteria for analysis of ecological hazards. *Rivista di Biologia* 96: 159–169.
3 Labie, D. (2007). Developmental neurotoxicity of industrial chemicals. *Medical Sciences (Paris)* 23: 868–872.
4 Robinson, T., McMullan, G., Marchant, R. et al. (2001). Remediation of dyes in textile effluent: a critical review on current treatment technologies with a proposed alternative. *Bioresource Technology* 77: 247–255.
5 Felsot, A.S., Racke, K.D., and Hamilton, D.J. (2003). Disposal and degradation of pesticide waste. *Reviews of Environmental contamination and Toxicology* 177: 123–200.
6 Lodolo, A., Gonzalez-Valencia, E., and Miertus, S. (2001). Overview of remediation technologies for persistent toxic substances. *Archives of Industrial Hygien and Toxicology* 52: 253–280.
7 Scullion, J. (2006). Remediating polluted soils. *Naturwissenschaften* 93: 51–65.
8 Shannon, M.J. and Unterman, R. (1993). Evaluating bioremediation: distinguishing fact from fiction. *Annual Review of Microbiology* 47: 715–738.
9 Snellinx, Z., Nepovim, A., Taghavi, S. et al. (2002). Biological remediation of explosives and related nitroaromatic compounds. *Environmental Science and Pollution Research International* 9: 48–61.
10 Lovley, D.R. (2003). Cleaning up with genomics: applying molecular biology to bioremediation. *Nature Reviews Microbiology* 1: 35–44.
11 Diaz, E. (2004). Bacterial degradation of aromatic pollutants: a paradigm of metabolic versatility. *International Microbiology* 7: 173–180.
12 Parales, R.E. and Haddock, J.D. (2004). Biocatalytic degradation of pollutants. *Current Opinion in Biotechnology* 15: 374–379.
13 Nojiri, H. and Tsuda, M. (2005). Functional evolution of bacteria in degradation of environmental pollutants. *Tanpakushitsu Kakusan Koso* 50: 1505–1509.

14 Janssen, D.B., Dinkla, I.J., Poelarends, G.J. et al. (2005). Bacterial degradation of xenobiotic compounds: evolution and distribution of novel enzyme activities. *Environmental Microbiology* 7: 1868–1882.

15 Zhang, J., Zhang, H., Li, X. et al. (2006). Soil microbial ecological process and microbial functional gene diversity. *Ying Yong Sheng Tai Xue Bao* 17: 1129–1132.

16 Arai, H., Ohishi, T., Chang, M.Y. et al. (2000). Arrangement and regulation of the genes for meta-pathway enzymes required for degradation of phenol in *Comamonas testosteroni* TA441. *Microbiology* 146: 1707–1715.

17 Solyanikova, I.P. and Golovleva, L.A. (2004). Bacterial degradation of chlorophenols: pathways, biochemical and genetic aspects. *Journal of Environmental Science and Health* 39: 333–351.

18 Symons, Z.C. and Bruce, N.C. (2006). Bacterial pathways for degradation of nitroaromatics. *Natural Product Reports* 23: 845–850.

19 Heidelberg, J.F., Paulsen, I.T., and Nelson, K.E. (2002). Genome sequence of the dissimilatory metal ion-reducing bacterium *Shewanella oneidensis*. *Nature Biotechnology* 20: 1118–1123.

20 Golyshin, P.N., Martins Dos Santos, V.A., and Kaiser, O. (2003). Genome sequence completed of *Alcanivorax borkumensis*, a hydrocarbon-degrading bacterium that plays a global role in oil removal from marine systems. *Journal of Biotechnology* 106: 215–220.

21 Rabus, R. (2005). Functional genomics of an anaerobic aromatic-degrading denitrifying bacterium, strain EbN1. *Applied Microbiology and Biotechnology* 68: 580–587.

22 Zhao, B. and Poh, C.L. (2008). Insights into environmental bioremediation by microorganisms through functional genomics and proteomics. *Proteomics* 8: 874–881.

23 Thompson, I.P., van der Gast, C.J., Ciric, L. et al. (2005). Bioaugmentation for bioremediation: the challenge of strain selection. *Environmental Microbiology* 7: 909–915.

24 Vinas, M., Sabate, J., Guasp, C. et al. (2005). Culture-dependent and -independent approaches establish the complexity of a PAH-degrading microbial consortium. *Canadian Journal of Microbiology* 51: 897–909.

25 Dinkla, I.J., Gabor, E.M., and Janssen, D.B. (2001). Effects of iron limitation on the degradation of toluene by Pseudomonas strains carrying the TOL (pWWO) plasmid. *Environmental Microbiology* 67: 3406–3412.

26 Kim, H.J. and Graham, D.W. (2003). Effects of oxygen and nitrogen conditions on the transformation kinetics of 1,2-dichloroethenes by Methylosinus trichosporium OB3b and its sMMOC mutant. *Biodegradation* 14: 407–414.

27 Lovanh, N., Hunt, C.S., and Alvarez, P.J. (2002). Effect of ethanol on BTEX biodegradation kinetics: aerobic continuous culture experiments. *Water Research* 36: 3739–3746.

28 Zhou, Q., Zhang, J., Fu, J. et al. (2008). Biomonitoring: an appealing tool for assessment of metal pollution in the aquatic ecosystem. *Analytica Chimica Acta* 606: 135–150.

29 Purohit, H.J., Raje, D.V., Kapley, A. et al. (2003). Genomics tools in environmental impact assessment. *Environmental Science and Technology* 37: 356A–363A.

30 Paul, D., Pandey, G., Meier, C. et al. (2006). Bacterial community structure of a pesticide-contaminated site and assessment of changes induced in community structure during bioremediation. *FEMS Microbiology Ecology* 57: 116–127.

31 Verma, J.P. and Jaiswal, D.K. (2016). Book review: advances in biodegradation and bioremediation of industrial waste. *Frontiers in Microbiology* 6: 1–2.

32 Frutos, F.J.G., Pérez, R., Escolano, O. et al. (2012). Remediation trials for hydrocarbon-contaminated sludge from a soil washing process: evaluation of bioremediation technologies. *Journall of Hazardous Materials* 199: 262–271.

33 Smith, E., Thavamani, P., Ramadass, K. et al. (2015). Remediation trials for hydrocarbon-contaminated soils in arid environments: evaluation of bioslurry and biopiling techniques. *International Biodeterioration and Biodegradation* 101: 56–65.

34 Fruchter, J. (2002). In situ treatment of chromium-contaminated groundwater. *Environmental Science and Technology* 36: 464A–472A.

35 Farhadian, M., Vachelard, C., Duchez, D. et al. (2007). In situ bioremediation of monoaromatic pollutants in groundwater: a review. *Bioresource Technology* 99: 5296–5308.

36 Jorgensen, K.S. (2007). In situ bioremediation. *Advances in Applied Microbiology* 61: 285–305.

37 Carberry, J.B. and Wik, J. (2001). Comparison of ex situ and in situ bioremediation of unsaturated soils contaminated by petroleum. *Journal of Environmental Science and Health* 36: 1491–1503.

38 Prpich, G.P., Adams, R.L., and Daugulis, A.J. (2006). Ex-situ bioremediation of phenol-contaminated soil using polymer beads. *Biotechnology Letters* 28: 2027–2031.

39 Kim, S.J., Park, J.Y., Lee, Y.J. et al. (2005). Application of a new electrolyte circulation method for the ex situ electrokinetic bioremediation of a laboratory-prepared pentadecane contaminated kaolinite. *Journal of Hazardous Materials* 118: 171–176.

40 Bouwer, E., Durant, N., Wilson, L. et al. (1994). Degradation of xenobiotic compounds in situ: capabilities and limits. *FEMS Microbiology Reviews* 15: 307–317.

41 Romantschuk, M., Sarand, I., and Petanen, T. (2000). Means to improve the effect of in-situ bioremediation of contaminated soil: an overview of novel approaches. *Environmental Pollution* 107: 179–185.

42 Mandelbaum, R.T., Shati, M.R., and Ronen, D. (1997). In situ microcosms in aquifer bioremediation studies. *FEMS Microbiology Reviews* 20: 489–502.

43 Schmidt, B.F., Chao, J., Zhu, Z. et al. (1997). Signal amplification in the detection of single-copy DNA and RNA by enzyme-catalyzed deposition (CARD) of the novel fluorescent reporter substrate Cy3.29-tyramide. *Journal of Histochemistry and Cytochemistry* 45: 365–373.

44 Scow, K.M. and Hicks, K.A. (2005). Natural attenuation and enhanced bioremediation of organic contaminants in groundwater. *Current Opinion in Biotechnology* 16: 246–253.

45 Janikowski, T.B., Velicogna, D., Punt, M. et al. (2002). Use of a two-phase partitioning bioreactor for degrading polycyclic aromatic hydrocarbons by a *Sphingomonas* sp. *Applied Microbiology and Biotechnology* 59: 368–376.

46 Di Gennaro, P., Collina, E., Franzetti, A. et al. (2005). Bioremediation of diethylhexyl phthalate contaminated soil: a feasibility study in slurry- and solid-phase reactors. *Environmental Science and Technology* 39: 325–330.

47 Omenn, G.S. (1992). Environmental biotechnology: biotechnology solutions for a global environmental problem, hazardous chemical wastes. *Asia Pacific Journal of Public Health* 6: 40–45.

48 Robles-Gonzalez, I.V., Fava, F., and Poggi-Varaldo, H.M. (2008). A review on slurry bioreactors for bioremediation of soils and sediments. *Microbial Cell Factories* 7: 5–17.

49 Kao, C.M., Chen, S.C., Wang, J.Y. et al. (2003). Remediation of PCE-contaminated aquifer by an in situ two-layer biobarrier: laboratory batch and column studies. *Water Resources* 37: 27–38.

50 Glover, K.C., Munakata-Marr, J., and Illangasekare, T.H. (2007). Biologically enhanced mass transfer of tetrachloroethene from DNAPL in source zones: experimental evaluation and influence of pool morphology. *Environmental Science and Technology* 41: 1384–1389.

51 El Fantroussi, S. and Agathos, S.N. (2005). Is bioaugmentation a feasible strategy for pollutant removal and site remediation? *Current Opinions in Microbiolgy* 8: 268–275.

52 van der Gast, C.J., Whiteley, A.S., and Thompson, I.P. (2004). Temporal dynamics and degradation activity of a bacterial inoculum for treating waste metal-working fluid. *Environmental Microbiology* 6: 254–263.

53 Roane, T.M., Josephson, K.L., and Pepper, I.L. (2001). Dual-bioaugmentation strategy to enhance remediation of co-contaminated soil. *Applied Environmental Microbiology* 67: 3208–3215.

54 Ledin, M. (2000). Accumulation of metals by microorganisms – processes and importance for soil systems. *Earth Science Reviews* 51: 1–31.

55 Rahman, K.S.M., Banat, I.M., Thahira, J. et al. (2002). Bioremediation of gasoline contaminated soil by a bacterial consortium amended with poultry litter, coir pith, and rhamnolipid biosurfactant. *Bioresource Technology* 81: 25–32.

56 Nyer, E.K., Payne, F., and Suthersan, S. (2002). Environment vs. bacteria or let's play "name that bacteria". *Ground Water Monitoring and Remediation* 23: 36–45.

57 Alisi, C., Musella, R., Tasso, F. et al. (2009). Bioremediation of diesel oil in a cocontaminated soil by bioaugmentation with a microbial formula tailored with native strains selected for heavy metals resistance. *Science of the Total Environment* 407: 3024–3032.

58 Li, X.J., Lin, X., Li, P.J. et al. (2009). Biodegradation of the low concentration of polycyclic aromatic hydrocarbons in soil by microbial consortium during incubation. *Journal of Hazardous Materials* 172: 601–605.

59 Gentry, T.J., Rensing, C., and Pepper, I.L. (2004). New approaches for bioaugmentation as a remediation technology. *Critical Reviews in Environmental Science and Technology* 34: 447–494.

60 Goldstein, R.M., Mallory, L.M., and Alexander, M. (1985). Reasons for possible failure of inoculation to enhance biodegradation. *Applied Environmental Microbiology* 50: 977–983.

61 Leahy, J.G. and Colwell, R.R. (1990). Microbial-degradation of hydrocarbons in the environment. *Microbiologial Reviews* 54: 305–315.

62 Mishra, S., Jyot, J., Kuhad, R.C. et al. (2001). In situ bioremediation potential of an oily sludge-degrading bacterial consortium. *Current Microbiology* 43: 328–335.

63 Moslemy, P., Neufeld, R.J., and Guiot, S.R. (2002). Biodegradation of gasoline by gellan gum-encapsulated bacterial cells. *Biotechnology and Bioengineering* 80: 175–184.

64 Obuekwe, C.O. and Al-Muttawa, E.M. (2001). Self-immobilized bacterial cultures with potential for application as ready-to-use seeds for petroleum bioremediation. *Biotechnology Letters* 23: 1025–1032.

65 McLoughlin, A.J. (1994). Controlled release of immobilized cells as a strategy to regulate ecological competence of inocula. In: *Biotechnics/Wastewater* (ed. T. Scheper), 1–45. Berlin: Springer.

66 Cassidy, M.B., Lee, H., and Trevors, J.T. (1996). Environmental applications of immobilized microbial cells: a review. *Journal of Industrial Microbiology and Biotechnology* 16: 79–101.

67 vanVeen, J.A., vanOverbeek, L.S., and vanElsas, J.D. (1997). Fate and activity of microorganisms introduced into soil. *Microbiology and Molecular Biology Reviews* 61: 121–135.

68 Bouchez, T., Patureau, D., Dabert, P. et al. (2000). Ecological study of a bioaugmentation failure. *Environmental Microbiology* 2: 179–190.

69 Dibble, J.T. and Bartha, R. (1979). Effect of environmental parameters on the biodegradation of oil sludge. *Applied Environmental Microbiology* 37: 729–739.

70 Atlas, R.M. (1995). Bioremediation of petroleum pollutants. *International Biodeterioration and Biodegradation* 35: 317–327.

71 Delille, D., Delille, B., and Pelletier, E. (2002). Effectiveness of bioremediation of crude oil contaminated subantarctic intertidal sediment: the microbial response. *Microbial Ecology* 44: 118–126.
72 Nikolopoulou, M. and Kalogerakis, N. (2009). Biostimulation strategies for fresh and chronically polluted marine environments with petroleum hydrocarbons. *Journal of Chemical Technology and Biotechnology* 84: 802–807.
73 Sarkar, D., Ferguson, M., Datta, R. et al. (2005). Bioremediation of petroleum hydrocarbons in contaminated soils: comparison of biosolids addition, carbon supplementation, and monitored natural attenuation. *Environmental Pollution* 136: 187–195.
74 Sugai, S.F., Lindstrom, J.E., and Braddock, J.F. (1997). Environmental influences on the microbial degradation of Exxon Valdez oil on the shorelines of Prince William Sound, Alaska. *Environmental Science and Technology* 31: 1564–1572.
75 Mulkins-Phillips, G.J. and Stewart, J.E. (1974). Effect of environmental parameters on bacterial-degradation of bunker-C oil, crude oils, and hydrocarbons. *Applied Microbiology* 28: 915–922.
76 Horel, A. and Schiewer, S. (2009). Investigation of the physical and chemical parameters affecting biodegradation of diesel and synthetic diesel fuel contaminating Alaskan soils. *Cold Regions Science and Technology* 58: 113–119.
77 Bordoloi, N.K. and Konwar, B.K. (2009). Bacterial biosurfactant in enhancing solubility and metabolism of petroleum hydrocarbons. *Journal of Hazardous Materials* 170: 495–505.
78 Ron, E.Z. and Rosenberg, E. (2002). Biosurfactants and oil bioremediation. *Current Opinion in Biotechnology* 13: 249–252.
79 Baek, K.H., Yoon, B.D., Kim, B.H. et al. (2007). Monitoring of microbial diversity and activity during bioremediation of crude OH-contaminated soil with different treatments. *Journal of Microbiology and Biotechnology* 17: 67–73.
80 Hamdi, H., Benzarti, S., Manusadzianas, L. et al. (2007). Bioaugmentation and biostimulation effects on PAH dissipation and soil ecotoxicity under controlled conditions. *Soil Biology and Biochemistry* 39: 1926–1935.
81 Hankard, P.K., Svendsen, C., Wright, J. et al. (2004). Biological assessment of contaminated land using earthworm biomarkers in support of chemical analysis. *Science of the Total Environment* 330: 9–20.
82 Bento, F.M., Camargo, F.A.O., Okeke, B.C. et al. (2005). Comparative bioremediation of soils contaminated with diesel oil by natural attenuation, biostimulation and bioaugmentation. *Bioresoures Technology* 96: 1049–1055.
83 Thomassin-Lacroix, E.J.M., Eriksson, M., Reimer, K.J. et al. (2002). Biostimulation and bioaugmentation for on-site treatment of weathered diesel fuel in Arctic soil. *Applied Microbiology and Biotechnology* 59: 551–556.

84 Simon, M.A., Bonner, J.S., McDonald, T.J. et al. (1999). Bioaugmentation for the enhanced bioremediation of petroleum in a wetland. *Polycyclic Aromatic Compounds* 14: 231–239.

85 Lendvay, J.M., Loffler, F.E., Dollhopf, M. et al. (2003). Bioreactive barriers: a comparison of bioaugmentation and biostimulation for chlorinated solvent remediation. *Environmental Science and Technology* 37: 1422–1431.

2

Bioremediation

A Green Solution to avoid Pollution of the Environment

Muhammad Mahroz Hussain[1], *Zia Ur Rahman Farooqi*[1], *Junaid Latif*[2], *Muhammad Umair Mubarak*[1], *and Fazila Younas*[1]

[1] *Institute of Soil and Environmental Sciences, University of Agriculture Faisalabad, Faisalabad, Pakistan*
[2] *College of Natural Resources and Environment, Northwest Agriculture and Forestry University, Yangling, China*

2.1 Introduction

Phytoremediation is a leading technology that helps to resolve the issues related to toxic metal removal with the help of a green revolutionary technique. In a green revolutionary technique the best genetically appropriate plants assist removal of toxic metals from contaminated soil. These plants can degrade, remove, immobilize, or metabolize toxic metal in a wide range of areas, e.g., wetland or a terrestrial land system. Phytoremediation is not a new technology. Almost 300 years ago different types of plants species were utilized to clean-up wastewater [1]. At the end of the nineteenth century, *Thlaspi caerulescens* and *Viola calaminaria* were the first plant species that were documented to accumulate high levels of metals in their leaves [2]. In 1935, Byers reported that plants of the genus *Astragalus* could accumulate up to 0.6% selenium in dry shoot biomass. One decade later, it was identified that plants could accumulate up to 1% Ni in shoots [3]. In the last decade, extensive research has been conducted to investigate the biology of metal phytoextraction. Metal hyperaccumulation is a phenomenon generally associated with species endemic to metalliferous soils, and it is found in only a very small proportion of such metallophytes. Furthermost, but not all, hyperaccumulators are strictly endemic to metalliferous soils. More than 430 taxa are described to date in all continents in temperate and tropical environments. Notable centers of distribution are: Ni – New

Soil Bioremediation: An Approach Towards Sustainable Technology, First Edition.
Edited by Javid A. Parray, Abeer Hashem Abd Elkhalek Mahmoud, and Riyaz Sayyed.

Caledonia, Cuba, South East Asia, Brazil, Southern Europe, and Asia Minor; Zn and Pb – Europe; Co and Cu – Southcentral Africa. Some families and genera are particularly well represented, i.e., for Ni: Brassicaceae (*Alyssum* and *Thlaspi*), Euphorbiaceae (*Phyllanthus*, *Leucocroton*), and Asteraceae (*Senecio, Pentacalia*); Zn: Brassicaceae (*Thlaspi*); Cu and Co: Lamiaceae and Scrophulariaceae. Phytoremediation can be practiced to scavenge both organic and inorganic pollutants present in solid substrates (soil), liquid substrates (water), and the air.

Phytoremediation is a comprehensive technique, which offers heavy metal (HM) contamination remediation with an innovative and cost-effective option. The use of plants to bring back contaminated sites is termed as phytoremediation as it uses the plant's natural characteristics to up take, accumulate, store, degrade, and remediate heavy metals [4]. With the green revolution, use of pesticides and fertilizers has polluted the soil with HMs like Cd, Pb, Ni, and Hg. Pesticides, beside their biocidal and fertilizing effects, contain considerable concentrations of HMs. Undeniably, pesticides can be very toxic and are responsible for farming diseases such as cancers and neurodegenerative diseases. However, in developed countries, there is a rapid change from subsistence farming to intensive farming, in order to feed more people. There are also some issues like lack of selectivity, overuse, and over exploitation that leads to risk for living organisms and humans by contaminating drinking water, food, and soils. Their presence in soil, water, plants, and even the atmosphere, together with their potential pharmacodynamic properties, can have harmful effects on the environment and on human health [4, 5]. This problem can be overcome by phytoremediation, which can reduce HM pollution and decrease their impact on the environment [5]. These techniques are exciting prospects for reducing environmental pollution. Plants can bioaccumulate, biotransform, and bioremediate HMs [6]. Unlike organic compounds, heavy metals cannot be degraded. Hence, an effective clean-up strategy via immobilization is required to reduce or remove toxicity. In recent years, scientists have commenced generation of cost-effective technologies that comprise use of microorganisms/biomass or live plants to clean polluted areas. These technologies are best applied at sites with shallow contamination of organic, nutrient, or metal pollutants that are acquiescent to one of the five applications, i.e., phytotransformation, rhizosphere bioremediation, phytostabilization, phytoextraction, and rhizo-filtration. The technology involves efficient use of plants to remove, detoxify, or immobilize environmental contaminants in a soil, water, or sediment mix through the natural, biological, chemical, or physical activities or processes of the plants [7]. The exploitation of plants to remediate soils contaminated with trace elements could offer a cheap and sustainable technology for bioremediation. Many modern tools and analytical devices have provided insight into the selection and optimization of the remediation process by plant species. Metal-hyperaccumulating plants, desirable for heavily polluted environments, can be established by the insertion of novel traits into high biomass plants in a transgenic approach, which is an encouraging strategy for the development of effective phytoremediation technology. The inherited manipulation of a phytoremediator plant needs many optimization processes, including mobilization of trace HM ions, their uptake into the root, stem, and other viable parts of the plant and their detoxification and allocation within the plant [8, 9].

2.2 Sources of Heavy Metals

HMs are the natural elements that have higher atomic weights and density above $5\,g/cm^3$. These are inevitable and cannot be avoided as they are found naturally, through weathering [10], volcanic eruptions [11], and fossil fuels [12]. Then they were used raw and as processing materials such as, Pt in hydrogenation [13], As in pesticides [14] Cd in fertilizers [15], and in a range of other industrial, domestic, agricultural, and medical applications [16]. It is assumed that almost all the heavy metal concentrations are higher and widespread in the environment due to road dusts [17–19].

2.2.1 Natural Sources

Heavy metals are found naturally in the environment as result of volcanic eruptions, and sedimentary and metamorphic rock deposits and their releases during weathering and pedogenic processes. These HMs are introduced into soil and groundwater and reaches the human food chain [20, 21]. All HMs in the environment originate in natural phenomena and human activities distribute them to other parts of the ecosystem [22, 23]. In addition, gases and fluid emissions from the earth's surface, atmosphere, sea floor, and volcanoes are additional important sources of HMs.

2.2.2 Anthropogenic Sources

HMs are released into the environment by industrial activities, ore mining, and through other product uses. Further anthropogenic sources are agricultural activities such as: fertilizer use, animal manures, and pesticides; metallurgical activities, smelting, metal finishing; dyes; energy production; transportation; and microelectronic products [24]. Fertilizers are used to provide essential nutrients to soil and crops for sustainable production and improved quality and pesticides are applied to protect crops from pest and diseases. Both products contain HM. Moreover, soil amendments, derived from sewage sludge also contain HM, which is mobilized during crop growth due to irrigation [25, 26]. It is believed that Pb, Hg, As, Cr, Cu, and Ni mainly came from anthropogenic sources with complex distribution and exposures [27]. Fossil fuel combustion is again one of the major sources of HM and there is a need for the reduction of this source by adopting other fuel technologies [28]. Urban areas are generally contaminated with Pb, Zn, Cd, and Cu due to fossil fuels emissions by traffic and the paint industry [29].

2.3 Impacts of Heavy Metals on Soil and Microbial Activity

2.3.1 Soil Microbial Community

Different HMs have different effects on soil microbes and their processes however, some of these microbes might have evolved tolerance and adaptation mechanisms against HM in the soil environment [30]. Soil microbes are also biomonitors of HM pollution in soil. To check HM effects on soil microbes, microbes under study are divided into three groups, i.e., sensitive, tolerant, and resistant. It was established that resistance is evolved in microbes for HMs by the passage of time after repeated exposure [31]. Microbial colonies in Zn and Ti contaminated soils have decreased C contents, Zn shows biocidal effect and decreased microbial activity [32]. Soil microbial properties, microbial biomass, respiration, N-mineralization, soil enzymes involved in cycling of C, N, P, and S were proven in response to different levels of HM pollution and it was revealed that microbial biomass and enzyme activities were reduced with increasing HM concentration. This implies that HM concentration in soils can severely decrease the functions of the soil microbial community and impair specific pathways of nutrient cycling [33]. In another study, results showed that microbial biomass in terms of C was depressingly affected by different levels of HMs. Enzyme activity was significantly dejected by HM stressed conditions. Soil phosphatase enzyme activities were found in the soils 200 m away from the HM [34, 35].

2.3.2 Soil Organic Matter

Heavy metals also have negative effects on Soil Organic Matter (SOM) as it decreased up to 1% in those soils contaminated with Zn and Cu by hindering the microbial activity and plant growth [36, 37]. In another study, HMs showed a positive relationship with SOM because an adverse HM effect on soil microbes appeared to increase the accumulation of organic matter but the relationship was negative in terms of soil respiration and microbial biomass since the bacteria were less effective in mineralizing SOM under these conditions [38, 39]. Microbial activities such as respiration, C and N mineralization, biological N2 fixation, and soil enzymes were assessed for this [40]. SOM degradation leads to changes in the soil, these are assessed by their change with regards to Zn, Pb, and Cu that caused a decrease in SOM after soil has received long-term application of sewage sludge [41].

2.3.3 Plants

HM's availability in high concentrations from mining, industrial, and disposal of industrial sewage sludge is increasing the pollution problem in agriculture and in the crops grown on these soils that receive HM contaminated materials. HMs cause numerous negative effects on plants that limit their growth and yield [42, 43]. Generally, HMs affect plant growth by causing cytological disorders, disturbing metabolic processes and physiological growth [44]. Heavy metals also involved in the production of ROS (reactive oxygen species), cause blocking of essential functional groups in biomolecules and displacement of essential metal ions [45, 46].

2.3.4 Water

HM contamination issue is increasing all over the world in every component of the environment including groundwater resources. Heavy metals, such as Cd, Cu, Pb, Cr, and Hg are major pollutants that contaminate water used for drinking and agricultural purposes. Plants growing on HM-polluted sites take in this contaminated water and accumulate HM in them and when these plants are used as food, these HM are transferred to the organism's bodies accumulating again [43]. Once fresh water organisms are exposed to HMs, like Cr, Ni, Cd, and Pb, accumulation is increased with increased exposure time, harming the consumer's kidneys and other body organs [47]. Food crops consumption that are irrigated with HM-contaminated water is also a major route of human exposure to HM [48, 49]. Metals are being utilized in a range of ways in industries and agriculture; particularly heavy metals such as Hg, Cd, Pb, and As. These constitute a significant potential threat to human health via contaminated water [50]. Water pollution remains a major source of morbidity and mortality in different parts of the world. Water environment treatment has led to improved health outcomes among Chinese people. Reduced water pollution mediated the associations between water environment treatment and health outcomes [51].

2.3.5 Humans

HMs are among one of the widespread pollutants in the world, which cause severe threats to almost every organism including humans. In agricultural soils, HMs like Cd, Cu, Pb, and Zn concentrations are 0.097, 22.6, 26.0, and 74.2 mg/

kg, respectively and can even reach 3.16, 99.3, 84.1, and 147 mg/kg, respectively. In water, Cd, Cu, Pb, and Zn concentrations are 0.080, 7.91, 15.7, and 18.7 ug/l respectively and cause contamination in drinking water and food that threatens human health [52]. Occupational exposures to Mn, Cu, Fe, Hg, Zn, and Al pose a risk factor for Parkinson's disease [53, 54]. Rice grown on Cd contaminated soils causes Cd loaded grains and poses kidney failure in humans consuming the contaminated rice [55–57]. Table 2.1 defines HM effects on plants and humans.

2.4 Fate of Pesticides and Its Biodegradation in Soil

Organophosphate pesticide primarily catalyzes the degradation of the neurotransmitter acetylcholine in the synapse. This enzyme rapidly hydrolyzes the acetylcholine, a neurotransmitter into choline by stopping the stimulation of nerves. These compounds block the normal activity of acetylcholinesterase by binding covalently to the enzyme, thereby changing the activity and function. The accumulation of neurotransmitter acetylcholine (Ach) in the synapse leads to blockage of nerves because regeneration of acetylcholine esterase is a slow process and takes hours or even days. This blockage of nerves causes permanent paralysis and finally the death of insects and pests. *Solanum lycopersicum* L. being the important human diet constituent is grown throughout the world. It is commonly used in a salad, consumed as a sauce and juice. Different types of pesticides like cypermethrin, deltamethrin, profenofos, and chlorpyriphos, are used to protect the tomato crop from damage caused by these insects and pests [87].

Bioaccumulation is the primary cause of toxicity of pesticides as their higher concentrations in biological systems leads to major health problems. The persistence of organophosphate pesticides in soil has also been related to the organic matter, clay content, and iron or aluminum oxy content of the soil. These have a higher affinity to absorbing the pesticides and act as a sink for such hydrophobic compounds affecting plants, humans, and animals. Pesticides undergo various changes in the environment comprising their adsorption transmission and degradation, which depends on the nature of soil, pesticide type, and its physico-chemical properties. The predominant process involved in transformation of such molecules is facilitated by microbes [78] followed by photolysis or photo-degradation and chemical transformations [43]. Chemical and microbial degradation are difficult to distinguish as both processes go side by side. Further the physical properties of the soil also play an integral part, as the clay content in soil leads to an increase in surface area, which enhances hydrolytic conversion [55].

Table 2.1 Effect of different heavy metals on plants and humans.

Heavy metals	Effects on plants	Citations	Effects on Humans	Citations
As	Growth, leaf gas-exchange, water potential, protein content, and biomass.	[58–61]	Carcinogen, cyto, and genotoxic, cause diabetes, cardiovascular diseases, and disrupts DNA repair.	[62–66]
Al	Decreases in biomass, photosynthesis, protein contents, disrupts NPK, Ca, Mg uptake	[67–70]	Neuro, geno, and cytotoxic, damages membrane and DNA, cause chromosomal and lymphocyte aberrations.	[71, 72]
Cr	Damages rhizobia, production of ROS and oxidative stress, low seed germination and reduced nutrient uptake	[73–76]	Carcinogen, damages DNA and reproductive system with birth and growth defects.	[73–77]
Ni	Reduces growth and photosynthetic pigment contents, low crop growth, increases imbalance between K and Ca	[78–80]	Cardiovascular diseases, haemo, immuno, neuro, geno, nephron, and hepatotoxic, carcinogen and cause reproductive system illness.	[81, 82]
Cu	Low yield, failure in making seed, less photosynthetic rate, ROS production, cell damage, disturbs photosynthesis and electron transport.	[83, 84]	Cardiovascular system and disrupts enzymes activity.	[85, 88]
Pb	Stunted growth, chlorosis and blackening of root system, inhibit photosynthesis, upset water balance, changes affects membrane structure.	[89, 90]	Mental retardation, growth impairment, comma, convulsions, anemia, hypertension, and immunotoxicity.	[91, 92]
Cd	Reduction in soil biota, stomata opening, transpiration, and photosynthesis.	[93–95]	Carcinogen, negative impact on cardiovascular, immune and reproductive system, nephrotoxic.	[96–98]
Zn	Inhibition of several enzymes lowers photosynthesis rate.	[99, 100]	Nausea, vomiting, diarrhea, kidney, and stomach damage, burning, stinging, itching	[86, 101, 102]
Hg	Reduces soil microbes, poor plant growth, reduced nutrient uptake, and reduced seed germination, decreased biomass.	[103–105]	Neurobehavioral defects, walking, vision, hearing, and speech is affected	[65, 91, 106]

Bioremediation is considered an environment friendly, green, and economical technology for degradation of persistent organic pollutants such as pesticides. *Pseudomonas putida*, *Bacillus subtilis*, *Burkholderia gladioli*, and *Pseudomonas aeruginosa* have been reported as efficient microbial species for the degradation of profenofos pesticide. These microbial species degrade profenofos by hydrolysis to yield 4-bromo-2-chlorophenol as metabolite [73]. Hydrolysis being the most significant step plays an important role in detoxifying the organophosphate compound that makes it vulnerable to further degradation. Esterase or phosphortriesterase enzymes are responsible for this reaction. Numerous microorganisms can hydrolyze the organophosphate pesticide. Illustrations include *Pseudomonas diminuta* MG and *Flavobacterium* ATCC 27551 possessing the organophosphate hydrolase enzyme [51]. Bacterial strains isolated from pesticide-contaminated soil and these strains were tested for their degradation capability and it was found that JCp4 and FCp1 degraded 84.4 and 78.6% the chlorpyrifos pesticide respectively. Moreover, these strains also showed plant growth promoting traits, which includes phytohormone production, phosphate solubilization, and N_2 fixation, etc. [49]. In Pakistan, pesticide residues were found above permissible limits in different vegetables analyzed for pesticide residues. It was found that all samples contained pesticide residues of carbofuran and chlorpyrifos and limits with concentrations ranging from 0.01–0.39 and 0.05–0.96 mg/kg, respectively were observed [81].

2.5 Strategies of Bioremediation

Bioremediation is the process of using organisms to neutralize or remove contaminants from waste. It is very important to understand that this form of waste remediation uses no toxic chemicals, although it may use an organism that can be detrimental under certain circumstances. A gory, but simple description of bioremediation is the use of maggots in wound care. Wounds that have contamination can have maggots introduced to them. The maggots then eat the contamination allowing the wound to heal correctly – a form of medical bioremediation. There are many other types that are used to control different waste contamination, which are now described.

2.5.1 Microbial Remediation

The use of microbes such as bacteria and fungi for soil rejuvenation is a form of environmental remediation. The objective of microbial remediation is to remove soil contaminants and pollutants [107]. Though the industrial use of microbes for removing contaminants goes back only three decades, microbes, aerobes, anaerobes, and facultative anaerobes have been contributing to soil improvement for billions of years. They help with N-fixation, limiting growth of plant pathogens, and the decomposition of heavy metals, pesticides, and hydrocarbons in the soil. The microbial flora is nourished by the contaminants, degrading them for energy and reproduction. Microbial remediation can be divided into three grades:

Natural attenuation: The process takes place naturally with indigenous soil microorganisms.

Biostimulation: The natural process receives external help in the form of nutrients, moisture, and an ideal pH for the microorganisms.

Bioaugmentation: This involves the use of externally introduced microorganisms, which is the case in situations such as oil spills where the naturally occurring microbes may die out because of the intensity of the contamination [108].

Rhizofiltration: Phytofiltration is used to inhibit organic pollutants in wastewater and surface water from mixing with water streams or groundwater using plants for filtration purpose, as they can absorb or adsorb the pollutants [109]. Phytofiltration can also be: rhizofiltration in which plant roots are used; blastofiltration in which seedlings are utilized; and caulofiltration that uses excised plant shoots) [110]. Due to phytofiltration, the movement of contaminants in the soil is minimized [111]. Phytofiltration is defined as the use of plants, both terrestrial and aquatic to absorb, concentrate, and precipitate contaminants from polluted aqueous sources with low contaminant concentration in their roots. Rhizofiltration can partially treat industrial discharge, agricultural runoff, or acid mine drainage. It can be used for lead, cadmium, copper, nickel, zinc, and chromium that are primarily retained within the roots [4]. The advantages of rhizofiltration include its ability to be used in *in-situ* or *ex-situ* applications and species other than hyperaccumulators can also be used. Plants like sunflower, Indian mustard, tobacco, rye, spinach, and corn have been studied for their ability to remove lead from effluent, with sunflowers having the greatest ability. Indian mustard has proven to be effective in removing a wide concentration range of lead (4–500 mg/l). This technology has been tested in the field with Uranium contaminated water at concentrations of 21–874 μg/l; the treated U-concentration reported by studies was <20 μg/l before discharge into the environment. The use of some metal accumulator aquatic plants species, both living and dead, and constructed wetlands for the removal of heavy metals from industrial wastewater has gained considerable interest [112]. Aquatic plants and microorganisms can remove metals from water through processes of biosorption and metabolism-dependent bioaccumulation. The use of dead or dried aquatic plants, for metal removal as a simple biosorbent material has advantages in its high efficiency including minimization of the volume of chemical and/or biological sludge, no nutrient requirements, low cost, conservation, transport, handling, and metal recovery [113–116].

2.5.2 Phytovolatilization

Phytovolatization is the process in which pollutants are up taken by the plants from the soil and then converted into a volatile form and then released in the atmosphere. Phytostabilization can be used for organic pollutants and other heavy metals like Se and Hg. But, as explained earlier, phytostabilization transfers the pollutants into the atmosphere, from one medium to another, and does not remove the pollutants permanently. The pollutants in the atmosphere can also be redeposited into the soil at a later time. In phytoremediation of organics, plant metabolism contributes to the contaminant reduction by transformation breakdown, stabilization, or volatilizing contaminant compounds from soil and groundwater. Phytodegradation is the breakdown of organics, taken up by the plant to simpler molecules that are incorporated into plant tissues. Plants contain enzymes that can breakdown and convert ammunition wastes, chlorinated solvents such as trichloroethylene, and other herbicides. The enzymes are usually dehalogenases, oxygenases, and reductases. Rhizodegradation is the breakdown of organics in the soil through microbial activity of the root zone and is a much slower process than phytodegradation. Yeast, fungi, bacteria, and other microorganisms consume and digest organic substances like fuels and solvents. All phytoremediation technologies are not exclusive and may be used concurrently, but the metal extraction depends on its bioavailable fraction in soil [117, 118].

2.5.3 Phytodegradation

Phytodegradation, also known as phytotransformation, is the use of plants and microorganisms to uptake, metabolize, and degrade the organic contaminant. In this method, plant roots are used in association with microorganisms to detoxify soil contaminated with organic compounds [119]. Some plants can decontaminate soil, sludge, sediment, and ground and surface water by producing enzymes. It involves organic compounds, including herbicides, insecticides, chlorinated solvents, and inorganic contaminants [120].

2.6 Adaptive Mechanism of Bioremediation for Heavy Metals, Pesticides, and Herbicides

Plants adopt different strategies and complex mechanisms for survival under critical conditions, which are mostly caused by biotic or abiotic stresses. Plant stress occurs through damage by certain living or nonliving species. Living organisms such as viruses, bacteria, parasites, fungi, beneficial or harmful insects, and native or cultivated plants mostly initiate biotic stresses. Abiotic stresses such as, mineral toxicity, excessive watering, drought, extreme temperatures, salinity negatively impact development, growth, yield, and seed quality of plants and other crops. Differentiation between the damage caused by living or nonliving agents is a very difficult task even with the help of accurate diagnosis and close examinations. Relative oxygen species generated in the plants that accumulate in cells under environmental stress results in the oxidation of carbohydrates, proteins, lipid, chlorophyll, nucleic acids, etc. [121–123].

2.6.1 Defense System

Plant cells have evolved intricate defense systems including: (i) Plant hormones (phytohormones), such as, ethylene, jasmonic acid [JA], salicylic acid [SA], abscisic acid [ABA], and brassino-steroids. These phytohormones are mostly required by plants for their growth and development and occasionally act as defense mechanisms. (ii) Enzymatic systems that comprise superoxide dismutase [SOD], ascorbate peroxidase [APX], catalase [CAT], glutathione reductases [GR], dehydro-ascorbate reductases [DHAR], mono dehydro-ascorbate reductases [MSHAR], glutathione peroxidase [GPX], glutathione-S-transferase [GST] and guaicol peroxidase [GOPX]. (iii) Nonenzymatic systems include glutathione [GSH], ascorbic acid, nonprotein amino acids, α-tocopherol, phenolic compounds and alkaloids, which can hunt primordial generated ROS [124, 125].

Different research has documented different mechanisms for the degradation of HMs and pesticides as follows.

2.6.1.1 Adsorption

The adsorption of orthophosphate-pesticides in soil and the resulting decrease in fluidity are important factors that influence their behavior in nature. The degree of adsorption as well as the rate and extent of final degradation are influenced by several factors including solubility, volatility, charge, polarity, molecular structure, and pesticide size. The process of soil particle adsorption can prevent degradation of pesticides by separating the pesticide from the enzyme that degrades it or by enhancing the degradation process. Abiotic hydrolysis degradation improves the adsorption process. Conversely, volatilization or leaching following adsorption leads to a reduction in the loss of pesticides. Various physical and chemical forces in the process of soil particle adsorption include van der Waals forces, dipole–dipole interactions, hydrogen bonds, and ions replacement. However, there is less information available for the adsorption of ionizable pesticides and extensive studies are needed to analyze the background mechanisms to predict the nature of pesticide interactions with the soil, as these phenomena may affect other processes [126, 127].

2.6.1.2 Photodegradation

Photolysis of organophosphorus pesticides can be a very important degradation pathway in aqueous environments as well as in the gaseous phase. The degradation of chlorpyrifos under environmental conditions has been studied and around 200 American crabs were killed by about 20 μg/l in the Ebre Delta Irrigation ditch in Spain. The content of chlorpyrifos and its conversion products was recorded four days after application. The chlorpyrifos transformation product is 3-methyl-4-nitrophenol, Acaricion, and S-methyl isomer. The half-life of chlorpyrifos is 13 hours and the degradation rate constant 0.053/hour, mainly by photolysis. The degradation of chlorpyrifos and the formation of its transformation products are closely related to environmental factors such as wind [126–128].

2.6.1.3 Hydrolysis

Hydrolysis is the most thoroughly studied degradation pathway for organophosphorus pesticides. The organophosphorus pesticide can be diverse and usually involves the cleavage of bonds, which produce the best product. A good example of bond cleavage can be found in the hydrolysis of diazinon, where the oxygen attached to the pyrimidine ring can most effectively stabilize the negative charge and similar behavior can be found for other phosphorothioates as well. During the hydrolysis of dichlorvos, the possible initial hydrolytic cleavage lies between the P and O atoms attached to the carbon atom of the double bond. Alkaline hydrolysis is the major pathway for malathion and is consistent with previous laboratory studies. It has been observed that only the biological and photochemical degradation of malathion is slow, further the biodegradation is important for parathion. Alkaline hydrolysis and photolysis are only minor ways of degrading parathion. Digestion mechanisms include copper-catalyzed chlorpyrifos hydrolysis. *P. putida* can use parathionmethyl as the sole source for C and/or P. Bacteria producing enzyme, the organophosphoric anhydrase, which hydrolyzes parathionmethyl to p-nitrophenol, further degrades to hydroquinone and 1, 2, 4-benzenetriol and then cleaves to acetic acid by glycerol oxygenase [126, 129, 130].

2.6.1.3.1 Enzymatic Degradation Enzymes known to hydrolyze many organophosphorus pesticides are made from a variety of aquatic species. These enzymes are known as organophosphoric anhydrases, although they are also known as paraoxonase, esterase, phosphotriesterase, diisopropylfluorophosphatase, and parathion hydrolase. However, these enzymes can hydrolyze a variety of organophosphorus acetylcholinesterase inhibitors. Among the aquatic species, enzymes have been identified that are partially characterized by salmon, invertebrates such as *Rangia cuneata*, protozoa *Tetrahymena thermophila*, and various thermophilic and other bacteria. These enzymes evolved in response to the metabolism of naturally occurring organophosphates and halogenated organic compounds. Several researchers have highlighted the presence of organophosphorus hydrolase (Aryldialkylphosphatase) genes in microbial cells that degrade organic phosphorus compounds and hydrolases as the main enzyme behind the process. An enzyme derived from an over-produced *P. diminuta* undergoes hydrolysis of a phosphate bond in an organophosphorus pesticide molecule resulting in an up-to 100-fold reduction in toxicity. The use of hydrolases and related genes to understand the complex interactions between microorganisms and pesticides can significantly improve understanding of biodegradation processes and facilitate bioremediation [126, 129, 131].

2.7 Behavior of Inorganic and Organic Pollutants in Soil

Translocation, mobility, uptake, and accumulation of contaminants in plants are noticeably caused by main two factors: (i) environmental factors, and (ii) genetic makeup between plant species to uptake contaminants, their translocation and storage to different organs and plant resistance against specific contaminants. These mechanisms of contaminants in different parts of plants are known to differ and are also imbalanced. The uptake mechanisms for organic or inorganic complexes are also diverse. The movement of inorganic compounds within plants for example, nutrients, metals, and metalloids is generally carried out by through active transport and passive diffusion. An example, using Ni, showed that passive diffusion was influenced at a high concentration of nickel whereas active transport of Ni^{2+} was revealed to play a vital role in its uptake from medium to low concentrations of Ni. Inorganic contaminants, usually transportation of metals, are carried from root membranes with the support of membrane transport proteins that belong to the CDF proteins family (cation diffusion facilitators). It is remarkable that a binding domain of protein identifies only specific ions and is responsible for its transportation [132]. In contrast to the inorganic translocation system, organic contaminants don't have any specific carrier to carry them through the plant. They simply move into the symplast and xylem apoplast due to their hydrophobicity and also pass into the leaves by simple diffusion [133]. Moreover, translocation of both organic and inorganic pollutants is influenced by soil rhizosphere microorganisms, which are in symbiosis with the roots [129, 134]. Microorganisms also have diverse functions, for example, they excrete organic compounds to the soil, which increases the bioavailability in the soil, and the transit of metals to the plant from the roots, they may also alter their chemical properties, which could make them harmless, harmful, mobile, or immobile [117].

Due to the distinct ability of HM accumulation in plants, they are divided into three classes: accumulators – obtain high levels of metals on the surface of the ground, easy to reap organs, and self-reliance of metal concentrations in the soil; excluders – they have very confined transit of metals to the shoots from the roots even if the soil is enrich with the metal contaminants; indicators – they manifest the levels of metal contamination in the rhizospheric soil [135]. Furthermost the plants belong to the class of excluder, as in the case of Pb, its translocation to the shoots is very confined but it accumulates in the cell walls of the cortex due to the weak transportation of metal ions to the shoots [136], or due to the prolonged distance between the shots/roots [125]. The degree of metal uptake and its translocation varies from plant to plant spices and it is one of the most notable features of plant resistance [137]. Whereas the ability of HM accumulation to one or another level, they are recognized as hyperaccumulators, accumulators, and nonaccumulators [138]. Hyperaccumulators have 100 times higher concentrations in their shoots than nonaccumulators. Until now, about 400 species have been characterized as metal hyperaccumulators as they genetically have great capacity to accumulate high levels of HM in their shoots. At the time of uptake from the roots, the major fraction of metal is present in the plant rhizodermis and cortex [134, 139, 140]. The distribution of metals in hyperaccumulator plants is truly systemized through proficiency of many detoxification mechanisms and is determined by the features of metal transport. Ni is easily transported to the tissues of stele, while the distribution of Cd and Pb is limited to the central cylinder by the endodermis. The accumulation and distribution of organic compounds in plants includes three phases: enzymatic modification and degradation, conjugation, and sequestration in cell walls [124]. To date, it is anticipated that using some biotechnological approaches and genetic engineering utilizing the knowledge of the metabolic and genetic processes regulating a metal tolerance will add to the plant resistance and accumulation of HMs. Solubility of metal and other contaminants and their bioavailability to the plants is chiefly effected by the chemical properties of soil for example, cation exchange capacity, loading rate, soil pH, redox potential, organic matter, soil texture, and clay content [141, 142]. Normally, higher levels of organic matter or clay and soil pH enhance chelation effect and the metals will be strongly bound to soil for longer periods and will be less bioavailable to plants. Soil temperature is also a crucial parameter accounting for variations in metal accumulation by crops [143, 144]. However, the introduction of various genetic changes in plants can enhance their survival at high concentration of contaminants and it advances their ability to bind or remove toxicants and to influence the synthesis of enzymes thus lessening the toxic effect of HMs [145, 146]. The plants with genetic potential for uptake, extraction, degradation, destabilization, and immobilization of contaminants are the elite implements for cleaning up contaminated soils by phytoremediation processes.

2.8 Environmental Implications of Bioremediation

Bioremediation and phytoremediation, like other remediation technologies do possess both positive and negative impacts.

2.8.1 Advantages of Phytoremediation

a) It is a clean-up technology, cost-effective, esthetically pleasing, and environmentally friendly.
b) It has a high probability of public acceptance.
c) It may reduce the entry of contaminants into the environment by preventing their leakage into groundwater systems.
d) It may be used on a larger scale to clean-up a diversity of contaminants, which is possible with other approaches.
e) Environmental disruption is negligible, and it preserves topsoil in *in-situ* treatment.
f) Plants act as soil stabilizers, which minimizes the grasshopper effect, and prevents contaminants from spreading in their surrounding environment.
g) It has the potential to treat a wide-range of hazardous pollutants in the environment.
h) Sites can be monitored easily with the naked eye.
i) Additional advantages of phytoremediation over bioremediation, physio-chemical and engineering methods include the production of useful byproducts, such as bioenergy or wood pulp.
j) Phytoremediation also supports bioremediation because plants supply nutrients and provide protection for rhizospheric microorganisms, which promotes remediation of pollutants. Additionally, the plants that are grown during phytoremediation provide stabilization of the soil and could potentially be used for green energy purposes.
k) Lessens the amount of landfill waste further (up to 90%), which can be further used as bio-ore of heavy metals.

2.8.2 Disadvantages of Phytoremediation

a) It is usually slower than other common treatment technologies and depends upon climatic conditions.
b) For better results spots must be large enough to cultivate and utilize agricultural machinery for planting and harvesting.
c) Contaminants collected in leaves and trunks can be released again to the environment during litterfall.

d) The contaminants should be present within reach of the root zone and should not be bound to the organic portion of soil to be accessible for the plants; typically 10–15 ft for trees and 3–6 ft for herbaceous plants.
e) It is a time-consuming and slow process – it may take several growing seasons to fully clean-up a site.
f) Very often the introduction of nonnative species may affect the whole biodiversity

2.9 Conclusion

Bioremediation is a remarkable approach to mitigate contaminants from polluted environments like water and soil, but it is not a permanent solution to an overall contamination problem. We must acknowledge that once pollutants are released into the environment, they cannot be completely degraded because of their movement between various environmental elements and food chains. For that reason, as a foremost strategy, we must stop or reduce the production of those pollutants, which could accumulate in the environment and cause environmental degradation. Second, we must implement advanced environmentally friendly approaches of bioremediation to overcome this problem. Bioremediation reduces capital and operational costs making this approach more economic than others, i.e., *ex-situ* and *in-situ* cleaning methods. In contrast to traditional methods that degrade soil structure and diminish its fertility bioremediation and phytoremediation enhance soil quality and fertility.

References

1 Qasim, S.R. (2017). *Wastewater Treatment Plants: Planning, Design, and Operation*. London: Routledge.

2 Carr, S.A., Liu, J., and Tesoro, A.G. (2016). Transport and fate of microplastic particles in wastewater treatment plants. *Water Research* 91: 174–182.

3 Zhang, Q., Yang, W.N., Ngo, H.H. et al. (2016). Current status of urban wastewater treatment plants in China. *Environment International* 92: 11–22.

4 McIntyre, T. (2003). Phytoremediation of heavy metals from soils. In: *Phytoremediation* (ed. D.T. Tsao), 97–123. Berlin, Heidelberg: Springer.

5 Pascal-Lorber, S. and Laurent, F. (2011). Phytoremediation techniques for pesticide contaminations. In: *Alternative Farming Systems, Biotechnology, Drought Stress and Ecological Fertilisation* (ed. E. Lichtfouse), 77–105. Dordrecht: Springer.

6 He, C., Li, L., and Gu, C. (2003). Phytoremediation techniques of heavy metal in sewage sludge. *Chinese Journal of Ecology* 22 (5): 78–81.

7 Jadia, C.D. and Fulekar, M. (2009). Phytoremediation of heavy metals: recent techniques. *African Journal of Biotechnology* 8 (6): 921–928.

8 Singh, O.V., Labana, S., Pandey, G. et al. (2003). Phytoremediation: an overview of metallic ion decontamination from soil. *Applied Microbiology and Biotechnology* 61 (5): 405–412.

9 Liu, P., Qiu, G., and Shang, L. (2007). Phytoremediation of mercury contaminated soil: a review. *Chinese Journal of Ecology* 6: 27.

10 Alloway, B.J. (2013). Sources of heavy metals and metalloids in soils. In: *Heavy Metals in Soils* (ed. B.J. Alloway), 11–50. Dordrecht: Springer.

11 Adamo, P., Denaix, L., Terribile, F. et al. (2003). Characterization of heavy metals in contaminated volcanic soils of the Solofrana river valley (southern Italy). *Geoderma* 117 (3–4): 347–366.

12 Callender, E. (2003). Heavy metals in the environment-historical trends. *Treatise on Geochemistry* 9: 612.

13 Rylander, P.N. (1967). Platinum metals in catalytic hydrogenation. *Annals of the New York; Academy of Sciences* 145 (1): 46–51.

14 Jayasumana, C., Gajanayake, R., and Siribaddana, S. (2014). Importance of arsenic and pesticides in epidemic chronic kidney disease in Sri Lanka. *BMC Nephrology* 15 (1): 124.

15 Roberts, T.L. (2014). Cadmium and phosphorous fertilizers: the issues and the science. *Procedia Engineering* 83: 52–59.

16 Tchounwou, P.B., Yedjou, C.G., Patlolla, A.K. et al. (2012). Heavy metal toxicity and the environment. In: *Molecular, Clinical and Environmental Toxicology: Volume 3: Environmental Toxicology* (ed. A. Luch), 133–164. Basel: Springer.

17 Wei, B. and Yang, L. (2010). A review of heavy metal contaminations in urban soils, urban road dusts and agricultural soils from China. *Microchemical Journal* 94 (2): 99–107.

18 Tang, W., Shan, B., Zhang, H. et al. (2010). Heavy metal sources and associated risk in response to agricultural intensification in the estuarine sediments of Chaohu Lake Valley, East China. *Journal of Hazardous Materials* 176 (1): 945–951.

19 Arora, M., Kiran, B., Rani, S. et al. (2008). Heavy metal accumulation in vegetables irrigated with water from different sources. *Food Chemistry* 111 (4): 811–815.

20 Bradl, H.B. (2005). Sources and origins of heavy metals. In: *Interface Science and Technology* (ed. H.B. Bradl), 1–27. Amsterdam: Elsevier.

21 Buccolieri, A., Buccolieri, G., Dell'Atti, A. et al. (2006). Natural sources and heavy metals. *Annali di Chimica* 96 (3–4): 167–181.

22 Garrett, R.G. (2000). Natural sources of metals to the environment. *Human and Ecological Risk Assessment: An International Journal* 6 (6): 945–963.

23 Dias, G.M. and Edwards, G.C. (2003). Differentiating natural and anthropogenic sources of metals to the environment. *Human and Ecological Risk Assessment: An International Journal* 9 (4): 699–721.

24 Ochieng, E.Z., Lalah, J.O., and Wandiga, S.O. (2009). Anthropogenic sources of heavy metals in the Indian Ocean Coast of Kenya. *Bulletin of Environmental Contamination and Toxicology* 83 (4): 600–607.

25 Zhang, C., Qiao, Q., Appel, E. et al. (2012). Discriminating sources of anthropogenic heavy metals in urban street dusts using magnetic and chemical methods. *Journal of Geochemical Exploration* 119–120: 60–75.

26 Xu, Y., Sun, Q., Yi, L. et al. (2014). The source of natural and anthropogenic heavy metals in the sediments of the Minjiang River Estuary (SE China): implications for historical pollution. *Science of the Total Environment* 493: 729–736.

27 Davis, H.T., Aelion, C., McDermott, S. et al. (2009). Identifying natural and anthropogenic sources of metals in urban and rural soils using GIS-based data, PCA, and spatial interpolation. *Environmental Pollution (Barking, Essex: 1987)* 157 (8–9): 2378–2385.

28 Pacyna, E.G., Pacyna, J.M., Fudala, J. et al. (2007). Current and future emissions of selected heavy metals to the atmosphere from anthropogenic sources in Europe. *Atmospheric Environment* 41 (38): 8557–8566.

29 Alloway, B.J. (2013). Sources of heavy metals and metalloids in soils. In: *Heavy Metals in Soils: Trace Metals and Metalloids in Soils and Their Bioavailability* (ed. B.J. Alloway), 11–50. Dordrecht: Springer.

30 Bååth, E. (1989). Effects of heavy metals in soil on microbial processes and populations (a review). *Water, Air, and Soil Pollution* 47 (3): 335–379.

31 Jansen, E., Michels, M., van Til, M. et al. (1994). Effects of heavy metals in soil on microbial diversity and activity as shown by the sensitivity-resistance index, an ecologically relevant parameter. *Biology and Fertility of Soils* 17 (3): 177–184.

32 Leita, L., De Nobili, M., Muhbachova, G. et al. (1995). Bioavailability and effects of heavy metals on soil microbial biomass survival during laboratory incubation. *Biology and Fertility of Soils* 19 (2): 103–108.

33 Kandeler, F., Kampichler, C., and Horak, O. (1996). Influence of heavy metals on the functional diversity of soil microbial communities. *Biology and Fertility of Soils* 23 (3): 299–306.

34 Wang, Y., Shi, J., Wang, H. et al. (2007). The influence of soil heavy metals pollution on soil microbial biomass, enzyme activity, and community composition near a copper smelter. *Ecotoxicology and Environmental Safety* 67 (1): 75–81.

35 Fließbach, A., Martens, R., and Reber, H.H. (1994). Soil microbial biomass and microbial activity in soils treated with heavy metal contaminated sewage sludge. *Soil Biology and Biochemistry* 26 (9): 1201–1205.

36 Chander, K. and Brookes, P.C. (1991). Effects of heavy metals from past applications of sewage sludge on microbial biomass and organic matter accumulation in a sandy loam and silty loam U.K. soil. *Soil Biology and Biochemistry* 23 (10): 927–932.

37 Weng, L., Temminghoff, E., Lofts, S. et al. (2002). Complexation with dissolved organic matter and solubility control of heavy metals in a Sandy soil. *Environmental Science & Technology* 36 (22): 4804–4810.

38 Valsecchi, G., Gigliotti, C., and Farini, A. (1995). Microbial biomass, activity, and organic matter accumulation in soils contaminated with heavy metals. *Biology and Fertility of Soils* 20 (4): 253–259.

39 Reuter, J.H. and Perdue, E.M. (1977). Importance of heavy metal-organic matter interactions in natural waters. *Geochimica et Cosmochimica Acta* 41 (2): 325–334.

40 Brookes, P.C. (1995). The use of microbial parameters in monitoring soil pollution by heavy metals. *Biology and Fertility of Soils* 19 (4): 269–279.

41 Clemente, R., Escolar, Á., and Bernal, M.P. (2006). Heavy metals fractionation and organic matter mineralisation in contaminated calcareous soil amended with organic materials. *Bioresource Technology* 97 (15): 1894–1901.

42 Barceló, J. and Poschenrieder, C. (1990). Plant water relations as affected by heavy metal stress: a review. *Journal of Plant Nutrition* 13 (1): 1–37.

43 Nagajyoti, P.C., Lee, K.D., and Sreekanth, T.V.M. (2010). Heavy metals, occurrence and toxicity for plants: a review. *Environmental Chemistry Letters* 8 (3): 199–216.

44 Påhlsson, A.-M.B. (1989). Toxicity of heavy metals (Zn, Cu, Cd, Pb) to vascular plants. *Water, Air, and Soil Pollution* 47 (3): 287–319.

45 Schützendübel, A. and Polle, A. (2002). Plant responses to abiotic stresses: heavy metal-induced oxidative stress and protection by mycorrhization. *Journal of Experimental Botany* 53 (372): 1351–1365.

46 Ernst, W.H.O. (1996). Bioavailability of heavy metals and decontamination of soils by plants. *Applied Geochemistry* 11 (1): 163–167.

47 Vinodhini, R. and Narayanan, M. (2008). Bioaccumulation of heavy metals in organs of fresh water fish *Cyprinus carpio* (common carp). *International Journal of Environmental Science and Technology* 5 (2): 179–182.

48 Khan, S., Cao, Q., Zheng, Y.M. et al. (2008). Health risks of heavy metals in contaminated soils and food crops irrigated with wastewater in Beijing, China. *Environmental Pollution* 152 (3): 686–692.

49 Ma, J., Ding, Z., Wei, G. et al. (2009). Sources of water pollution and evolution of water quality in the Wuwei basin of Shiyang river, Northwest China. *Journal of Environmental Management* 90 (2): 1168–1177.

50 Castro-González, M.I. and Méndez-Armenta, M. (2008). Heavy metals: implications associated to fish consumption. *Environmental Toxicology and Pharmacology* 26 (3): 263–271.

51 Wang, Q. and Yang, Z. (2016). Industrial water pollution, water environment treatment, and health risks in China. *Environmental Pollution* 218: 358–365.
52 Cheng, S. (2003). Heavy metal pollution in China: origin, pattern and control. *Environmental Science and Pollution Research* 10 (3): 192–198.
53 Uversky, V.N., Li, J., and Fink, A.L. (2001). Metal-triggered structural transformations, aggregation, and fibrillation of human α-Synuclein: a possible molecular link between Parkinson's disease and heavy metal exposure. *Journal of Biological Chemistry* 276 (47): 44284–44296.
54 Muchuweti, M., Birkett, J.W., Chinyanga, E. et al. (2006). Heavy metal content of vegetables irrigated with mixtures of wastewater and sewage sludge in Zimbabwe: implications for human health. *Agriculture, Ecosystems & Environment* 112 (1): 41–48.
55 Zhuang, P., McBride, M.B., Xia, H. et al. (2009). Health risk from heavy metals via consumption of food crops in the vicinity of Dabaoshan mine, South China. *Science of the Total Environment* 407 (5): 1551–1561.
56 Singh, A., Sharma, R.K., Agrawal, M. et al. (2010). Health risk assessment of heavy metals via dietary intake of foodstuffs from the wastewater irrigated site of a dry tropical area of India. *Food and Chemical Toxicology* 48 (2): 611–619.
57 Mahmood, A. and Malik, R.N. (2014). Human health risk assessment of heavy metals via consumption of contaminated vegetables collected from different irrigation sources in Lahore, Pakistan. *Arabian Journal of Chemistry* 7 (1): 91–99.
58 Fowler, B.A. (2013). *Biological and Environmental Effects of Arsenic*, vol. 6. Amsterdam: Elsevier.
59 Marin, A., Pezeshki, S.R., Masschelen, P.H. et al. (1993). Effect of dimethylarsenic acid (DMAA) on growth, tissue arsenic, and photosynthesis of rice plants. *Journal of Plant Nutrition* 16 (5): 865–880.
60 Burlo, F., Guyarro, I., Carbonell-Barrachina, A.A. et al. (1999). Arsenic species: effects on and accumulation by tomato plants. *Journal of Agricultural and Food Chemistry* 47 (3): 1247–1253.
61 Stoeva, N., Berova, M., and Zlatev, Z. (2005). Effect of arsenic on some physiological parameters in bean plants. *Biologia Plantarum* 49 (2): 293–296.
62 Sun, H.-J., Rathinasabapathi, B., Wu, B. et al. (2014). Arsenic and selenium toxicity and their interactive effects in humans. *Environment International* 69: 148–158.
63 Banerjee, M., Banerjee, N., Bhattacharjee, P. et al. (2013). High arsenic in rice is associated with elevated genotoxic effects in humans. *Scientific Reports* 3: 2195.
64 Abernathy, C.O., Liu, Y.P., Longfellow, D. et al. (1999). Arsenic: health effects, mechanisms of actions, and research issues. *Environmental Health Perspectives* 107 (7): 593.
65 Mergler, D., Anderson, H.A., Chan, L.H. et al. (2007). Methylmercury exposure and health effects in humans: a worldwide concern. *Ambio: A Journal of the Human Environment* 36 (1): 3–11.

66 Sun, H.J., Xiang, P., Luo, J. et al. (2016). Mechanisms of arsenic disruption on gonadal, adrenal and thyroid endocrine systems in humans: a review. *Environment International* 95: 61–68.

67 Guo, P., Qi, Y.-P., Cai, Y.,.-T. et al. (2018). Aluminum effects on photosynthesis, reactive oxygen species and methylglyoxal detoxification in two Citrus species differing in aluminum tolerance. *Tree Physiology* 38 (10): 1548–1565.

68 Khan, A., Khan, S., Khan, M.A. et al. (2015). The uptake and bioaccumulation of heavy metals by food plants, their effects on plants nutrients, and associated health risk: a review. *Environmental Science and Pollution Research* 22 (18): 13772–13799.

69 Mišík, M., Burke, I.T., Reismüller, M. et al. (2014). Red mud a byproduct of aluminum production contains soluble vanadium that causes genotoxic and cytotoxic effects in higher plants. *Science of the Total Environment* 493: 883–890.

70 Sade, H., Meriga, B., Surapu, V. et al. (2016). Toxicity and tolerance of aluminum in plants: tailoring plants to suit to acid soils. *Biometals* 29 (2): 187–210.

71 Dórea, J. (2015). Exposure to mercury and aluminum in early life: developmental vulnerability as a modifying factor in neurologic and immunologic effects. *International Journal of Environmental Research and Public Health* 12 (2): 1295.

72 Rajiv, S., Jerobin, J., Saranya, V. et al. (2016). Comparative cytotoxicity and genotoxicity of cobalt (II, III) oxide, iron (III) oxide, silicon dioxide, and aluminum oxide nanoparticles on human lymphocytes in vitro;. *Human & Experimental Toxicology* 35 (2): 170–183.

73 Annangi, B., Bonassi, S., Marcos, R. et al. (2016). Biomonitoring of humans exposed to arsenic, chromium, nickel, vanadium, and complex mixtures of metals by using the micronucleus test in lymphocytes. *Mutation Research/Reviews in Mutation Research* 770: 140–161.

74 Stambulska, U.Y., Bayliak, M.M., and Lushchak, V.I. (2018). Chromium (VI) toxicity in legume plants: modulation effects of rhizobial symbiosis. *BioMed Research International* 3: 1–18.

75 Heer, M. and Egert, S. (2015). Nutrients other than carbohydrates: their effects on glucose homeostasis in humans. *Diabetes/Metabolism Research and Reviews* 31 (1): 14–35.

76 Mishra, S. and Bharagava, R.N. (2016). Toxic and genotoxic effects of hexavalent chromium in environment and its bioremediation strategies. *Journal of Environmental Science and Health, Part C* 34 (1): 1–32.

77 Kolahian, S., Sadri, H., Larijani, A. et al. (2016). Supplementation of diabetic rats with leucine, zinc, and chromium: effects on function and histological structure of testes. *International Journal for Vitamin and Nutrition Research* 85 (5–6): 311–321.

78 Shabani, L. and Sabzalian, M.R. (2016). Arbuscular mycorrhiza affects nickel translocation and expression of ABC transporter and metallothionein genes in *Festuca arundinacea*. *Mycorrhiza* 26 (1): 67–76.

79 Shanying, H., Yang, X., He, Z. et al. (2017). Morphological and physiological responses of plants to cadmium toxicity: a review. *Pedosphere* 27 (3): 421–438.

80 Amari, T., Ghnaya, T., and Abdelly, C. (2017). Nickel, cadmium and lead phytotoxicity and potential of halophytic plants in heavy metal extraction. *South African Journal of Botany* 111: 99–110.

81 Das, K., Das, S., and Dhundasi, S. (2008). Nickel, its adverse health effects & oxidative stress. *Indian Journal of Medical Research* 128 (4): 412.

82 Lippmann, M., Ito, K., Hwang, J.S. et al. (2006). Cardiovascular effects of nickel in ambient air. *Environmental Health Perspectives* 114 (11): 1662.

83 Vinit-Dunand, F., Epron, D., Alaoui-Sossé, B. et al. (2002). Effects of copper on growth and on photosynthesis of mature and expanding leaves in cucumber plants. *Plant Science* 163 (1): 53–58.

84 Yruela, I. (2005). Copper in plants. *Brazilian Journal of Plant Physiology* 17 (1): 145–156.

85 Araya, M., Epron, D., and Alaoui-Sossé, B. (2006). Understanding copper homeostasis in humans and copper effects on health. *Biological Research* 39 (1): 183–187.

86 Valberg, L., Flanagan, P.R., and Chamberlain, M.J. (1984). Effects of iron, tin, and copper on zinc absorption in humans. *The American Journal of Clinical Nutrition* 40 (3): 536–541.

87 Vuori, E., Makinen, S.M., Kara, R. et al. (1980). The effects of the dietary intakes of copper, iron, manganese, and zinc on the trace element content of human milk. *The American Journal of Clinical Nutrition* 33 (2): 227–231.

88 Scuderi, P. (1990). Differential effects of copper and zinc on human peripheral blood monocyte cytokine secretion. *Cellular Immunology* 126 (2): 391–405.

89 Sharma, P. and Dubey, R.S. (2005). Lead toxicity in plants. *Brazilian Journal of Plant Physiology* 17 (1): 35–52.

90 Seregin, I. and Ivanov, V. (2001). Physiological aspects of cadmium and lead toxic effects on higher plants. *Russian Journal of Plant Physiology* 48 (4): 523–544.

91 de Vries, W., Römkens, P.F.A.M., and Schütze, G. (2007). Critical soil concentrations of cadmium, lead, and mercury in view of health effects on humans and animals. In: *Reviews of Environmental Contamination and Toxicology* (ed. P. de Voogt), 91–130. New York;, NY: Springer New York;.

92 James, H.M., Hilburn, M.E., and Blair, J.A. (1985). Effects of meals and meal times on uptake of lead from the gastrointestinal tract in humans. *Human Toxicology* 4 (4): 401–407.

93 Das, P., Samantaray, S., and Rout, G. (1997). Studies on cadmium toxicity in plants: a review. *Environmental Pollution* 98 (1): 29–36.

94 Di Toppi, L.S. and Gabbrielli, R. (1999). Response to cadmium in higher plants. *Environmental and Experimental Botany* 41 (2): 105–130.

95 Benavides, M.P., Gallego, S.M., and Tomaro, M.L. (2005). Cadmium toxicity in plants. *Brazilian Journal of Plant Physiology* 17 (1): 21–34.

96 Fowler, B.A. (2009). Monitoring of human populations for early markers of cadmium toxicity: a review. *Toxicology and Applied Pharmacology* 238 (3): 294–300.

97 Klaassen, C.D., Liu, J., and Diwan, B.A. (2009). Metallothionein protection of cadmium toxicity. *Toxicology and Applied Pharmacology* 238 (3): 215–220.

98 Godt, J., Scheidig, F., Grosse-Siestrup, C. et al. (2006). The toxicity of cadmium and resulting hazards for human health. *Journal of Occupational Medicine and Toxicology* 1 (1): 22.

99 Assche, F.V. and Clijsters, H. (1990). Effects of metals on enzyme activity in plants. *Plant, Cell & Environment* 13 (3): 195–206.

100 Banks, M., Schwab, A.P., Fleming, G.R. et al. (1994). Effects of plants and soil microflora on leaching of zinc from mine tailings. *Chemosphere* 29 (8): 1691–1699.

101 Solomons, N.W. and Jacob, R. (1981). Studies on the bioavailability of zinc in humans: effects of heme and nonheme iron on the absorption of zinc. *The American Journal of Clinical Nutrition* 34 (4): 475–482.

102 Sandström, B. and Sandberg, A.-S. (1992). Inhibitory effects of isolated inositol phosphates on zinc absorption in humans. *Journal of Trace Elements and Electrolytes in Health and Disease* 6 (2): 99–103.

103 Boening, D.W. (2000). Ecological effects, transport, and fate of mercury: a general review. *Chemosphere* 40 (12): 1335–1351.

104 Mishra, A. and Choudhuri, M. (1998). Amelioration of lead and mercury effects on germination and rice seedling growth by antioxidants. *Biologia Plantarum* 41 (3): 469–473.

105 Munzuroglu, O. and Geckil, H. (2002). Effects of metals on seed germination, root elongation, and coleoptile and hypocotyl growth in *Triticum aestivum and Cucumis sativus. Archives of Environmental Contamination and Toxicology* 43 (2): 203–213.

106 Burbacher, T.M., Rodier, P.M., and Weiss, B. (1990). Methylmercury developmental neurotoxicity: a comparison of effects in humans and animals. *Neurotoxicology and Teratology* 12 (3): 191–202.

107 Singh, H. (2006). *Mycoremediation: Fungal Bioremediation*. Chichester: Wiley.

108 Kulkarni, M. and Chaudhari, A. (2007). Microbial remediation of nitro-aromatic compounds: an overview. *Journal of Environmental Management* 85 (2): 496–512.

109 Cunningham, S.D. and Berti, W.R. (1993). Remediation of contaminated soils with green plants: an overview. *in vitro; Cellular & Developmental Biology: Plant* 29 (4): 207–212.

110 Cataldo, D. and Wildung, R. (1978). Soil and plant factors influencing the accumulation of heavy metals by plants. *Environmental Health Perspectives* 27: 149.

111 Reeves, R.D. (2003). Tropical hyperaccumulators of metals and their potential for phytoextraction. *Plant and Soil* 249 (1): 57–65.

112 Wu, H., Zhang, J., Ngo, H.H. et al. (2015). A review on the sustainability of constructed wetlands for wastewater treatment: design and operation. *Bioresource Technology* 175: 594–601.

113 Bratby, J. (2016). *Coagulation and Flocculation in Water and Wastewater Treatment*. London: IWA Publishing.

114 Giuliano, A., Sabia, G., Cabez, J.F. et al. (2017). Assessment of performance and advantages related to the use of a natural coagulant in the industrial wastewater treatment. *Environmental Engineering & Management Journal (EEMJ)* 16 (8): 1709–1719.

115 Goncalves, A.L., Pires, J.C., and Simões, M. (2017). A review on the use of microalgal consortia for wastewater treatment. *Algal Research* 24: 403–415.

116 Wang, Q., Wei, W., Gong, Y. et al. (2017). Technologies for reducing sludge production in wastewater treatment plants: state of the art. *Science of the Total Environment* 587: 510–521.

117 Limmer, M. and Burken, J. (2016). Phytovolatilization of organic contaminants. *Environmental Science & Technology* 50 (13): 6632–6643.

118 Ijaz, A., Imran, A., ul Haq, M.A. et al. (2016). Phytoremediation: recent advances in plant-endophytic synergistic interactions. *Plant and Soil* 405 (1–2): 179–195.

119 He, Y., Langenhoff, A.M., Sutton, N.B. et al. (2017). Metabolism of ibuprofen by *Phragmites australis*: uptake and phytodegradation. *Environmental Science & Technology* 51 (8): 4576–4584.

120 Al-Baldawi, I.A., Sheikh Abdullah, S.R., Anuar, N. et al. (2015). Phytodegradation of total petroleum hydrocarbon (TPH) in diesel-contaminated water using *Scirpus grossus*. *Ecological Engineering* 74: 463–473.

121 Vera-Estrella, R., Gómez-Méndez, M.F., Amezcua-Romero, J.C. et al. (2017). Cadmium and zinc activate adaptive mechanisms in *Nicotiana tabacum* similar to those observed in metal tolerant plants. *Planta* 246 (3): 433–451.

122 Hasan, M., Cheng, Y., Kanwar, M.K. et al. (2017). Responses of plant proteins to heavy metal stress – a review. *Frontiers in Plant Science* 8: 1492.

123 Van Oosten, M.J. and Maggio, A. (2015). Functional biology of halophytes in the phytoremediation of heavy metal contaminated soils. *Environmental and Experimental Botany* 111: 135–146.

124 Malar, S., Vikram, S.S., Favas, P.J.C. et al. (2016). Lead heavy metal toxicity induced changes on growth and antioxidative enzymes level in water hyacinths [*Eichhornia crassipes (Mart.)*]. *Botanical Studies* 55 (1): 54.

125 Banerjee, G., Pandy, S., Ray, A.K. et al. (2015). Bioremediation of heavy metals by a novel bacterial strain *Enterobacter cloacae* and its antioxidant enzyme activity, flocculant production, and protein expression in presence of lead, cadmium, and nickel. *Water, Air, & Soil Pollution* 226 (4): 91.

126 Kumar, S., Kaushik, G., Dar, M.A. et al. (2018). Microbial degradation of organophosphate pesticides: a review. *Pedosphere* 28 (2): 190–208.

127 Goyal, N., Jain, S., and Banerjee, U. (2003). Comparative studies on the microbial adsorption of heavy metals. *Advances in Environmental Research* 7 (2): 311–319.

128 Day, T.A., Bliss, M.S., Tomes, A.R. et al. (2018). Desert leaf litter decay: coupling of microbial respiration, water-soluble fractions and photodegradation. *Global Change Biology* 24 (11): 5454–5470.

129 John, E.M. and Shaike, J.M. (2015). Chlorpyrifos: pollution and remediation. *Environmental Chemistry Letters* 13 (3): 269–291.

130 Alneyadi, A.H. and Ashraf, S.S. (2016). Differential enzymatic degradation of thiazole pollutants by two different peroxidases – a comparative study. *Chemical Engineering Journal* 303: 529–538.

131 Silverman, A.I., Sedlak, D.L., and Nelson, K.L. (2018). Simplified process to determine rate constants for sunlight-mediated removal of trace organic and microbial contaminants in unit process open-water treatment wetlands. *Environmental Engineering Science* 36 (1): 43–59.

132 Dar, M.I., Naikoo, M.I., Green, I.D. et al. (2018). Heavy metal hyperaccumulation and hypertolerance in Brassicaceae. In: *Plants under Metal and Metalloid Stress* (eds. M. Hasanuzzaman, K. Nahar and M. Fujita), 263–276. Springer.

133 Clemens, S. and Ma, J.F. (2016). Toxic heavy metal and metalloid accumulation in crop plants and foods. *Annual Review of Plant Biology* 67: 489–512.

134 Epelde, L., Lanzén, A., Blanco, F. et al. (2015). Adaptation of soil microbial community structure and function to chronic metal contamination at an abandoned Pb-Zn mine. *FEMS Microbiology Ecology* 91 (1): 1–11.

135 Yin, H., Niu, J., Ren, Y. et al. (2015). An integrated insight into the response of sedimentary microbial communities to heavy metal contamination. *Scientific Reports* 5: 14266.

136 Pishchik, V., Oliveira, R.S., Ren, Y. et al. (2016). Mechanisms of plant and microbial adaptation to heavy metals in plant–microbial systems. *Microbiology* 85 (3): 257–271.

137 Richter, J., Ploderer, M., Mongelard, G. et al. (2017). Role of CrRLK1L Cell Wall sensors HERCULES1 and 2, THESEUS1, and FERONIA in growth adaptation triggered by heavy metals and trace elements. *Frontiers in Plant Science* 8: 1554.

138 Mesa, J., Mateos-Naranjo, E., Caviedes, M.A. et al. (2015). Scouting contaminated estuaries: heavy metal resistant and plant growth promoting rhizobacteria in the native metal rhizoaccumulator *Spartina maritima*. *Marine Pollution Bulletin* 90 (1–2): 150–159.

139 Krämer, U. (2018). The plants that suck up metal. *German Research* 40 (3): 18–23.

140 de la Torre, V.S.G., Majorel-Loulergue, C., Gonzalez, D.A. et al., Wide cross-species RNA-seq comparison reveals convergent molecular mechanisms involved in nickel hyperaccumulation across angiosperms. *CellPress*. Available at: https://papers.ssrn.com/sol3/papers.cfm?abstract_id=3272237 (2018).

141 Deng, T.-H.-B., Van Der Ent, A., Tang, Y.-T. et al. (2018). Nickel hyperaccumulation mechanisms: a review on the current state of knowledge. *Plant and Soil* 423 (1–2): 1–11.

142 Pilon-Smits, E.A. (2017). Mechanisms of plant selenium hyperaccumulation. In: *Selenium in Plants* (eds. E.A.H. Pilon-Smits, L.H.E. Winkel and Z.-Q. Lin), 53–66. Springer.

143 Im, J., Yang, K., Jho, E.H. et al. (2015). Effect of different soil washing solutions on bioavailability of residual arsenic in soils and soil properties. *Chemosphere* 138: 253–258.

144 Xian, Y., Wang, M., and Chen, W. (2015). Quantitative assessment on soil enzyme activities of heavy metal contaminated soils with various soil properties. *Chemosphere* 139: 604–608.

145 Dias, D., de Castro Moreira, M.E., Contin Gomes, M.J. et al. (2015). Rice and bean targets for biofortification combined with high carotenoid content crops regulate transcriptional mechanisms increasing iron bioavailability. *Nutrients* 7 (11): 9683–9696.

146 Ullah, A., Fahad, S., Munis, H.F. et al. (2015). Phytoremediation of heavy metals assisted by plant growth promoting (PGP) bacteria: a review. *Environmental and Experimental Botany* 117: 28–40.

3

Laccase

The Blue Copper Oxidase

Deepa Thomas and A.K. Gangawane

Faculty of Applied Sciences, Parul University, Vadodara, Gujarat, India

3.1 What Is Laccase?

Laccase is an important oxidase enzyme, which is widely used industry. The multiple potential applications of laccase in these industries define this particular enzyme as a promising alternative to existing costly and polluting processes [1–4].

Laccase is one of a small group of enzymes called the large blue copper proteins or blue copper oxidases that have been the subject of study since the end of the last century. The blue oxidases have been intensively studied because they have the ability to reduce molecular oxygen to water [5, 6]. Laccase (benzenediol: oxygen oxidoreductase, EC 1.10.3.2) contain 15–30% carbohydrates and have a molecular mass of 60–90 kDa. These are glycosylated polyphenol oxidases that contain four copper ions per molecule that carry out one electron oxidation of phenolic and its related compound and reduce oxygen to water [4]. Laccases oxidize a surprisingly wide variety of organic and inorganic compounds, including diphenols, polyphenols, substituted phenols, diamines, and aromatic amines, with concomitant reduction of molecular oxygen to water [7]. Laccases exhibit a broad substrate range, which varies from one laccase to another. Although it is known that diphenol oxidase, monophenols like 2, 6-dimethoxy phenol or guaiacol may also serve as substrates. Laccases catalyze monoelectronic oxidation of molecules to corresponding reactive radicals with the help of four copper atoms, which form the main catalytic core of the laccase, accompanied with the diminution of oxygen to water molecules and simultaneous oxidation of substrate to produce radicals (Figure 3.1). All substrates cannot be directly oxidized by

Soil Bioremediation: An Approach Towards Sustainable Technology, First Edition.
Edited by Javid A. Parray, Abeer Hashem Abd Elkhalek Mahmoud, and Riyaz Sayyed.

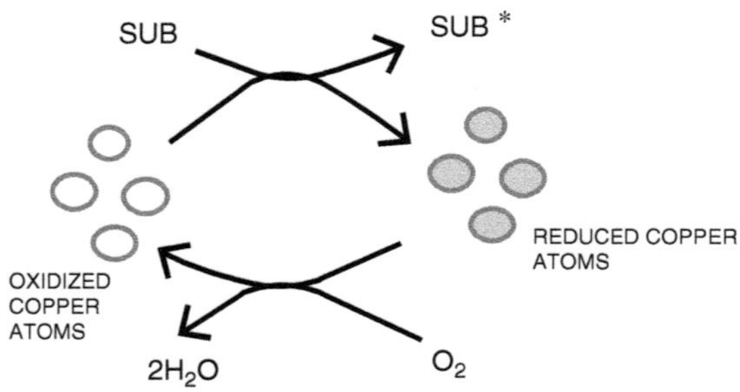

Figure 3.1 Oxidation of substrate with laccase enzyme.

laccases, either because of their large size, which restricts their penetration into the enzyme active site, or because of their particular high redox potential. To overcome this hindrance, suitable chemical mediators are used, which are oxidized by the laccase and their oxidized forms are then able to interact with high redox potential substrate targets [8]

3.2 Distribution of Laccases

Laccases are widely distributed in higher plants, bacteria, fungi, and insects. In plants, laccases are found in cabbages, turnip, potatoes, pears, apples, and other vegetables. Laccases have also been present in various Ascomyceteous, Basidiomycetous, and Deuteromycteous fungi. Laccase has been found in the cuticles of many insect species, and is involved in cuticle sclerotization. In several holometabolous insects, laccase has been identified as the principal enzyme associated with cuticular hardening [9]. Plant laccases have not been characterized or used extensively despite their wide occurrence, because their detection and purification is often difficult, as the crude plant extracts contain a large number of oxidative enzymes with broad substrate specificities. The majority of laccases characterized so far have been derived from fungi especially from white-rot basidiomycetes that are efficient lignin degraders [9]. Fungal laccases have higher redox potential than bacterial or plant laccases (up to +800 mV), and their action seems to be relevant in nature, they are also finding important applications in biotechonology. Thus, fungal laccases are involved in the degradation of lignin or in the removal of potentially toxic phenols arising during lignin degradation. In addition, fungal laccases are hypothesized to take part in the synthesis of dihydroxynaphthalene melanins, darkly pigmented polymers that organisms produce against environmental stress or in fungal morphogenesis by catalyzing the formation of extracellular pigments. Concerning their use in the biotechnology area, fungal laccases have widespread applications, ranging from effluent decoloration and detoxification to pulp bleaching, removal of phenolics from wines, organic synthesis, biosensors, synthesis of complex medical compounds, and dye transfer blocking functions in detergents and washing powders, many of which have been patented. The biotechnological use of laccase has been expanded by the introduction of laccase-mediator systems, which are able to oxidize nonphenolic compounds that are otherwise barely or not oxidized by the enzyme alone. However large-scale production of fungus is a still challenging [10]. Further, very little is known about the potential of bacterial laccases for bioremediation applications. Wastewater treatment involving bacteria is, however, considered to be more stable, as bacteria generally tolerate a broader range of habitats and grow faster than fungi. Moreover, in contrast to fungal laccases, some bacterial laccases can be highly active and much more stable at high temperatures, at high pH as well as at high chloride concentrations [11–13].

3.3 Application of Laccase

Laccases play an important role in: the food industry, the paper and pulp industry, the textile industry, soil bioremediation, and biodegradation of environmental phenolic pollutants. These enzymes are used for pulp delignification, pesticide or insecticide degradation, waste detoxification, textile dye transformation, food technological uses, and biosensor and analytical applications (Figure 3.2).

Recently laccases have been efficiently applied to nanobiotechnology due to their ability to catalyze electron transfer reactions without additional cofactor.

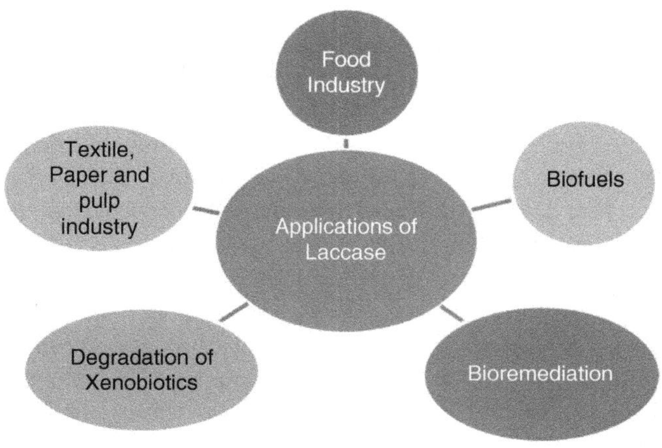

Figure 3.2 Applications of laccase.

The technique for the immobilization of biomolecule such as layer-by-layer, micropatterning, and self-assembled monolayer technique can be used for preserving the enzymatic activity of laccases [14]. Many micropollutants present in municipal wastewater, such as pharmaceuticals or biocides, are not easily removed in conventional biological treatments, resulting in a constant input into the aquatic environment. As these compounds are designed to be biologically active, they can affect sensitive aquatic organisms even at low concentrations. One potential means to reduce the amounts released to the environment is to improve their biodegradation in a post-treatment step using microorganisms that produce oxidative enzymes such as laccases. Due to their wide range of substrates and the sole requirement of oxygen as the cosubstrate, laccases appear to be a promising biocatalyst to enhance the biodegradation of micropollutants in wastewater in a complementary treatment step [14].

3.3.1 Laccase in Bioremediation

One of the main environmental problems, faced by the world today, is the contamination of soil, water, and air by toxic chemicals. With the extensive use of pesticides in agriculture and industrialization, the pollution of the environment with mandate organic compounds has become a serious problem. Laccases have many possible applications in bioremediation. Laccases may be applied to degrade various substances such as undesirable contaminants, byproducts, or discarded waste materials.

3.3.1.1 Degradation of Xenobiotics

One of the potential environmental applications for laccases is the bioremediation of contaminated soils, as laccases are able to oxidize toxic organic pollutants, such as various xenobiotics, Polycyclic Aromatic Hydrocarbons (PAHs), chlorophenols, and other contaminants [15, 16]. Phenolic compounds are present in wastes from several industrial processes: coal conversion, petroleum refining, production of organic chemicals, and olive oil production among others. Immobilized laccase was found to be useful in removing phenolic and chlorinated phenolic pollutants [17]. Laccase-Mediator-Systems (LMS) has been extensively studied in the oxidation of recalcitrant PAHs, main components of several ship spills. In this sense, laccase is being included in several enzymatic bioremediation programs [18].

Laccase may be applied to degrade plastic waste having olefin units and can be used to degrade polyurethanes. LMS facilitated the degradation of phenolic compounds (environmental hormones) from biphenol and alkylphenol derivatives and also the decomposition of fluorescent brighteners [18].

3.3.1.2 Decolorization of Dyes

The textile industry accounts for most of the dyestuff market and utilizes large volumes of water and chemicals. The chemical reagents used are very diverse in chemical composition, ranging from inorganic compounds to polymers and organic products. Due to their chemical structure, dyes are resistant to fading on exposure to light, water, and different chemicals and most of them are difficult to decolorize due to their synthetic origin [19, 20].

Several dyes are made from known carcinogens like benzidine and other aromatic compounds. Most currently existing processes to treat dye wastewater are ineffective and not economical. Therefore, the use of enzyme laccases seems like an attractive solution due to their potential in degrading dyes of various chemical structure including synthetic dyes currently used in many industries. The use of laccase in the textile industry is growing very fast since, besides decolorizing textile effluents, laccase is used to bleach textiles and even to synthesize dyes [21, 22]. They could also be used for decolorizing the dye-house effluents that are hardly decolorized by conventional sewage treatment plants.

3.3.2 Effluent Treatment

Laccase enzyme offers several advantages of great interest to biotechnological applications of industrial effluent treatment. In addition to dye-house effluents, laccases can decolorize wastewaters from pulp mills by removing colored phenolic compounds. This enzyme may also be used to eliminate odors emitted from garbage disposal sites, livestock farms, or pulp mills. In addition to the potential role of such degradation processes for natural attenuation processes in freshwater environments, this enzyme laccase also offers new perspectives for biotechnological applications such as wastewater treatment [18].

3.3.3 Pulp and Paper Industry

In the paper industry, the separation and degradation of lignin in wood pulp are conventionally obtained using chlorine- or oxygen-based chemical oxidants. Laccases have shown to be applicable to the bioremediation of pulp and paper industry waste by effecting direct dechlorination [23]. The oxygen delignification process has been introduced in recent years to replace the polluting chlorine-based methods. In spite of this new method the pretreatments of wood pulp with laccase can provide milder and cleaner strategies of delignification that also respect the integrity of cellulose. Other uses of laccases for the pulp and paper industry include an improvement in the paper-making properties of pulp and can be even used for deinking and decolorizing printed papers [23].

3.3.4 Laccases: Biosensors and Biofuel Cells

A biosensor is an integrated biological-component probe with an electronic transducer, thus converting a biochemical signal into a quantifiable electrical response that detects, transmits, and records information regarding a physiological or biochemical change. Some of the major attributes of a good biosensing system are its specificity, sensitivity, reliability, portability, real-time analysis, and operation simplicity. Thus, laccases can be applied as biosensors or bioreporters [24, 25]. A number of biosensors containing laccase have been developed for immunoassays, and for determination of glucose, aromatic amines, and phenolic compounds. Laccase catalysis can be used to assay other enzymes. Laccase enzyme covalently conjugated to a biobinding molecule can be used as a reporter for immunochemical (ELISA (enzyme-linked immunosorbent assay), Western blotting), cytochemical, histochemical, or nucleic acid-detection assays. The bioreporter applications are of interest in the high-sensitivity diagnostic field [25].

In addition to biosensors, laccases could be immobilized on the cathode of biofuel cells that could provide power, for example, for small transmitter systems. Fuel cells are very attractive energy sources, particularly at micro-, mini-, portable-, or mobile-scale, that potentially have higher energy conversion/usage efficiency and lower pollution effect than any of the existing or emerging energy sources [14].

3.3.5 Laccase in Food Industries

3.3.5.1 Wine Stabilization

Laccase is used to improve the quality of drinks and for the stabilization of certain perishable products containing plant oils. In the food industry, wine stabilization is the main application of laccase. Polyphenols have undesirable effects on wine production and on its organoleptic characteristics, so their removal from the wine is very important. Many innovative treatments, such as enzyme inhibitors, complexing agents, and sulfate compounds, have been proposed for the removal of phenolics responsible for discoloration, haze, and flavor changes but the possibility of using enzymatic laccase treatments as a specific and mild technology for stabilizing beverages against discoloration and clouding represents an attractive alternative. Since such an enzyme is not yet permitted as a food additive, the use of immobilized laccase might be a suitable method to overcome such legal barriers as in this form it may be classified as technological aid. So, laccase could find an application in preparation of must and wine, and in fruit juice stabilization [8].

3.3.5.2 Baking Industry

In the bread-making process the laccase enzyme affixes bread and dough enhancement additives to the bread dough, these result in improved freshness of the bread texture and flavor. Laccase, when added in the baking process, results in an oxidizing effect, and improves the strength of structures in baked products. Laccase imparts many characteristics to the baked products including an improved crumb structure, increased softness, and improved volume [26].

3.4 Conclusion

The laccase enzymes are receiving much attention from researchers around the globe due to their specific nature. The interest in utilizing laccases for many biotechnological and industrial applications has increased rapidly since the discovery of this enzyme in white-rot fungi. Emerging technologies include selective delignification in the production of cellulosics in pulp bleaching, biosensors, and biofuel, and in the treatment of environmental pollutants and toxicants generated in various industrial processes. Therefore, laccases have been widely studied for various applications, including the bioremediation of soil and contaminated effluents as well as their use in biosensors. This chapter shows that laccase has a great potential application in environmental protection. However, much more research is required to make use of laccases to protect the environment and in other industrial applications.

References

1 Illanes, A., Cauerhff, A., Wilson, L. et al. (2012). Recent trends in biocatalysis engineering. *Bioresource Technology* 115: 48–57. https://doi.org/10.1016/j.biortech.2011.12.050.

2 Nigam, P.S. (2013). Microbial enzymes with special characteristics for biotechnological applications. *Biomolecules* 3 (3): 597–611.

3 Gurung, N., Ray, S., Bose, S. et al. (2013). A broader view: microbial enzymes and their relevance in industries, medicine, and beyond. *Biomed Research International* 2013: 329121.

4 Rodríguez Couto, S. and Herrera, J.L.T. (2006). Industrial and biotechnological applications of laccases: a review. *Biotechnology Advances* 24 (5): 500–513.

5 Gianfreda, L., Xu, F., and Bollag, J.M. (1999). Laccases: a useful group of oxidoreductive enzymes. *Bioremediation Journal* 3 (1): 1–25.

6 Thurston, C.F. (1994). The structure and function of fungal laccases. *Microbiology* 140: 19–26.
7 Arora, D.S. and Sharma, R.K. (2010). Ligninolytic fungal laccases and their biotechnological applications. *Applied Biochemistry and Biotechnology* 160: 1760–1788.
8 Imran, M., Asad, M.J., Hadri, S.H. et al. (2012). Production and industrial applications of laccase enzyme. *Journal of Cell and Mollecular Biology* 10 (1): 1–11.
9 Sharma, P., Goel, R., and Caplash, N. (2007). Bacterial laccases. *World Journal of Microbiology and Biotechnology* 23: 823–832.
10 Brijwani, K., Rigdon, A., and Vadlani, P.V. (2010). Fungal laccases: production, function, and applications in food processing. *Enzyme Research* 2010: 149748.
11 Bugg, T.D., Ahmad, M., Hardiman, E.M. et al. (2011). The emerging role for bacteria in lignin degradation and bio-product formation. *Current Opinion in Biotechnology* 22 (3): 394–400.
12 Harms, H., Schlosser, D., and Wick, L.Y. (2011). Untapped potential: exploiting fungi in bioremediation of hazardous chemicals. *Nature Reviews in Microbiology* 9 (3): 177–192.
13 Reiss, R., Ihssen, J., and Thöny-Meyer, L. (2011). Bacillus pumilus laccase: a heat stable enzyme with a wide substrate spectrum. *BMC Biotechnology* 11: 9.
14 Kunamneni, A., Plou, F.J., Ballesteros, A. et al. (2007). Laccases and their applications: a patent review. *Recent Patents on Biotechnology* 2008 (2): 10–12.
15 Bressler, D.C., Fedorak, P.M., and Pickard, M.A. (2000). Oxidation of carbazole, N-ethylcarbazole, fluorene, and dibenzothiophene by the laccase of Coriolopsis gallica. *Biotechnology Letters* 22 (14): 1119–1125.
16 Pointing, S.B. (2001). Feasibility of bioremediation by white-rot fungi. *Applied Microbiology and Biotechnology* 57 (1–2): 20–33.
17 Niku-Paavola, M.-L. and Viikari, L. (2000). Enzymatic oxidation of alkenes. *Journal of Molecular Catalysis B* 10 (4): 435–444.
18 Viswanath, B., Rajesh, B., Janardhan, A. et al. (2014). Fungal laccases and their applications in bioremediation. *Enzyme Research* 2014: 1–21.
19 Banat, I.M., Nigam, P., Singh, D. et al. (1996). Microbial decolorization of textile-dye-containing effluents: a review. *Bioresource Technology* 58: 217–227.
20 Zollinger, H. (2002). *Synthesis, Properties and Applications of Organic Dyes and Pigments*. New York;: Wiley.
21 Setti, L., Giuliani, S., Spinozzi, G. et al. (1999). Laccase catalyzed-oxidative coupling of 3-methyl 2-benzothiazolinone hydrazone and methoxyphenols. *Enzyme and Microbial Technology* 25 (3–5): 285–289.
22 Raghukumar, C. (2000). Fungi from marine habitats: an application in bioremediation. *Mycological Research* 104 (10): 1222–1226.
23 Bajpai, P. (1999). Application of enzymes in the pulp and paper industry. *Biotechnology Progress* 15 (2): 147–157.

24 Kuznetsov, B.A., Shumakovich, G.P., Koroleva, O.V. et al. (2001). On applicability of laccase as label in the mediated and mediatorless electroimmunoassay: effect of distance on the direct electron transfer between laccase and electrode. *Biosensors and Bioelectronics* 16 (1–2): 73–84.

25 Fogel, R. and Limson, J.C. (2013). Electrochemically predicting phenolic substrates suitability for detection by amperometric laccase biosensors. *Electroanalysis* 25: 1237–1246.

26 Minussi, R.C., Pastore, G.M., and Duran, N. (2002). Potential applications of laccase in the food industry. *Trends in Food Science and Technology* 13: 205–216.

4

Genome Assessment

Functional Gene Identification Involved in Heavy Metal Tolerance and Detoxification

Uttara Mahapatra[1], *Ayantika Pal*[2], *Ajay Kumar Manna*[1], *and Dijendra Nath Roy*[3]

[1] *Department of Chemical Engineering, National Institute of Technology, Agartala, Tripura, India*
[2] *Department of Zoology, Women's College, Tripura University, Agartala, Tripura, India*
[3] *Department of Bio Engineering, National Institute of Technology, Agartala, Tripura, India*

4.1 Introduction

Bioremediation is a novel technique for the treatment of heavy metal polluted environments (soil and water), which involves using living organisms (microorganisms and plants) to reduce the effects of toxic heavy metals turning them into less harmful residues. This method has a huge scope to substitute the conventional methods because of its cost effectiveness and sustained ecological balance. Bioremediation processes can work in natural process or be encouraged by the addition of electron acceptor, nutrient, or other factors in the process [1]. In this framework, it is worth mentioning that sulfate-reducing bacteria assist the conversion of sulfate to sulfide with the sulfides reacting with heavy metals to precipitate toxic metals as metal sulfide. These metal sulfides are stable and can easily be isolated from the waste [2]. Further, bacteria used for bioremediation are selected to target a specific form and oxidation state of toxic pollutants, such as reduction of soluble hexavalent uranium or degradation of a specific hydrocarbon. Significantly, soil and groundwater rarely possess a unique chemical as contamination [3]. Bioremediation through microbial resources facilitates reduction of heavy metal pollution, otherwise becomes a major concern in aquatic and soil ecosystems due to its nonbiodegradability, bioaccumulation, and biomagnifications through the ecological food chain, resulting in ecological and environmental adverse effects. Various management procedures to control heavy metal pollution, such as chemical precipitation, oxidation, or reduction, filtration, reverse osmosis,

Soil Bioremediation: An Approach Towards Sustainable Technology, First Edition.
Edited by Javid A. Parray, Abeer Hashem Abd Elkhalek Mahmoud, and Riyaz Sayyed.

membrane technology, ion exchange, evaporation, and electrochemical treatment, are used to diminish or eradicate heavy metals in the environment. However, bioremediation is also a cost-effective and environmentally friendly approach with less formation of byproducts compared with earlier mentioned physicochemical methods. Thus, bioremediation is considered and accepted as an alternative treatment technology of heavy metal waste. Microbial bioremediation is being investigated as a preferred method for treating this contamination, where current technologies are inadequate. In order for bioremediation to be effective, the operative microorganisms must grow, function, and compete with other bacteria in the presence of all contaminants. The contamination of toxic metals to soil affecting human health directly or indirectly leads to great economic losses [4]. The activities of metals in soil always make them challenging materials to decontaminate as they may create stable complexes with naturally available substances, including binding to the soil components, and become precipitated as insoluble mineral forms. Any kind of soils naturally possess trace levels of metals; however, when this level exceeds the tolerable limit of living organisms, it registers as pollution. In soils, metals may be present in a dissolved status or as insoluble particles or precipitate as pure or mixed substances in soil solution [5].

As a result, bioremediation is an ecofriendly and resourceful technique to decontaminate environments from heavy metals by bringing into play of the intrinsic biological approaches of microorganisms and plants to eliminate harmful contaminants. This chapter highlights microbial remediation of toxic heavy metal pollution in the environment. It also emphasizes the magnitude of biotechnological practice applied in improving the capability of microbial resources to effectively breakdown heavy metals at a faster rate. In addition to this, genetic regulation is also addressed to identify gene involvement in bioremediation process as well as opportunities for its future use and its limitations. However, strict observance of biosafety regulations should be pursued in the exploitation of biotechnological practice to ensure safety of the environment.

4.2 Tolerance and Degradation Mechanisms of Toxic Metals by Microorganisms

4.2.1 Cadmium (Cd)

Rhodobacter sphaeroides was applied for the biodegradation of Cd-contaminated wastewater [6]. The authors studied microbial tolerance capacity on pH, temperature, cadmium, and inoculum mass for removal rate. An EC_{50} value of 5.34 mg/l for Cd was reported in this study. The optimal pH was 7.0 and the optimal temperature was 35–40 °C. The removal rate for Cd by *R. sphaeroides* was 97.92% and it was found that precipitation of heavy metal and biosorption were responsible for bioremediation. In another experiment on Cd(II) biosorption, living and nonliving cells of *Bacillus cereus* RC-1 were used [7]. Nonliving cells have higher uptake efficiency as compared to living cells. The maximum uptake capacities obtained from Langmuir isotherm for living cells (24.01 mg/g) and nonliving cells (31.95 mg/g) were determined under specific initial Cd concentration, pH, and contact time. It was also indicated that the mechanism of bioaccumulation was more dependent on extracellular accumulation than intracellular biosorption.

Galedar and Younesi [8] investigated Cd removal using ethanol pretreated *Saccharomyces cerevisiae*. These cells were immobilized by means of polysulfone in batch and continuous methods [8]. The results showed that maximum adsorption capacity of Cd by batch system (optimum conditions pH 8.0 with initial ion and biomass concentrations of 100 and 8 mg/l, respectively) was 3.1 mg/g and for a continuous system (pH 8.0 with initial ion dose of 100 mg/l, flow rate of 1 ml/min, and column height of 25 cm) was 3.74 mg/g. Moreover, cell immobilization increases the removal rate by 48% in comparison to nonimmobilized ethanol pretreated cells. Hasan et al. [9] successfully utilized free and immobilized forms of *Arthrobacter* sp. as biosorbent for Cd(II) removal in batch and continuous processes from water [9]. Bioadsorption capacity by employing the former was 270.27 mg/g while with latter 188.67 mg/g, respectively. A maximum Cd(II) removal percentage of 88.9% was achieved at 1.0 ml/min of flow rate and 19 cm of bed height using a column system. Additionally, optimization in the same process conditions was performed for maximum percent removal by using a central composite design. The values for experimental data (88.9%) and optimization plots (89.64%) were approximately similar.

In another study, biosorption of Cd(II) on *Geobacillus toebii* subsp. *Decanicus* (G1) and *Geobacillus thermoleovorans* subsp. *Stromboliensis* (G2) in a batch system was examined [10]. For G1 and G2, equilibrium uptake capacity was determined by variable solution pH 6.0 and 4.0, respectively and rest of the parameters such as temperature 70 °C, contact time 60 minutes, biomass dosage 50 mg, initial metal dosage 50 mg/l, remained the same. The maximum adsorption capacity for G1 and G2 were found to be 29.2, mg/g and 38.8, mg/g, respectively. The same authors Özdemir et al. [11] employed the aforementioned thermophilic bacteria, G1 and G2 to study the resistance of heavy metals such as Cd(II), Cu(II), Ni(II), Zn(II), and Mn(II). The order of metal resistance from most sensitive to most resistance was found to be Cd˂Zn˂Cu˂Ni˂Mn for both the microorganisms. They further concluded that dead cells adsorb more than live cells. The Cd adsorption efficiency by live and dead cells was found to be 17.44 mg/g and 46.2 mg/g of cells, respectively in 7.32 mg/l Cd solution. In another experiment, Özdemir et al. [12] used two thermophilic strains of bacteria, *Anoxybacillus amylolyticus and Geobacillus thermantarcticus* to adsorb Cd(II) using different variables. Percent Cd(II) removal at optimum parameters of 50 mg/l of Cd(II) in 60 minutes by 50 mg of *A. amylolyticus* (pH 5.0) and *G. thermantarcticus* (pH 4.0) were 74.1 and 85.4, respectively at 60 °C [12].

Bravo et al. [13] measured the total content of Cd soil and isolated a microorganism that showed tolerance toward cadmium namely *Enterobacter* sp. CdDB41 from the cacao soils obtained from Colombia (Northeastern). The maximum growth uptake as shown by this strain CdDB41 was found to be 2.21 and 2.32 J at a $CdCl_2$ concentration of 6 mg/l and 24 mg/l [13].

Oves et al. [14] examined metal sorption ability of *Bacillus thuringiensis* strain OSM29 which was recovered from cauliflower grown in industrial wastewater. The metal removal was detected by using different concentrations of Cd from 25 to 150 mg/l. The optimum physico-chemical parameters shown by the strain OSM29 was at pH 6.0, temperature 32 °C, and contact time 30 minutes. It was found that strain OSM29 has the potential to adsorb 87%, 25 mg/l of initial Cd concentration. Liu et al. [15] isolated a Cd-resistant microbial strain, *Delftia* sp. B9 from soil contaminated with Cd and applied it in the remediation of paddy soils. Soil and pot experiments were conducted with bacterial inoculum to see its effect on Cd soil speciation, Cd accumulation, and rice grain biomass. It was demonstrated by TEM-EDS (Transmission Electron Microscopy-Energy-Dispersive Spectroscopy) that Cd bioaccumulation took place within the cell. It was found that concentration of Cd in rice was below the 0.2 mg/kg permissible limit as set by the Food Safety of People's Republic of China. Therefore, *Delftia* sp. B9 successfully stabilized Cd in soil and reduced accumulation of Cd in rice grain.

In a very recent study by Vishan et al. [16] employed a batch system for investigating Cd(II) biosorption of *Bacillus badius* AK. The optimum physico-chemical parameters for biosorption were measured at pH (7.0), contact time (30 minutes), initial biomass dosage (2 g/l), temperature (40 °C), initial Cd (II) concentration (100 mg/l), and agitation (150 rpm). An uptake capacity of 131.58 mg Cd/g of biomass and monolayer adsorption by *B. badius* was attained. Hence, these bacteria can be applied fruitfully for the treatment of wastewater commercially. *Burkholderia cepacia* GYP1 was collected from farmland contaminated with various heavy metals and was used to study its Cd accumulation under oligotrophic conditions processes [17]. The authors performed a detailed cellular study on the biosorption mechanism for 7 days by using flow cytometry, ATR-FTIR (Attenuated total reflectance - Fourier-transform infrared spectroscopy), and electron microscopy. It was demonstrated that on day one, Cd accumulation takes place on the outer membrane, and after the second day intracellular Cd concentration continued increasing and then attained a stabilizing point. An increasing Cd amount was mostly present extracellularly resulting in the release of extracellular polymeric substances responsible for protecting the bacteria against heavy metals. A Cd accumulation of 116 mg/g dry weight of biomass was achieved in this process. Sakpirom et al. [18] explored *Rhodopseudomonas palustris* TN110, a nonsulfur purple bacterium, having ability for Cd bioremediation as well as nitrogen fixation. Bioremediation through strain TN110 was done by synthesizing a spherical form of around 4.85 nm size cadmium sulfide (CdS) nanoparticles. Under an optimal pH of 7.5, temperature at 30 °C, and illuminance of 3000 lx, this strain has the ability to convert 25.61% of 0.2 mM $CdCl_2$ into CdS nanoparticles, which have an IC_{50} value of 1.76 mM. Additionally, these nanoparticles also upregulate two genes, i.e., V-Fe (*vnfG*) and Mo-Fe (*nifH*) nitrogenase gene, which are responsible for nitrogen fixation but they also downregulate Fe-Fe nitrogenase genes (*anfG*) [18].

4.2.2 Chromium (Cr)

Sediment samples from Xiamen China containing 55 strains and one isolated strain that had an ability to tolerate hexavalent chromium were collected. One strain, *Sporosarcina saromensis* M52 has a huge potential for Cr(VI) bioremediation. Zhao et al. [19] elucidated the removal results of strain M52 at two different concentrations of Cr(VI) in 24 hours at 35 °C: 50–200 mg/l at pH 7.0–8.5 and 100 mg/l at pH 8.0 (i.e., removal efficiency of 100%). Therefore, *Sp. saromensis* M52 can be a useful tool for the remediation of toxic and harmful metals [19]. In another experiment, a novel actinobacterial isolate was discovered from the feces of chicken and goat and characterized as *Streptomyces werraensis* LD22. This strain can detoxify, resist, and adsorb Cr(VI). The results revealed that 95% resistance was showed by LD22 at a maximum level of 250 mg/l of Cr(VI) (added in the form of $K_2Cr_2O_7$). Furthermore, the maximum Cr(VI) uptake capacity at pH 7.0 by *St. werraensis* LD22 was found at an initial metal concentration of 100 mg/l with biomass dose of 3 g/l [20].

Yahya et al. [21] made use of a nonviable bacterium, *Acinetobacter haemolyticus* for treating Cr(III) present in synthetic and industrial effluent. For synthetic wastewater, under maintaining specific conditions (pH-5.0, initial Cr(III)-100 mg/l, biomass dose-15 mg of dried weight biomass, and contact time-30 minutes), maximum removal of 198.80 mg/g of Cr(III) was achieved. R^2 values for both equilibrium models (Langmuir and Freundlich) and kinetic model (pseudo second) were greater than 0.95. A microbial biomass recovery of 90% was obtained from desorption studies using 1 M of H_2SO_4, HNO_3, or CH_3COOH. For raw tannery wastewater, *A. haemolyticus* has a removal efficacy of 79.87 mg/g of Cr(III) [21]. Sandana Mala et al. [22] isolated a novel bacterium, *Bacillus methylotrophicus* from tannery sludge for the reduction of chromate (transforming Cr(VI) to Cr(III)), which is facilitated by enzyme chromate reductase. Enzyme activity of 212.84 U/mg (initial) and 312.99 U/mg (inducible) at 48 hours was attained by a low-cost medium with the addition of 0.25 mM chromate. Further, it was investigated that the activity of enzyme increases to 356.48 U/mg in the presence of reduced glutathione. Optimal conditions influencing the enzyme activity were pH 7.0 and temperature 30 °C. The V_{max} and K_m values for this enzyme were 59.89 M/min/mg protein and 86.5 M, respectively [22].

The biosorption of Cr(VI) by a novel exopolysaccharide purified from *Arthrobacter* ps-5 was studied. The effect of pH, initial EPS concentration, and ionic strength of different salts were simultaneously studied on metal adsorption. At pH 6.0 there was an uptake capacity of 84.47 mg/g Cr(VI), increasing the exopolysaccharide dosage from 0.4 to 1.5 g/l resulted in a biosorption capacity of 31.57–70.49 mg/g, and adding $CaCl_2$ and KCl salts of 1.0–4.0 mmol/l decreased the biosorption potential in the order of $K^+ < Ca^{2+}$ [23]. Soil samples from the electroplating industry at Coimbatore, India were exploited for isolating *Bacillus subtilis* SS-1, a bacterium resistant to chromium. The optimum parameters at which batch tests were conducted were: pH at 2.0, time 6 hours, initial metal concentration of 100 mg/l, and biomass dose of 0.1 g/l – 98.7% Cr(VI) removal was achieved by this process [24].

An optimization experiment had been carried out indirectly on a biosorption system by employing *Arthrobacter viscosus* anchored on NaY zeolite [25]. The aim of this study was to optimize the removal and reduction of Cr(VI). The results suggested that higher pH values of the solution favored removal of total chromium whereas lower pH enhanced reduction of Cr(VI). After optimizing the studied parameters it was found that zeolite has a maximum chromium content of 0.9% and highest uptake of 13.0 mg Cr/g of zeolite was attained at pH 4.0, Cr(VI) concentration of 100 mg/l, and biomass dosage of 5 g/l. Chromium removal from wastewater is an important issue of ecological concern [26]. This investigation shows the utilization of *Kocuria rhizophila* 14ASP (endophytic bacterium) isolated from *Oxalis corniculate* (hyperaccumulator) for Cr(III) bioadsorption from aqueous solution. This strain shows minimum inhibitory concentration at 8 mM of Cr. Effects of specific parameters like pH, initial metal concentration, and contact time on biosorption was assessed at around 35 °C. The ideal pH was 4.0 and contact time achieved after 60 minutes for Cr(III) biosorption. Through the Langmuir model, maximum uptake capacity for Cr was 14.4 mg/g. *Mesorhizobium amorphae* CCNWGS0123 was explored as a biosorbent to remove chromium. The highest adsorption capacity at optimal pH, for Cr(III)-4.0 and Cr(VI)-2.0, initial chromium ions of 100 mg/l, and biomass 1.0 g/l were 53.52 mg/l for Cr(III) and 47.67 mg/l for Cr(VI). Thus, Cr(III) has higher adsorption efficiency than hexavalent chromium *M. amorphae* [27].

4.2.3 Lead (Pb)

The purified forms of extracellular polymeric substances specifically polysaccharide, protein, biosurfactant, and DNA (ssDNA and dsDNA) from *Pseudomonas aeruginosa* were studied for their Pb(II) biosorption limits [28]. For optimization of lead adsorption, Pb(II) concentration and extracellular polymeric components were mainly focused and upgraded. The uptake efficiency by polysaccharide, protein, biosurfactant, ssDNA, and dsDNA were 0.48, 4.8, 0.041, 0.42, and 0.30 mg/mg of the respective substances. From the results, it was concluded that protein has the highest adsorption capacity and thus can be used for the biosorption of Pb(II). In a very recent experiment performed by Xing et al. [29] on bioaccumulation of Pb(II) from *Bacillus coagulans*, strain R11 collected from a lead mine have been investigated. This study was specifically performed to check the initial Pb(II) adsorption effect on three morphologically different structures of *B. coagulans* R11 namely spores, vegetative, and decay cells. The Pb(II) removal capacity for spores, vegetative, and decay cells in mg/g were 14.86, 17.53, and 17.41, respectively [29].

Kalita and Joshi [30] isolated a thermophilic bacteria *Pseudomonas* sp. W6 from a hot water source located at Northeast India for evaluating the biosorption capacity of lead. This bacterium can resist up to 1.0 mM of lead. Mass spectroscopy outcome revealed that percentage biosorption capacity for batch and column was 65 and 61.2, respectively [30]. *B. subtilis* FZUL-33 was isolated for the mineralization of Pb(II) from aqueous solutions [31]. An equilibrium uptake capacity of 381.31 ± 11.46 mg/g for *B. subtilis* FZUL-33 was determined at 25 °C, pH 5.5, contact time 10 minutes, and initial Pb(II) concentration 1300 mg/l isolated *R. sphaeroides* from water injected in the oil field and utilized it for bioremediation purpose for the removal of lead from soil [32]. Optimum pH, temperature, and inoculum size for this bacterial species were 7.0, 30–35 °C, and 2×10^8/ml, respectively. Moreover, growing wheat seeds in this contaminated soil showed phytoavailability of Pb was lessened in the soil. The efficiency for Pb bioremediation on wheat root and leaf was 14.78 and 24.01%, respectively. It was suggested that precipitation of inert compounds such as lead sulfide and lead sulfate is the main mechanism by which *R. sphaeroides* treats the lead-contaminated soil.

Many researchers have used optimization as a mean to support their experiment [33]. One such study has been done on *Klebsiella* sp. 3S1 for the remediation of Pb. The authors examined the bioremediation process in aqueous solution by applying Rotatable Central Composite Experimental Design. The ideal conditions for attaining this experiment were pH 5.0, temperature 34 °C, and a biosorbent dose of 0.4 g/l. Through the Langmuir model, an uptake rate of 140.19 mg/g was achieved. Jin et al. [34] carried out optimization procedures for *Arthrobacter* sp. Of strain 25, which has the potential to tolerate Pb(II). Response surface methodology based on the Box–Behnken model was used for optimizing the parameters. The adsorption rate obtained from this experiment was 9.6 mg/g (at pH 5.75, biosorbent dose 9.9 g/l, and initial Pb(II) concentration of 108.79 mg/l). This outcome signifies that *Arthrobacter* sp. 25 can be employed for treating a low level of Pb(II) concentration in wastewater [34]. Abu Hasan et al. [35] also performed a similar study based on the Box–Behnken design for lead removal from aqueous solutions using *B. cereus* I6. The best adsorption capacity was at pH 7.0 and by increasing the Pb level from 10 to 50 mg/l and biomass amount from 0.3 to 1.5 g/l, the metal uptake rate also increased significantly [35].

Mwandira et al. [36] novelly employed *Pararhodobacter* sp. Along with microbially induced calcium carbonate precipitation for the remediation of lead. A removal rate of 1036 mg/l Pb(II) was achieved at an incubation period of 6 hours and an increase in pH from 8.0 to 9.1. This method also proved to be an environmentally friendly and sustainable process for removing lead [36]. Wei et al. [37] adsorbed Pb(II) indirectly on extracellular polymeric substances purified from *Klebsiella* sp. J1. Pb(II) biosorption at pH 6.0 and extracellular polymeric substances dose of 0.2 g/l was 99.5 mg/g. These data significantly fitted both the Langmuir isotherm equilibrium model and the pseudo-second kinetic model [37].

4.2.4 Zinc (Zn)

Li et al. [6] employed *R. sphaeroides* anaerobically for the bioremoval of Zn from polluted wastewater. Zn was added in the form of $Zn(NO_3)_2$ under various concentrations from 0, 20, 50, 100, and 150 mg/l. For Zn(II), EC_{50} value of 69.79 mg/l was reported. Various parameters with respect to pH of 5.0, 6.0, 7.0, 8.0, and 9.0, temperature of 25, 30, 35, 40, and 45 °C, and inoculum size of (1, 2, 3, 4, and 5)$\times 10^8$ m/l were studied during the removal process. pH and temperature have a greater impact on the adsorption capacity than inoculum size. An optimal pH and temperature were found to be 7.0 and 35–40 °C, respectively. This process achieved a 97.76% removal of Zn. Özdemir et al. [10] used a batch stirring reactor for the biosorption of Zn(II) in the form of $ZnSO_4$ on two *Geobacillus* subsp. Namely, *G. decanicus* (G1) and *G. stromboliensis* (G2). Various dosages of powdered biomass (0.25–10 g/l) was exposed for 15–120 minutes on different initial Zn(II) concentrations (0.01–0.3 g/l) at 30–80 °C. As per the Langmuir isotherm, adsorption capacity in mg/g of Zn(II) for G1 and G2 were 29 and 21.1, respectively. The mechanism of biosorption by dried cells of G1 and G2 was examined by FT-IR spectra.

A comparative study has been made by Magnin et al. [38], studying Zn biosorption by employing two *Rhodobacter capsulatus* strains, i.e., B10 (wild-type strain) and RC220 (strain lacking endogenous plasmid) under different chemical, physical, and biological parameters. These bacterial consortia were grown in synthetic media utilizing lactate as its carbon source. At pH 7.0, initial Zn dose 10 mg/l, and contact time 30–120 minutes, the zinc adsorption by this bacterial biomass was 17 and 16 mg Zn/g dry weight of biomass, respectively. The Zn uptake ability of B10 and RC220 attained a maximum value of 164 ± 8 and 73.9 mg/g dry weight, respectively. Thus, it was concluded that *R. capsulatus* B10, having a higher uptake value could be used in the remediation of zinc from the environment. Furthermore, future studies are needed to stury the process of *R. capsulatus* RC220, which was undergoing adsorption without having any genetically embedded zinc resistance plasmid located in it. Limcharoensuk et al. [39] isolated three microbial consortiums from a zinc mine located in Tak Province of Thailand. These bacteria were named as *P. aeruginosa* B237, *Cupriavidus taiwanensis* E324, and *Tsukamurella paurometabola* A155. Bioadsorption (physical adsorption) rather than bioaccumulation was found to be a suitable mechanism for the adsorption of zinc on the cell walls of bacteria. A maximal adsorption ability of 16.89 and 16.75 mg/g was attained by *P. aeruginosa* B237 and *T. paurometabola* A155, respectively.

4.2.5 Nickel (Ni)

Gheethi et al. [41] isolated two microorganisms namely *Bacillus megaterium*1295S and *Sporosarcina pasteurii* 586S from sewage sludge for removing Ni from aqueous solution. The removal efficiency process was examined with respect to biomass concentrations (0.1–3 mg/l), time (1–10 hours), pH (2.0–8.0), temperature (20–55 °C), and Ni (0.1–0.8 mg/l). The uptake capacities of living and dead cell biomass were compared. Percentage removal capacities for dead cells were greater than live cells for both strains. Dead biomass of *B. megaterium*1295S showed more efficiency than *Sp. Pasteurii* 586S for removing Ni(II) at 196.4 and Ni(II) 200.2 mg/g [40]. Ni(II) biosorption from the industrial effluent by an economical, greener, and dead strain *Streptomyces roseorubens* SY was studied. The optimum parameters for the adsorption of Ni(II) were evaluated at pH 5.0, contact time 1 hour, agitation speed 150 rpm, biomass dosage 1.0 g/l, and Ni(II) concentration 100 mg/l. A highest uptake capacity of 208.39 mg of Ni(II)/g was obtained by Langmuir isotherm. Sathyavathi et al. [42] isolated a bacterial stain from electroplating industrial wastewater comprising of nickel. This strain has been named as *Microbacterium* sp. MRS-1 and has the ability to transform $NiSO_4$ (soluble) to NiO (insoluble) nanoparticles, light green in color. When 2000 mg/l of $NiSO_4$ is incubated for a prolonged time, a green color precipitate of NiO nanoparticles is observed in the bottom flask, which has a distinctive floral structure and a size of 100–500 nm. This strain, MRS-1, is applied to the effluent by the industry itself. A 95% removal of Ni was observed. Thus, this strain has a bioremediation potential and can be explored as an effective tool for removing nickel from wastewater as well as synthesizing nanoparticles. Calcite precipitation induced by microorganisms, i.e., MICP (in present study *B. cereus* NS4) is a new and productive technology in the area of heavy metals remediation because of its large surface area and porosity whereas char produced from the Cinnamomum leaves can be utilized as an effective adsorbent [43]. In this first ever study, a combinatorial experiment has been performed by the addition of biochar in a Ni(II) solution and simultaneously treating that Ni(II) using MICP method. The results suggested that at 50 and 100 mg/l of initial concentrations, the percent immobilizations of Ni(II) were about 89 and 66, respectively using this MICP process. Conversely, biochar inhibits the biocalcite by making the adsorption bonds weaker and thus leads to suppression of the nickel remediation process.

Another researcher carried out experiments on synthetic solutions resembling actual wastewater containing nickel [44]. The use of *Rhodococcus opacus* for the removal of Ni(II) was investigated in this study under various conditions. It was revealed that *R. opacus* at pH 5.0 has an uptake capacity and removal efficiency of 7.63 mg/g and 92% for 5 mg/l of Ni(II). Moreover, activation energy for this biosorption was around 12.56 kJ/mol. In another study biosorption of Ni(II) on *G. toebii* subsp. *Decanicus* (G1) and *G. thermoleovorans* subsp. *Stromboliensis* (G2) in a batch system was examined by Özdemir et al. [10]. For G1 and G2, equilibrium uptake capacity was determined by variable solution pH 6.0 and 4.0, respectively and the rest of the parameters such as temperature 70 °C, contact time 60 minutes, biomass dosage 50 mg, initial metal dosage 50 mg/l, remained the same. The maximum Ni(II) adsorption capacity for G1 and G2 were found to be 21 mg/g and 42 mg/g, respectively.

4.2.6 Copper (Cu)

In a very recent work performed on the samples collected from a copper mining site, the bioadsorption property of *Rhodococcus erythropolis* for Cu(II) was investigated [45]. Under various metal and biomass concentration, and pH, the uptake capacity for Cu(II) was analyzed by this bacterium. A total adsorption capacity by *R. erythropolis* of 68.03 mg Cu(II)/g biomass was achieved in this process by considering the optimal pH at 6.0. Letnik et al. [46] has exploited the capacity of *Micrococcus luteus* to sequester copper. At 370 mg/l of Cu(II), only 10% of these cells survived. At pH 6.0 a sorption ability of 59 mg Cu(II)/g dry cells was obtained. The mechanism by which *M. luteus binds* copper has been tested through three arrangements: nonimmobilized cells, and immobilized cells in electrospun polymer composites and immobilized cells in alginate. It was found that after immobilization, the Cu(II) uptake capacity achieved was up to 76 mg/g [46].

Four copper resistant bacterial strains of *B. cereus* (CURB-2, CURB-4, CURB-6, and CURB-8) have been isolated, characterized, and screened from water and residual samples collected from a coastal area of Tamil Nadu and tested in batch culture under laboratory conditions [47]. These strains were subjected to 100 mg/l of Cu (II) solution at pH 7.0 for 24 hours. The removal elucidated by strains CURB-2, 6, and 8 and CURB-4 were 80% and 99.9%, respectively. Liu et al. [48] studied the applicability of immobilization on *B. subtilis* for the removal of Cu(II). The bacteria were immobilized inside chitosan beads. Batch experiments were performed under suitable and specific conditions for comparing the functionality of chitosan beads with and without microbial biomass. In all respects, the data concluded that immobilized biochitosan beads have a better performance and Cu(II) adsorption efficiency due to the presence of functional groups that permits the cell wall to tightly bind with the metal ions than compared to only chitosan beads. The optimum pH attained was at 6.0 and adsorption efficiency was 100.70 mg/g.

In a unique work carried out by Pan et al. [49] biosorption of Cu(II) on extracellular polymeric substances (EPS) of *Synechocystis* sp. Was examined under fluorescence spectroscopy. Three groups of fluorophores were observed, i.e., one protein and two humic-like fluorophores. The result revealed that Cu(II) quenches these three fluorophores specifying Cu(II)-EPS formation. Moreover, humic substances of EPS have more binding capability with Cu(II) than the protein one. Toxic wastewater released from the mining industry contains a huge quantity of pollutants such as heavy metals [50]. In current work, the authors have given attention to finding a cleaner and greener approach for the removal of the heavy metal copper. The biosorbent used for the sorption of Cu(II) in this research was dead cells of *M. amorphae* CCNWGS0123. The maximum removal of Cu(II) was at optimum conditions pH 5.0, temperature 28 °C, and initial Cu(II) dosage of 100 mg/l. Highest sorption was seen at 0.5 g/l Cu(II). Özdemir et al. [10] used a batch stirring reactor for the biosorption of Cu(II) in the form of $CuCl_2 \cdot 2H_2O$ on two *Geobacillus* subsp. namely, *G. decanicus* (G1) and *G. stromboliensis* (G2). The various dosage of powdered biomass (0.25–10 g/l) was exposed for 15–120 minutes on different initial Cu(II) concentrations (0.01–0.3 g/l) at 30–80 °C. As per the Langmuir isotherm, adsorption capacity in mg/g of Cu(II) for G1 and G2 were 48.5 and 41.5, respectively. The mechanism of biosorption by dried cells of G1 and G2 were examined by FT-IR (Fourier-transform infrared spectroscopy) spectra.

4.2.7 Aluminum (Al)

A microbial strain from an explosive contaminated site has been isolated [51]. This strain, *Bacillus safensis* KTSMBNL 26 has the ability to adsorb aluminum under optimum conditions of pH, contact time, and temperature. In this experiment, the researchers grew the cells in a batch for 24 hours at pH 6.0 and temperature 35 °C with initial Al dosage of 100 mg/l. The removal efficiency of 92 mg/l was achieved in the batch process. However, this efficiency decreases slightly with immobilization due to a longer lag phase for adsorption, 84 mg/l. However, for commercial treatment of wastewater the immobilization process is preferred over the batch system because of higher cell density, less expense in processing, and reuse.

Huang et al. [52] isolated a novel aluminum tolerant acidophilic strain, *Burkholderia* sp. SB1 (gram-negative, rod shape) from the red soil, which is acidic in nature [52]. This bacterium has the ability to survive in extreme acidic conditions of pH 2.2 and can tolerate Al(III) up to 270 mg/l. It was validated through laboratory experiments that at a concentration of 54 mg/l of Al(III) almost 97.7% can be removed. Moreover, this bacterium can function under variable temperatures from 25 to 37 °C and exhibit resistance to different antibiotics such as ampicillin, tetracycline, and streptomycin but was sensitive to chloramphenicol. As per the report, it is the first ever bacteria to have such excellent aluminum tolerance and resistance capacity. In the future, these bacteria can be exploited for bioremediation purpose.

Twenty-three heavy metal resistance microbes were isolated from water and soil contaminated with aluminum [53]. Based on the biochemical and morphological analysis only *Providencia rettgeri* MAM-4 has a strong ability to remove Al(III). At 200 mg/l of Al(III), the uptake efficiency of this bacteria increases four-fold in comparison to another strain, *B. cereus* ATCC 11778. It was also found that inoculating *P. MAM*-4 with clay added a significant enhancement in Al(III) removal and the mutant strain of this bacterium was able to remove more aluminum than the parent strain.

4.2.8 Arsenic (As)

A team of researchers from China utilized a chemoautotrophic and acidophilic bacterium strain, *Acidithiobacillus ferrooxidans* BY3 for studying the adsorption and transformation of arsenic biologically [54]. It was observed that the pH, Eh, Fe(II) oxidation rates, and density of bacteria expressively affected by the arsenic stress that is present in Fe(II) and a minute level of $NaAsO_2$ imparts energy for nourishment of *A. ferrooxidans* BY3. High As(III) also affected the structure of bacteria cell membrane composition by decreasing the volume, and vibration patterns of surface groups such as $—CH_2$, —OH, and —NH on the cell membrane. Statistical data showed a 10.07 fold increase in the conversion rate of As(III) to As(V) in the growth media comprising 1600 mg/l of $NaAsO_2$. Hence, it was proved that *A. ferrooxidans* BY3 is very much adapted and resistant to a high concentration of arsenic. In another research work, sites contaminated with industrial pollutants and paddy grasslands are a rich source of arsenic contamination [55]. From these sites located at Pakistan, two purple nonsulfur arsenic-resistant bacteria namely *Rhodospirillum sp.* Q3B *and Rhodospirillaceae sp.* Q3C. It was noticed that under anaerobic environment, these strains have the highest arsenic sorption capacity. The percentage arsenic reduction for *Rhodospirillum* sp. Q3B and *Rhodospirillaceae* sp. Q3C was 74.92 and 76.67 (v/v), respectively in anaerobic conditions. Moreover, these strains have a peculiar feature of respiring arsenic in the existence of citrate, oxalate, and lactate, which can lead to arsenic detoxification as well as producing hydrogen gas, which could be exploited in the future for biogas production.

Singh et al. [56] reported an arsenic tolerant microbial strain, *Bacillus aryabhattai* NBRI014. The study focuses on the intracellular bioaccumulation and biovolatilization of As(V) tolerance, this bacterium was grown in nutrient media containing 25 000 mg/l of As(V) for 48 hours. This investigation also provided an insight into the seven ars genes: *arsR, arsD, arsA, arsB, arsC, arsH,* and *arsAB.* During arsenic stress, these genes were found to be upregulated in the bacterial cells and thus can be employed for the remediation of arsenic. As arsenite is easily portable in soil and drains effectively into groundwater, in this investigation, an As(III) tolerant bacterium *Sporosarcina ginsengisoli* CR5 was isolated from soil contaminated with As [57]. Purposefully, the precipitation of microbial calcite by *Sp. ginsengisoli* CR5 for remediating As(III) contaminated soil was examined. These bacteria can grow at a high concentration of 50 mM As(III). Thus, induction of calcite precipitation biologically is an environmentally friendly approach for removing arsenic from the contaminated sites. Podder and Majumder [58] in an innovative study investigated the performance of *Bacillus arsenicus* MTCC 4380, biofilm was supported on Neem leaves/$MnFe_2O_4$ composite on scavenging As(III) and As(V). Ideal conditions for bioaccumulation and biosorption were found to be dependent on temperature and contact time. In the above study, equilibrium was attained at a biosorbent dose of 0.9 g/l, 30 °C, and 4 hours. Moreover, there was an improvement in initial sorption rate (h) with an increase in the amount of As(III) – 3.86–694.03 mg/g min and As(V) – 4.72–520.39 mg/g min. This experiment was favorable for future ecological applications.

4.2.9 Mercury (Hg)

A bacterial strain, *Pseudoxanthomonas* sp. SE1 has been isolated from soil contaminated with mercury [59]. This strain showed resistance to mercury and thus, has the ability to remove this metal. The bacterial capacity to resist Hg was proved experimentally by growing this microbe in presence of Hg, in nutrient broth and low phosphate media and expressed in terms of 72 hours MIC and EC_{50} values. These values for nutrient media were calculated to be 1.7 and 1.4 mg/l, respectively and for low phosphate media followed the order 23.1 and 22.6 mg/l, respectively. This strain can volatilize Hg photographic film and thus remove Hg from the solution. After this, adding 1.5 mg/l Hg in low phosphate media and growing the SE1 culture carried out bioremediation experiments. The results showed a volatilization rate of approximately 60% in 6 hours and adsorption rate around 40%. Furthermore, an enzyme mercuric reductase having an activity of 9.7 mU/mg of protein and encoded by *MerA* gene was detected in this isolate, which can convert Hg(II) to Hg.

Dash and Das [60] employed two marine microbes and constructed a transgenic bacterium by transforming a plasmid for a *mer* operon from *B. thuringiensis* PW-05 into a plasmid of *B. cereus* BW-03 a mercury resistant bacterium, for studying the mercury biosorption capacity. The transgenic was named as *B. cereus* BW-03 (PW-05). *In vitro*, this transformant has the ability to remove approximately greater than 99% of mercury (40% by bioadsorption and >53% by volatilization). Moreover, immobilizing the transformant increases the mercury reduction rate to approximately 100%. The characteristic properties of this transformant are that it can resist a wide range of pH 8.0, salinity (5–30 mg/l), and mercury (5–50 mg/l).

A separate study had been done by Muneer et al. [61] on the effluents discharged from tanneries. The investigators isolated bacteria, protozoa, and yeast, which were resistant to mercury and identified them as *Bacillus licheniformis*, *Tetrahymena rostrata*, and *Candida parapsilosis* based on 16S rRNA sequencing. Except in the lag phase, which was extended by the addition of mercury, these microorganisms demonstrated a typical growth pattern. The Hg(II) tolerating capability of these three microbial consortia was evaluated both individually as well as in various combinations. The individual Hg(II) removal efficiency for *B. licheniformis*, *Te. rostrata*, and *C. parapsilosis* was 73, 40, and 80%, respectively. Moreover, reduction in the Hg(II) level from the combination of the three was observed as follows *B. licheniformis* and *Te. Rostrata*, 73%, *Te. rostrata* and *C. parapsilosis*, 77%, and *C. parapsilosis* and *B. licheniformis*, 85%. The maximum Hg(II) removal of 88% was observed by the combination of these three microbial consortia at an incubation period of 96 hours.

Biosorption capacity of some of the microorganisms that bioremediate heavy metals along with their permissible limits in drinking water and soil are shown in Table 4.1.

Table 4.1 Information related to some of the toxic heavy metals and their respective bacterial biosorbents.

Heavy metal	Permissible limit[a]		Bacterial biosorbent	Uptake capacity (mg/g)	References
	Drinking water (mg/l)	Soil (mg/kg)			
Copper	1.3	2–100	*Arthrobacter* ps-5	169.15	[23]
			Geobacillus decanicus	48.5	[10]
			Geobacillus stromboliensis	41.5	
			Micrococcus luteus	59	[46]
			Rhodococcus erythropolis	68.03	[45]
			Enterobacter sp.	1.87	[62]
			Bacillus sp. FM1	78.125	[63]
			Pseudomonas aeruginosa AT18	86.95	[64]
Lead	0.015	2–200	*Arthrobacter* ps-5	216.09	[23]
			Arthrobacter sp. 25	9.6	[34]
			Enterobacter sp.	5.6	[62]
			Klebsiella sp. J1	99.5	[37]
			Klebsiella sp. 3S1	140.19	[33]
			Bacillus subtilis FZUL-33	381.31	[31]
			Bacillus coagulans R11 (Spore cells)	14.86	[29]
			Bacillus coagulans R11 (Vegetative cells)	17.53	
			Bacillus coagulans R11 (Decay cells)	17.41	

(Continued)

Table 4.1 (Continued)

Heavy metal	Permissible limit[a] Drinking water (mg/l)	Soil (mg/kg)	Bacterial biosorbent	Uptake capacity (mg/g)	References
Chromium	0.1	1–1000	*Arthrobacter* ps-5	84.47	[23]
			Pseudomonas aeruginosa AT18	200.00	[64]
			Arthrobacter viscosus	13.0	[25]
			Kocuria rhizophila 14ASP	14.4	[26]
			Bacillus sp. FM1	64.102	[63]
			Kocuria rhizophila	14.4	[26]
			Acinetobacter haemolyticus	198.80	[21]
Cadmium	0.005	0.01–0.7	*Kocuria rhizophila*	9.07	[26]
			Enterobacter sp.	2.69	[62]
			Tsukamurella paurometabola A155 *Pseudomonas aeruginosa* B237	16.89	[39]
			Bacillus cereus RC-1 (Live cells)	24.01	[7]
			Bacillus cereus RC-1 (Dead cells)	31.95	
			Saccharomyces cerevisiae (Batch system)	3.1	[8]
			Saccharomyces cerevisiae (Continuous system)	3.74	
			Arthrobacter sp. (Batch system)	270.27	[9]
			Arthrobacter sp. (Continuous system)	188.67	
			Geobacillus decanicus	29.2	[10]
			Geobacillus stromboliensis	38.8	

			Bacillus badius AK	131.58	[16]
			Burkholderia cepacia GYP1	116	[17]
Zinc	5.0	10–300	*Tsukamurella paurometabola* A155	16.75	[39]
			Pseudomonas aeruginosa B237	16.89	
			Geobacillus decanicus	29	[10]
			Geobacillus stromboliensis	21.1	
			Rhodobacter capsulatus B10 (wild strain)	164	[38]
			Rhodobacter capsulatus RC220 (mutant plasmid)	73.9	
			Acinetobacter sp.	36	[65]
			Enterobacter sp.	4.3	[62]
			Pseudomonas aeruginosa AT18	74.5	[64]
			Pseudomonas aeruginosa	46.1	[66]
Nickel	–	5–500	*Bacillus megaterium* 1295S (Dead cells)	196.4	[40]
			Sporosarcina pasteurii 586S (Dead cells)	200.2	
			Streptomyces roseorubens SY	208.39	[41]
			Geobacillus decanicus	21	[10]
			Geobacillus stromboliensis	42	
Arsenic	0.05	1–50	*Bacillus arsenicus* MTCC 4380 (As(III))	694.03	[58]
			Bacillus arsenicus MTCC 4380 (As(V))	520.39	

[a] Adapted from [126].

4.3 Genetic Engineering in Bioremediation Processes for Some Major Elements

Bioremediation is a procedure that utilizes microorganisms or their enzymes to stimulate the removal and degradation of heavy metal contamination present in the environment. The application of microbial metabolism for this purpose has proved to be economical and safe when compared to other physico-chemical processes. However, these processes are usually very slow, which allows the accumulation of toxic heavy metals in the system. Through the process of adsorption, microorganisms take up heavy metals and the process is known as biosorption. Moreover, to improve and enhance bioremediation, bioaugmentation is done either by including microorganisms that have catabolic genes or are genetically modified (GMOs). Ongoing advances in the field of molecular biology have been able to discover novel strains of bacteria or have been able to modify and manipulate existing strains. Chen et al. [67] reported *Enterobacter* sp. strain EG16 (plant-associated microbe) that had been isolated from soil contaminated with heavy metals. The authors investigated whether Cd exposure up-regulates *ahpC* and down-regulates *fur* genes, which are responsible for the simultaneous enhancement of siderophore production. There is also an increase in the indole-3-acetic acid (IAA) hormone in the plant and thus promoted plant growth [67]. Chang et al. [68] isolated novel bacteria, which have the capacity to oxidize arsenite from surface soil and groundwater, contaminated with arsenic in Vietnam. In one bacterium, *Citrobacter freundii* VTan4, *ArsH* gene is present, which has an oxidoreductase activity with an unidentified property and the other *Pseudomonas putida* VTw33, *ArsB, and ArsH* gene, which also has a similar function. The presence of these genes benefitted these bacteria to oxidize 1 mM of As(III) (toxic form) to As(V) (less toxic) within a time period of 30–50 hours. Thus, these two strains proved to be a very effective and meaningful tool that can be employed for the bioremediation of heavy metal. Sturm et al. [69] focused on heavy metal chromium in their research and investigated on *Leucobacter chromiiresistens* JG31, which has a *chrA* gene for providing these bacteria resistance to chromate. *L. chromiiresistens* JG31 can tolerate a chromate concentration of 400 mM and forms biofilm under a chromate stressed environment. Antonucci et al. [70] found that *Thermus thermophilus* HB27 showed effective tolerance to cadmium and arsenic. The presence of three genes namely *Tt*ArsC, *Tt*ArsX, and *Tt*SmtB are responsible for the encoding of arsenate reductase, membrane arsenic efflux transporter, and transcriptional repressor, respectively. Table 4.2 summarizes the functional genes that are present inside the bacteria for defense and are responsive toward heavy metal under stress conditions.

Table 4.2 Identification of functional genes involved in tolerance and detoxification.

Microorganism	Strain	Genes involved	Heavy metal	Brief functional aspect	References
Pseudomonas putida	CD2	*czcCBA1*, *cadA2R*, *colRS*	Cadmium	Metal efflux and signal transduction system	[71]
Enterobacter sp.	EG16	*ahpC*, *fur*	Cadmium	Elevates siderophore production	[67]
Escherichia coli	BL21(DE3)	*capB*	Cadmium	Resistance to cadmium	[72]
Desulfovibrio desulfuricans	DSM 1926	*smtAB*, *cadAC*, *cadD*	Cadmium	Provides heavy metal resistance	[73]
Desulfococcus multivorans	DSM 2059	*smtAB*, *cadAC*, *cadD*	Cadmium	Provides heavy metal resistance	[73]
Rhodopseudomonas palustris	TN110	*nifH*, *vnfG*	Cadmium	Expression of nitrogen-fixing genes	[18]
Burkholderia cepacia	GYP1	*czcC*	Cadmium	Heavy metal efflux system	[17]
Arthrobacter sp.	FB24	*chrA*	Chromium	Encodes for putative chromate efflux protein	[74]
Microbacterium sp.	A20, K19, K21, K22, K24, K27, K2B2, K30, K31, K33, K35, K36, K40, K41, K5D, PF3	*chrR*	Chromium	Helps in chromium reduction	[75]
Leucobacter chromiiresistens	JG31	*chrA*	Chromate	Thrive resistance at extreme chromium concentration	[69]
Paenibacillus yonginensis	DCY84T	*AtAIP*, *AtALS3*, *AtALMT1*	Aluminum	Show resistance to aluminum at stress condition	[76]
Ochrobactrum tritici	SCII24	*aoxAB*, *cytC*, *moeA*, *aoxRS*	Arsenic	Signal transduction	[77]
Bacillus aryabhattai	NBRI014	*ars*	Arsenic	Confer resistance to arsenic	[56]
Corynebacterium glutamicum	—	*arsR*, *arsB*, *arsC*	Arsenic	Regulatory protein, arsenite permease, and arsenate reduction enzyme	[78]

(Continued)

Table 4.2 (Continued)

Microorganism	Strain	Genes involved	Heavy metal	Brief functional aspect	References
Brevibacterium linens	AE038-8	*ACR3, ACR2, arsR, arsM*	Arsenic	Encodes for membrane efflux pump, cytosolic arsenate reductase, As(III)-repressor protein, putative SAM that methylate As(III)	[79]
Brevibacterium linens	AE038-8	*arsC, ACR3, arsR, arsO*	Arsenic	Proteins for detoxifying As, As(III) efflux pump, arsenate reductase, As(III) transcriptional repressor	[80]
Citrobacter freundii strain	VTan4	*arsH*	Arsenic	Organoarsenical oxidase	[68]
Pseudomonas putida strain	VTw33	*arsB, arsH*	Arsenic	Organoarsenical oxidase	[68]
Pseudomonas xanthomarina	S11	*arsC, arsR, arsH, arsO, acr3, arsRCDA, aioXRS, aioBA*	Arsenic	Pump extruding As[III], putative protein-tyrosine phosphatase, arsenite oxidase regulation, phosphorous metabolism	[81]
Sinorhizobium sp.	M14	*aioA, repA, traI, orf12*	Arsenic	Arsenite oxidation, replication proteins, transfer genes, putative proteins	[82]
Alkaliphilus oremlandii	OhILAs	*arrA*	Arsenic	Encodes respiratory arsenate reductase	[83]
Rhodococcus sp.	S43	*aioA*	Arsenite	Encodes arsenite oxidase	[84]
Pseudoxanthomonas	SE1	*merA*	Mercury	Encodes mercuric reductase enzyme	[59]
Cupriavidus metallidurans	MSR33	*merA, merB*	Mercury	Encodes enzyme mercuric reductase, organomercurial lyase protein	[85]
Stenotrophomonas sp.	TD3	*chrR, zupT, hmrE*	Chromium and zinc	NADPH-dependent FMN reductase, zinc transporter, and heavy metal resistance protein	[86]
Thermus thermophilus	HB27	*Tt*ArsC, *Tt*ArsX, *Tt*SmtB	Cadmium and arsenic	arsenate reductase, arsenic efflux protein, arsenic responsive transcriptional repressor	[70]

4.4 Biotechnological Intervention for Some Important Heavy Metals

Heavy metals are very toxic and lethal to mankind and the environment. They contribute to the majority of the wastewater pollution in groundwater and have thus become an issue for concern. Recent developments and emphasis on biotechnology and its application for the bioremediation of heavy metals will provide a sustainable, cost-effective, and an environmentally friendly approach. However, commercialization of such processes is required. Several researchers have investigated and isolated microbes showing specific and particular characteristics toward certain heavy metals. In this section, some of the captivated microorganisms having the highest removal rate have been expanded upon. Wu et al. [87] focused on one such heavy metal, arsenic, from the soils at the Kangjiawan lead-zinc mine in southern China. A huge number of novel bacteria, 170, showed resistance to arsenic. Two strains specifically *Bacillus* sp. Tw1 and *Pseudomonas* sp. Tw224 have the ability to resist As(III) and As(V) to ≥20 and 600 mM, respectively. The oxidation rate was obtained as *Bacillus* sp. Tw1 (42.48%) and *Pseudomonas* sp. Tw224 (52.01%). A reduction rate of 34.94% was achieved for *Pseudomonas* sp. Tw224. In another research Maizel et al. [80] isolated *Brevibacterium linens* AE038-8, from Argentina, which showed a tolerance ability of 75 mM for As(III) and 1 M for As(V) in complex media. Groundwater contaminated with arsenic has been found at Burdwan, West Bengal, India by Dey et al. [88]. From this groundwater, two bacteria have been isolated namely *Bacillus* sp. SW2 and *Aneurinibacillus aneurinilyticus* SW4. These microbial populations are tolerant to 4,500 and 550 ppm of As(III), respectively. The removal by *Bacillus* sp. SW2 and *An. aneurinilyticus* SW4 of As(III) was 51.45% and 51.99% and As(V) was 53.29% and 50.37%, respectively. In an innovative experiment performed by Kang et al. [89] the effect of four strains of bacteria synergistically on bioremediation of Cd, Cu, and Pb were studied. The combination of bacterial mixtures gives higher results in terms of heavy metal resistance, growth rate, and urease activity. These strains were isolated at a mine site in Korea and were identified as *Enterobacter cloacae* KJ-47, *E. cloacae* KJ-46, *Sporosarcina soli* B-22, and *Viridibacillus arenosi* B-21. The remediation rates of the combination of these four microbes for heavy metals (Cd, Cu, and Pb) were found to be 85.4%, 5.6%, and 98.3%, respectively. Tamindžija et al. [90] isolated *B. cereus* strains from different places. These bacteria have a high capacity toward Cr(VI). Removal rates for Cr(VI) from strains *B. cereus* PCr2a and *B. cereus* PCr2a were 22.9% and 98.5%, respectively. Other than the above-mentioned research many others have been carried out, details of which appear in Table 4.3, which tabulates the various biochemical processes with the applied microbial consortia's along with their subsequent rate of removal and resistance to different heavy metals.

Table 4.3 Summary of process, removal, and resistance to different heavy metals as investigated by various researchers.

Heavy metal	Bacteria	Increased Tolerance/ Resistance/Growth/ MIC	Removal/Reduction rate	Process	References
As	*Bacillus* sp. Tw1	600 mM As(V) and ≥20 mM As(III)	42.48% (oxidation)	Oxidation and reduction	[87]
	Pseudomonas sp. Tw224		34.94 and 52.01% (Redox)		
	Brevibacterium linens AE038–8	1 M As(V) and 75 mM As(III)	30–50 mM As(III)	Bioreduction	[80]
	Pseudomonas sp.	100 mg/l As(III/V)	—	Bioacclimation	[91]
	Pseudomonas xanthomarina S11	—	42.72 mM As[III]	Biooxidation	[81]
	Agrobacterium tumefaciens LBA288	1–20 mM As(III)	—	Recombinant DNA technology	[92]
	Stenotrophomonas sp. LM24R	5–20 mM As(III)			
	Pseudomonas sp. OS20R and *Alcaligenes* sp. LM16R	20 mM As(III)			
	Brevundimonas sp.GZS11	25 mM As(III)			
	Bacillus sp. KM02	4500 mg/l As(III)	51.45 As(III) and 53.29% As(V)	Bioremediation	[88]
	Aneurinibacillus aneurinilyticus BS-1	550 mg/l As(III)	51.99% As(III) and 50.37% As(V)		
	Azospirillum sp. MM17	60 mM As(III)	92% of 100 mM As(III)	Biooxidation	[93]
	Sporosarcina ginsengisoli CR5	50 mM As(III)	96.3% As	Biomineralization	[57]
	Pseudomonas sp. As7325	75 mg/l As(III) and 7500 mg/l As(V)	99% of 500 μg/l As(V) and 100% of 1000 μg/l As(V)	Biooxidation	[94]
	Bacillus sp., *Enterobacter* sp., *Ochrobactrum* sp.	—	11–38% As(V)	Biotransformation	[95]
	Sinorhizobium sp. M14	—	—	Biooxidation	[96]
	Alkaliphilus oremlandii OhILAs	40 mM As(V)	—	Biotransformation	[83]

Cd	*Serratia marcescens*	—	16% Cd(II)	Biomineralization	[97]
	Enterobacter cloacae EMB19		8% Cd(II)		
	Pseudomonas aeruginosa	7 mM Cd(II)	94.0% Cd(II)	Bioremediation	[98]
	Lysinibacillus sphaericus CH-5	2 g/l	99.95% Cd	Bioremediation	[99]
	Desulfovibrio desulfuricans	—	99.9% Cd	Bioremediation	[100]
	Bacillus sp. EB L14	1–15 mM Cd (II)	75.78% Cd (II)	Bioremediation	[101]
	Viridibacillus arenosi B-21, *Sporosarcina soli* B-22, *Enterobacter cloacae* KJ-46, and *Enterobacter cloacae* KJ-47	—	85.4% Cd	Bioremediation	[89]
	Enterobacter cloacae B1	900 mg/l Cd	64.17% Cd	Bioaccumulation	[102]
Pb	*Viridibacillus arenosi* B-21, *Sporosarcina soli* B-22, *Enterobacter cloacae* KJ-46, and *Enterobacter cloacae* KJ-47	—	98.3% Pb	Bioremediation	[89]
	Enterobacter cloacae B1	1100 mg/l Pb	95.25% Pb	Bioaccumulation	[102]
	Bacillus sp. EB L14	1–15 mM Pb (II)	80.48% Pb(II)	Bioremediation	[101]
	Arthrospira platensis	100 mg/l Pb(II)	50% Pb(II)	Optimization (Biosorption)	[103]
Cr	*Desulfovibrio desulfuricans*	—	74.2% Cr	Bioremediation	[100]
	Bacillus cereus group PCr2a	—	22.9%	Bioremoval	[90]
	Bacillus cereus NCr4		98.5%		
	Klebsiella sp. PB6	100 µg/ml Cr	89.5% Cr(VI)	Bioreduction	[104]
	Streptococcus spp. PZ4	—	85% Cr(VI)	Bioreduction	[105]
	Acenetobacter calcoaciticus	—	89.39% Cr(VI)	Bioreduction	[106]
	Kosakonia cowanii MKPF2, *Klebsiella pneumoniae* MKPF5, *Acinetobacter gerneri* MKPF7, *Klebsiella variicola* MKPF8, and *Serratia marcescens* MKPF12	2000, 1800, 1800, 1400, and 2000 mg/l Cr(VI)	20 mg/l Cr(VI)	Bioreduction	[107]

(Continued)

Table 4.3 (Continued)

Heavy metal	Bacteria	Increased Tolerance/ Resistance/Growth/ MIC	Removal/Reduction rate	Process	References
	Acinetobacter calcoaceticus	500 mg/l Cr(VI)	67.14% Cr(VI)	Bioreduction	[108]
	Acinetobacter AB1	400 mg/l of Cr(VI)	50 mg/l	Bioreduction and bioaccumulation	[109]
	Acinetobacter haemolyticus	90 mg/l Cr(VI)	0.52 μg Cr(VI) per mg of protein	Bioreduction	[110]
	Pseudomonas aeruginosa ZM130	—	76.6–98.7% Cr(VI)	Optimization	[111]
Ni	*Enterobacter cloacae* B1	700 mg/l Ni	36.77% Ni	Bioaccumulation	[102]
	Desulfovibrio desulfuricans	—	98.3% Ni	Bioremediation	[100]
Cu	*Pseudomonas* sp. NA	—	110 mg/l Cu(II)	Bioreduction and biosorption	[112]
	Bacillus sp. EB L14	1–15 mM Cu (II)	21.25% Cu(II)	Bioremediation	[101]
	Stenotrophomonas maltophilia PD2	50 mg/l Cu	90% Cu	Optimization (Bioremediation)	[113]
	Kocuria flava CR1	1000 mg/l Cu	97% Cu	Bioremediation	[114]
	Viridibacillus arenosi B-21, *Sporosarcina soli* B-22, *Enterobacter cloacae* KJ-46, and *Enterobacter cloacae* KJ-47	—	5.6% Cu	Bioremediation	[89]
Hg	*Cupriavidus metallidurans* MSR33	0.12 mM Hg(II)	—	Recombinant DNA technology	[85]

4.5 Future Perspective

Transgenic microbes may be considered important in the enhancement of degradation and detoxification of heavy metal contaminants, which is the major xenobiotic. Further, more research must be completed in order to increase the survival of microorganisms while in the toxic environment during bioremediation. Presently, the rate of survival is enough to combat heavy metal pollution effectively. Environmental factors such as temperature, low nutrient level, low pH, are not easily controlled and usually obstruct their utilization and the efficiency of the bioremediation process [115]. However, attempts have been carried out to avoid horizontal gene transfer to indigenous microorganisms from engineered microbes by using antisense RNA technology or by using suicidal genes. Interestingly, the use of selectable markers such as antibiotic genes must be withdrawn and substituted with other selectable markers to prevent unintentional propagation of genes resistant to antibiotics in other soil microorganisms. Furthermore, additional research is necessary for understanding the metabolic passageway of transgenic microbes or indigenous microbes used in bioremediation processes, so that their effectiveness and potential higher efficiency may be achieved in favor of bioremediation. The ability of microorganisms applied in bioremediation must compete with indigenous microbial population, which is essential for the accomplishment of bioremediation.

The purpose of organic loading including the addition of compost in soils rich in metal contamination could create a distinction in the soil microbial population by: altering pH; reducing the solubility of heavy metals; enhancing allochthonous biomass of microbes; or by making various nutritional supplements sufficiently accessible [116, 117]. Moreover, biochar is an end product of pyrolysis of biomass obtained from sources such as crop residue, manure, and solid wastes, which can be used to stimulate microorganisms for bioremediation by making the environment more favorable [118]. Comprehensive literature by several authors has highlighted the potential value of biochar as an efficient agent in immobilization of metals and organic pollutants [119–121]. Biochar is capable of donating, accepting, or transferring electrons within their surroundings abioticaly or through biological process [122, 123]. It is recommended by a few scientists that biochar also assists the microbial electron transport pathway since they show comparable functional uniqueness to soil redox-active organic matter [124]. Biochar actively plays an optimum role by enhancing the pH of contaminated soils to enhance the bioavailability of heavy metals to be consumed by plants. The rapidity and toxicity of many elements including Cr, Pb, As, Ni, Zn, and Cu are also dependent on their oxidation status, which is regulated by redox type reactions [125]. This ionic status of metal also creates important an aspect in bioremediation by microbial sources.

The progress of new genetic tools for a better perception of microbial gene structures and functions is required to promote progression in the pathway for improved unsafe-waste removal. This chapter has focused on the accomplishments of biotechnological applications and strategies for environmental protection, detoxification, and the removal of heavy metals from waste. It also examines recent developments and future prospects for the bio/phytoremediation of toxic pollutants from contaminated soil and water.

4.6 Conclusions

This chapter highlighted heavy metal contamination and its bioremediation by using microorganisms. Heavy metals are a major pollutant of the environment and pose a health risk along with involved enzymatic reactions and various aspects of mechanisms in the living system. Interestingly, microbes can effectively remediate heavy metal polluted environments. This method showed the effectiveness of bioremediation, which is potentially a better alternative for the elimination of heavy metals from polluted sites in comparison to physicochemical methods, which are not as effective and are costly due to the higher level of energy required. Microorganisms have their own intrinsic self-mechanisms that enable them to live in heavy metal stressed environments and subsequently eradicate the metals from the immediate environment. These microbes utilize different procedures such as precipitation, biosorption, and enzymatic transformation of metals complexes. Furthermore, these microbes also help in phytoremediation techniques, in which phytoextraction and phytostabilization are very effective. However, the environmental conditions also need to support optimally effective bioremediation. Synthetic biotechnology is a budding tool that may be functional to create microbial communities with the capability of degrading and eradicating heavy metals from soils and water using the metabolic machinery of such consortia of organisms. This knowledge must be highlighted for more efficient bioremediation of noxious wastes from the environment. Metagenomic advancement and metabolic study may also be emphasized to revise the purposeful structures of microbial population within contaminated sites. Further study of metal resistance genes could be used to enhance degradation of specific heavy metals by selective strains of particular microbes.

References

1 Lin, C.C. and Lin, H.L. (2005). Remediation of soil contaminated with the heavy metal (Cd^{2+}). *Journal of Hazardous Materials* 122 (1–2): 7–15.

2 Cohen, R.R.H. (2006). Use of microbes for cost reduction of metal removal from metals and mining industry waste streams. *Journal of Cleaner Production* 14: 1146–1157.

3 Riley, R.G. and Zachara, J.M. (1992). *Subsurface Science Program DOE/ER-0547T*. Washington, DC: US Department of Energy, Office of Energy Research.

4 Debnath, B., Majumdar, M., Bhowmik, M. et al. (2020). The effective adsorption of tetracycline onto zirconia nanoparticles synthesized by novel microbial green technology. *Journal of Environmental Management* 261: 110235.

5 Choudhury, R., Majumder, M., Roy, D.N. et al. (2016). Phytotoxicity of Ag nanoparticles prepared by biogenic and chemical methods. *International Nano Letters* 6 (3): 153–159.

6 Li, X., Peng, W., Jia, Y. et al. (2016). Removal of cadmium and zinc from contaminated wastewater using *Rhodobacter sphaeroides*. *Water Science and Technology* 75 (11): 2489–2498.

7 Huang, F., Dang, Z., Guo, C.L. et al. (2013). Biosorption of Cd(II) by live and dead cells of *Bacillus cereus* RC-1 isolated from cadmium-contaminated soil. *Colloids and Surfaces B: Biointerfaces* 107: 11–18.

8 Galedar, M. and Younesi, H. (2013). Biosorption of ternary cadmium, nickel and cobalt ions from aqueous solution onto *Saccharomyces cerevisiae* cells: batch and column studies. *American Journal of Biochemistry and Biotechnology* 9 (1): 47–60.

9 Hasan, S.H. and Srivastava, P. (2011). Biosorptive abatement of Cd^{2+} by water using immobilized biomass of *arthrobacter* sp.: response surface methodological approach. *Industrial and Engineering Chemistry Research* 50 (1): 247–258.

10 Özdemir, S., Kilinc, E., Poli, A. et al. (2009). Biosorption of Cd, Cu, Ni, Mn and Zn from aqueous solutions by thermophilic bacteria, *Geobacillus toebii* sub. sp. *decanicus* and *Geobacillus thermoleovorans* sub. sp. *stromboliensis*: equilibrium, kinetic and thermodynamic studies. *Chemical Engineering Journal* 152 (2): 195–206.

11 Özdemir, S., Kilinc, E., Poli, A. et al. (2012). Cd, Cu, Ni, Mn and Zn resistance and bioaccumulation by thermophilic bacteria, *Geobacillus toebii* subsp. *decanicus* and *Geobacillus thermoleovorans* subsp. *stromboliensis*. *World Journal of Microbiology and Biotechnology* 28 (1): 155–163.

12 Özdemir, S., Klnç, E., Poli, A. et al. (2013). Biosorption of heavy metals (Cd^{2+}, Cu^{2+}, Co^{2+}, and Mn^{2+}) by thermophilic bacteria, *Geobacillus thermantarcticus* and *Anoxybacillus amylolyticus*: equilibrium and kinetic studies. *Bioremediation Journal* 17 (2): 86–96.

13 Bravo, D., Pardo-Díaz, S., Benavides-Erazo, J. et al. (2016). Cadmium and cadmium-tolerant soil bacteria in cacao crops from northeastern Colombia. *Journal of Applied Microbiology* 124 (5): 1175–1194.

14 Oves, M., Khan, M.S., and Zaidi, A. (2013). Biosorption of heavy metals by *Bacillus thuringiensis* strain OSM29 originating from industrial effluent contaminated north Indian soil. *Saudi Journal of Biological Sciences* 20 (2): 121–129.

15 Liu, Y., Tie, B., Li, Y. et al. (2018). Inoculation of soil with cadmium-resistant bacterium *Delftia* sp. B9 reduces cadmium accumulation in rice (*Oryza sativa* L.) grains. *Ecotoxicology and Environmental Safety* 163: 223–229.

16 Vishan, I., Saha, B., Sivaprakasam, S. et al. (2019). Evaluation of Cd(II) biosorption in aqueous solution by using lyophilized biomass of novel bacterial strain *Bacillus badius* AK: biosorption kinetics, thermodynamics and mechanism. *Environmental Technology and Innovation* 14: 100323.

17 Zhang, J., Li, Q., Zeng, Y. et al. (2019). Bioaccumulation and distribution of cadmium by *Burkholderia cepacia* GYP1 under oligotrophic condition and mechanism analysis at proteome level. *Ecotoxicology and Environmental Safety* 176: 162–169.

18 Sakpirom, J., Kantachote, D., Siripattanakul-Ratpukdi, S. et al. (2019). Simultaneous bioprecipitation of cadmium to cadmium sulfide nanoparticles and nitrogen fixation by *Rhodopseudomonas palustris* TN110. *Chemosphere* 223: 455–464.

19 Zhao, R., Wang, B., Cai, Q.T. et al. (2016). Bioremediation of hexavalent chromium pollution by *Sporosarcina saromensis* M52 isolated from offshore sediments in Xiamen, China. *Biomedical and Environmental Sciences* 29 (2): 127–136.

20 Latha, S., Vinothini, G., and Dhanasekaran, D. (2015). Chromium [Cr(VI)] biosorption property of the newly isolated actinobacterial probiont *Streptomyces werraensis* LD22. *3 Biotech* 5 (4): 423–432.

21 Yahya, S.K., Zakaria, Z.A., Samin, J. et al. (2012). Isotherm kinetics of Cr(III) removal by non-viable cells of *Acinetobacter haemolyticus*. *Colloids and Surfaces B: Biointerfaces* 94: 362–368.

22 Sandana Mala, J.G., Sujatha, D., and Rose, C. (2015). Inducible chromate reductase exhibiting extracellular activity in *Bacillus methylotrophicus* for chromium bioremediation. *Microbiological Research* 170: 235–241.

23 Shuhong, Y., Meiping, Z., Hong, Y. et al. (2014). Biosorption of Cu^{2+}, Pb^{2+} and Cr^{2+} by a novel exopolysaccharide from *Arthrobacter* ps-5. *Carbohydrate Polymers* 101 (1): 50–56.

24 Sukumar, C., Janaki, V., Kamala-Kannan, S. et al. (2014). Biosorption of chromium(VI) using *Bacillus subtilis* SS-1 isolated from soil samples of electroplating industry. *Clean Technologies and Environmental Policy* 16 (2): 405–413.

25 Silva, B., Figueiredo, H., Quintelas, C. et al. (2012). Improved biosorption for Cr(VI) reduction and removal by *Arthrobacter viscosus* using zeolite. *International Biodeterioration and Biodegradation* 74: 116–123.

26 Haq, F., Butt, M., Ali, H. et al. (2015). Biosorption of cadmium and chromium from water by endophytic *Kocuria rhizophila* : equilibrium and kinetic studies. *Desalination and Water Treatment* 3: 1–13.

27 Xie, P., Hao, X., Mohamad, O.A. et al. (2012). Comparative study of chromium biosorption by *Mesorhizobium amorphae* strain CCNWGS0123 in single and binary mixtures. *Applied Biochemistry and Biotechnology* 2: 570–587.

28 Vimalnath, S. and Subramanian, S. (2018). Studies on the biosorption of Pb(II) ions from aqueous solution using extracellular polymeric substances (EPS) of *Pseudomonas aeruginosa*. *Colloids and Surfaces B: Biointerfaces* 40 (3): 307–332.

29 Xing, S.C., Chen, J.Y., Lv, N. et al. (2018). Biosorption of lead (Pb^{2+}) by the vegetative and decay cells and spores of *Bacillus coagulans* R11 isolated from lead mine soil. *Chemosphere* 211: 804–816.

30 Kalita, D. and Joshi, S.R. (2017). Study on bioremediation of Lead by exopolysaccharide producing metallophilic bacterium isolated from extreme habitat. *Biotechnology Reports* 16: 48–57.

31 Lin, W., Huang, Z., Li, X. et al. (2016). Bio-remediation of acephate–Pb(II) compound contaminants by *Bacillus subtilis* FZUL-33. *Journal of Environmental Sciences (China)* 45: 94–99.

32 Li, X., Peng, W., Jia, Y. et al. (2016). Bioremediation of lead contaminated soil with *Rhodobacter sphaeroides*. *Chemosphere* 40 (3): 307–332.

33 Muñoz, A.J., Espínola, F., Moya, M. et al. (2015). Biosorption of Pb(II) ions by *Klebsiella* sp. 3S1 isolated from a wastewater treatment plant: kinetics and mechanisms studies. *BioMed Research International* 156: 228–235.

34 Jin, Y., Wang, X., Zang, T. et al. (2016). Biosorption of Lead(II) by *Arthrobacter* sp. 25: process optimization and mechanism. *Journal of Microbiology and Biotechnology* 26 (8): 1428–1438.

35 Abu Hasan, H., Sheikh Abdullah, S.R., Tan Kofli, N. et al. (2016). Interaction of environmental factors on simultaneous biosorption of lead and manganese ions by locally isolated *Bacillus cereus*. *Journal of Industrial and Engineering Chemistry* 37: 295–305.

36 Mwandira, W., Nakashima, K., and Kawasaki, S. (2017). Bioremediation of lead-contaminated mine waste by *Pararhodobacter* sp. based on the microbially induced calcium carbonate precipitation technique and its effects on strength of coarse and fine grained sand. *Ecological Engineering* 109: 57–64.

37 Wei, W., Wang, Q., Li, A. et al. (2016). Biosorption of Pb (II) from aqueous solution by extracellular polymeric substances extracted from *Klebsiella* sp. J1: adsorption behavior and mechanism assessment. *Scientific Reports* 6: 1–10.

38 Magnin, J.-P., Gondrexon, N., and Willison, J.C. (2014). Zinc biosorption by the purple non-Sulphur bacterium *Rhodobacter capsulatus*. *Canadian Journal of Microbiology* 60 (12): 829–837.

39 Limcharoensuk, T., Sooksawat, N., Sumarnrote, A. et al. (2015). Bioaccumulation and biosorption of Cd^{2+} and Zn^{2+} by bacteria isolated from a zinc mine in Thailand. *Ecotoxicology and Environmental Safety* 122: 322–330.

40 Gheethi, A.A., Efaq, A.N., Mohamed, R.M. et al. (2017). Bio-removal of Nickel ions by *Sporosarcina pasteurii* and *Bacillus megaterium*, a comparative study. *IOP Conference Series: Materials Science and Engineering* 226 (1): 012044.

41 Long, J., Gao, X., Su, M. et al. (2018). Performance and mechanism of biosorption of nickel(II) from aqueous solution by non-living *Streptomyces roseorubens* SY. *Colloids and Surfaces A: Physicochemical and Engineering Aspects* 548: 125–133.

42 Sathyavathi, S., Manjula, A., Rajendhran, J. et al. (2014). Extracellular synthesis and characterization of nickel oxide nanoparticles from *Microbacterium* sp. MRS-1 towards bioremediation of nickel electroplating industrial effluent. *Bioresource Technology* 165 (C): 270–273.

43 Zhang, J., Kumari, D., Fang, C. et al. (2019). Combining the microbial calcite precipitation process with biochar in order to improve nickel remediation. *Applied Geochemistry* 103: 68–71.

44 Cayllahua, J.E.B., de Carvalho, R.J., and Torem, M.L. (2009). Evaluation of equilibrium, kinetic and thermodynamic parameters for biosorption of nickel(II) ions onto bacteria strain, *Rhodococcus opacus*. *Minerals Engineering* 22 (15): 1318–1325.

45 dos Baltazar, M., P.G., Gracioso, L.H., Avanzi, I.R. et al. (2018). Copper biosorption by *Rhodococcus erythropolis* isolated from the Sossego mine – PA – Brazil. *Journal of Materials Research and Technology* 8 (1): 475–483.

46 Letnik, I., Avrahami, R., Port, R. et al. (2017). Biosorption of copper from aqueous environments by *Micrococcus luteus* in cell suspension and when encapsulated. *International Biodeterioration and Biodegradation* 116: 64–72.

47 Rajaram, R., Banu, J.S., and Mathivanan, K. (2013). Biosorption of Cu (II) ions by indigenous copper-resistant bacteria isolated from polluted coastal environment. *Toxicological and Environmental Chemistry* 95 (4): 590–604.

48 Liu, Y.G., Liao, T., He, Z.B. et al. (2013). Biosorption of copper(II) from aqueous solution by *Bacillus subtilis* cells immobilized into chitosan beads. *Transactions of Nonferrous Metals Society of China (English Edition)* 23 (6): 1804–1814.

49 Pan, X., Liu, J., Song, W. et al. (2012). Biosorption of Cu(II) to extracellular polymeric substances (EPS) from *Synechoeystis* sp.: a fluorescence quenching study. *Frontiers of Environmental Science & Engineering* 6 (4): 493–497.

50 Mohamad, O.A., Hao, X., Xie, P. et al. (2012). Biosorption of Copper (II) from aqueous solution using non-living *Mesorhizobium amorphae* Strain CCNWGS0123. *Microbes and Environments* 27 (3): 234–241.

51 Dhanarani, S., Viswanathan, E., Piruthiviraj, P. et al. (2016). Comparative study on the biosorption of aluminum by free and immobilized cells of *Bacillus safensis* KTSMBNL 26 isolated from explosive contaminated soil. *Journal of the Taiwan Institute of Chemical Engineers* 69: 61–67.

52 Huang, S., Wang, X., Liu, X. et al. (2017). Isolation, identification, and characterization of an aluminum-tolerant bacterium *Burkholderia* sp. SB1 from an acidic red soil. *Pedosphere* 28 (6): 905–912.

53 Abo-Amer, A.E., Ramadan, A.B., Abo-State, M. et al. (2012). Biosorption of aluminum, cobalt, and copper ions by *Providencia rettgeri* isolated from wastewater. *Journal of Basic Microbiology* 53 (6): 477–488.

54 Gao, Q., Tang, D., Song, P. et al. (2018). Bio-adsorption and bio-transformation of Arsenic by *Acidithiobacillus ferrooxidans* BY3. *International Microbiology* 21 (4): 207–214.

55 Saleem, H., ul Ain Kokab, Q., and Rehman, Y. (2019). Arsenic respiration and detoxification by purple non-Sulphur bacteria under anaerobic conditions. *Comptes Rendus Biologies* 2018: 1–7.

56 Singh, N., Gupta, S., Marwa, N. et al. (2016). Arsenic mediated modifications in *Bacillus aryabhattai* and their biotechnological applications for arsenic bioremediation. *Chemosphere* 164: 524–534.

57 Achal, V., Pan, X., Fu, Q. et al. (2012). Biomineralization based remediation of As(III) contaminated soil by *Sporosarcina ginsengisoli*. *Journal of Hazardous Materials* 201–202: 178–184.

58 Podder, M.S. and Majumder, C.B. (2016). Study of the kinetics of arsenic removal from wastewater using *Bacillus arsenicus* biofilms supported on a Neem leaves/ $MnFe_2O_4$ composite. *Ecological Engineering* 88: 195–216.

59 Mahbub, K.R., Krishnan, K., Naidu, R. et al. (2016). Mercury resistance and volatilization by *Pseudoxanthomonas* sp. SE1 isolated from soil. *Environmental Technology and Innovation* 6: 94–104.

60 Dash, H.R. and Das, S. (2015). Bioremediation of inorganic mercury through volatilization and biosorption by transgenic *Bacillus cereus* BW-03. *International Biodeterioration and Biodegradation* 103: 179–185.

61 Muneer, B., Iqbal, M.J., Shakoori, F.R. et al. (2013). Tolerance and biosorption of mercury by microbial consortia: potential use in bioremediation of wastewater. *Pakistan Journal of Zoology* 45 (1): 247–254.

62 Abd, E.R.R.E.S., Sobhy, E.E., and Perihan, S.A. (2016). Uptake of some heavy metals by metal resistant *Enterobacter* sp. isolate from Egypt. *African Journal of Microbiology Research* 7 (23): 2875–2884.

63 Masood, F. and Malik, A. (2011). Biosorption of metal ions from aqueous solution and tannery effluent by *Bacillus* sp. FM1. *Journal of Environmental Science and Health – Part A Toxic/Hazardous Substances and Environmental Engineering* 46 (14): 1667–1674.

64 Pérez Silva, R.M., Ábalos Rodríguez, A., Gómez Montes De Oca, J.M. et al. (2009). Biosorption of chromium, copper, manganese and zinc by *Pseudomonas aeruginosa* AT18 isolated from a site contaminated with petroleum. *Bioresource Technology* 100 (4): 1533–1538.

65 Tabaraki, R., Ahmady-Asbchin, S., and Abdi, O. (2013). Biosorption of Zn(II) from aqueous solutions by *Acinetobacter* sp. isolated from petroleum spilled soil. *Journal of Environmental Chemical Engineering* 1 (3): 604–608.

66 Ahmady-Asbchin, S., Safari, M., and Tabaraki, R. (2015). Biosorption of Zn (II) by *Pseudomonas aeruginosa* isolated from a site contaminated with petroleum. *Desalination and Water Treatment* 54 (12): 604–608.

67 Chen, Y., Chao, Y., Li, Y. et al. (2016). Survival strategies of the plant-associated bacterium *Enterobacter* sp. Strain EG16 under cadmium stress. *Applied and Environmental Microbiology* 82 (6): 1734–1744.

68 Chang, J.S., Yoon, I.H., and Kim, K.W. (2018). Arsenic biotransformation potential of microbial arsH responses in the biogeochemical cycling of arsenic-contaminated groundwater. *Chemosphere* 191: 729–737.

69 Sturm, G., Brunner, S., Suvorova, E. et al. (2018). Chromate resistance mechanisms in *Leucobacter chromiiresistens*. *Applied and Environmental Microbiology* 84 (23) https://doi.org/10.1128/AEM.02208-18.

70 Antonucci, I., Gallo, G., Limauro, D. et al. (2018). Characterization of a promiscuous cadmium and arsenic resistance mechanism in *Thermus thermophilus* HB27 and potential application of a novel bioreporter system. *Microbial Cell Factories* 17 (1): 1–9.

71 Hu, N. and Zhao, B. (2007). Key genes involved in heavy-metal resistance in *Pseudomonas putida* CD2. *FEMS Microbiology Letters* 267 (1): 17–22.

72 Qin, W., Liu, X., Yu, X. et al. (2017). Identification of cadmium resistance and adsorption gene from *Escherichia coli* BL21 (DE3). *RSC Advances* 7 (81): 51460–51465.

73 Naz, N., Young, H.K., Ahmed, N. et al. (2005). Cadmium accumulation and DNA homology with metal resistance genes in sulfate-reducing bacteria. *Applied and Environmental Microbiology* 71 (8): 4610–4618.

74 Henne, K.L., Nakatsu, C.H., Thompson, D.K. et al. (2009). High-level chromate resistance in *Arthrobacter* sp. strain FB24 requires previously uncharacterized accessory genes. *BMC Microbiology* 9: 199.

75 Learman, D.R., Ahmad, Z., Brookshier, A. et al. (2019). Comparative genomics of 16 *Micro* 3 *bacterium* spp. that tolerate multiple heavy metals and antibiotics. *PeerJ* 6: e6258.

76 Sukweenadhi, J., Kim, Y.-J., Choi, E.-S. et al. (2015). *Paenibacillus yonginensis* DCY84T induces changes in *Arabidopsis thaliana* gene expression against aluminum, drought, and salt stress. *Microbiological Research* 172: 7–15.

77 Branco, R., Francisco, R., Chung, A.P. et al. (2009). Identification of an aox system that requires cytochrome c in the highly arsenic-resistant bacterium *Ochrobactrum tritici* SCII24. *Applied and Environmental Microbiology* 75 (15): 5141–5147.

78 Mateos, L.M., Ordóñez, E., Letek, M. et al. (2006). *Corynebacterium glutamicum* as a model bacterium for the bioremediation of arsenic. *International Microbiology* 9 (3): 207–215.

79 Maizel, D., Utturkar, S.M., Brown, S.D. et al. (2015). Draft genome sequence of *Brevibacterium linens* AE038-8, an extremely arsenic-resistant bacterium. *Genome Announcements* 30 (2): 4–5.

80 Maizel, D., Blum, J.S., Ferrero, M.A. et al. (2016). Characterization of the extremely arsenic-resistant *Brevibacterium linens* strain AE038-8 isolated from contaminated groundwater in Tucumán, Argentina. *International Biodeterioration and Biodegradation* 107: 147–153.

81 Koechler, S., Arsène-Ploetze, F., Brochier-Armanet, C. et al. (2015). Constitutive arsenite oxidase expression detected in arsenic-hypertolerant *Pseudomonas xanthomarina* S11. *Research in Microbiology* 166 (3): 205–214.

82 Drewniak, L., Dziewit, L., Ciezkowska, M. et al. (2013). Structural and functional genomics of plasmid pSinA of *Sinorhizobium* sp. M14 encoding genes for the arsenite oxidation and arsenic resistance. *Journal of Biotechnology* 164 (4): 479–488.

83 Fisher, E., Dawson, A.M., Polshyna, G. et al. (2008). Transformation of inorganic and organic arsenic by *Alkaliphilus oremlandii* sp. nov. strain OhILAs. *Annals of the New York; Academy of Sciences* 1125: 237–241.

84 Sultana, M., Vogler, S., Zargar, K. et al. (2012). New clusters of arsenite oxidase and unusual bacterial groups in enrichments from arsenic-contaminated soil. *Archives of Microbiology* 194 (7): 623–635.

85 Rojas, L.A., Yáñez, C., González, M. et al. (2014). Characterization of the metabolically modified heavy metal-resistant *Cupriavidus metallidurans* strain MSR33 generated for mercury bioremediation. *Heavy Metal Contamination of Water and Soil: Analysis, Assessment, and Remediation Strategies* 6 (3): 305–3272.

86 Ge, S. and Ge, S.C. (2016). Simultaneous Cr(VI) reduction and Zn(II) biosorption by *Stenotrophomonas* sp. and constitutive expression of related genes. *Biotechnology Letters* 38 (5): 877–884.

87 Wu, D., Zhang, Z., Gao, Q. et al. (2018). Isolation and characterization of aerobic, culturable, arsenic-tolerant bacteria from lead–zinc mine tailing in southern China. *World Journal of Microbiology and Biotechnology* 34 (12): 177.

88 Dey, U., Chatterjee, S., and Mondal, N.K. (2016). Isolation and characterization of arsenic-resistant bacteria and possible application in bioremediation. *Biotechnology Reports* 10: 1–7.

89 Kang, C.H., Kwon, Y.J., and So, J.S. (2016). Bioremediation of heavy metals by using bacterial mixtures. *Ecological Engineering* 89: 64–69.

90 Tamindžija, D., Chromikova, Z., Spaić, A. et al. (2019). Chromate tolerance and removal of bacterial strains isolated from uncontaminated and chromium-polluted environments. *World Journal of Microbiology and Biotechnology* 40 (3): 307–332.

91 Liu, Q., Guo, H., Li, Y. et al. (2013). Acclimation of arsenic-resistant Fe(II)-oxidizing bacteria in aqueous environment. *International Biodeterioration and Biodegradation* 35: 56.

92 Drewniak, L., Ciezkowska, M., Radlinska, M. et al. (2015). Construction of the recombinant broad-host-range plasmids providing their bacterial hosts arsenic resistance and arsenite oxidation ability. *Journal of Biotechnology* 196–197: 42–51.

93 Bahar, M.M., Megharaj, M., and Naidu, R. (2016). Oxidation of arsenite to arsenate in growth medium and groundwater using a novel arsenite-oxidizing diazotrophic bacterium isolated from soil. *International Biodeterioration and Biodegradation* 106: 178–182.

94 Kao, A.C., Chu, Y.J., Hsu, F.L. et al. (2013). Removal of arsenic from groundwater by using a native isolated arsenite-oxidizing bacterium. *Journal of Contaminant Hydrology* 155: 1–8.

95 Jareonmit, P., Mehta, M., Sadowsky, M.J. et al. (2012). Phylogenetic and phenotypic analyses of arsenic-reducing bacteria isolated from an old tin mine area in Thailand. *World Journal of Microbiology and Biotechnology* 28 (5): 2287–2292.

96 Debiec, K., Krzysztoforski, J., Uhrynowski, W. et al. (2017). Kinetics of arsenite oxidation by *Sinorhizobium* sp. M14 under changing environmental conditions. *International Biodeterioration and Biodegradation* 119: 476–485.

97 Bhattacharya, A., Naik, S.N., and Khare, S.K. (2018). Harnessing the bio-mineralization ability of urease producing *Serratia marcescens* and *Enterobacter cloacae* EMB19 for remediation of heavy metal cadmium (II). *Journal of Environmental Management* 215: 143–152.

98 Kermani, J.N.A., Ghasemi, F.M., Khosravan, A. et al. (2010). Cadmium bioremediation by metal-resistant mutated bacteria isolated from active sludge of industrial effluent. *Iranian Journal of Environmental Health Science & Engineering* 7 (4): 279–286.

99 Kang, C.H., Han, S.H., Shin, Y. et al. (2014). Bioremediation of Cd by microbially induced calcite precipitation. *Applied Biochemistry and Biotechnology* 172 (4): 1929–1937.

100 Joo, J., Choi, J.H., Kim, I.H. et al. (2015). Effective bioremediation of cadmium (II), nickel (II), and chromium (VI) in a marine environment by using *Desulfovibrio desulfuricans*. *Biotechnology and Bioprocess Engineering* 20 (5): 937–941.

101 Guo, H., Luo, S., Chen, L. et al. (2010). Bioremediation of heavy metals by growing hyperaccumulaor endophytic bacterium *Bacillus* sp. L14. *Bioresource Technology* 101 (22): 8599–8605.

102 Banerjee, G., Pandey, S., Ray, A.K. et al. (2015). Bioremediation of heavy metals by a novel bacterial strain *Enterobacter cloacae* and its antioxidant enzyme activity, flocculant production, and protein expression in presence of lead, cadmium, and nickel. *Water, Air, and Soil Pollution* 226: 91.

103 Wajda, Ł., Duda-Chodak, A., Tarko, T. et al. (2017). Application of principal component analysis for the optimisation of lead(II) biosorption. *World Journal of Microbiology and Biotechnology* 33 (10): 1–10.

104 Wani, P.A. and Omozele, A.B. (2015). CR (VI) removal by indigenous *Klebsiella* species PB6 isolated from contaminated soil under the influence of various factors. *Current Research in Bacteriology* 8 (3): 62–69.

105 Ahmad Wani, P., Zainab, I.O., Wasiu, I.A. et al. (2015). Chromium (VI) reduction by *Streptococcus Species* isolated from the industrial area of Abeokuta, Ogun State, Nigeria. *Research Journal of Microbiology* 10 (2): 66–75.

106 Samantaray, D.P. and Misra, B.B. (2012). Effect of metal on hexavalent chromium reduction by *Acenetobacter Calcoaciticus* effect of metal on hexavalent chromium reduction. *The Bioscan* 7 (4): 627–629.

107 Kabir, M.M., Fakhruddin, A.N.M., Chowdhury, M.A.Z. et al. (2018). Isolation and characterization of chromium(VI)-reducing bacteria from tannery effluents and solid wastes. *World Journal of Microbiology and Biotechnology* 37: 126.

108 Mishra, V., Samantaray, D.P., Dash, S.K. et al. (2010). Study on hexavalent chromium reduction by chromium resistant bacterial isolates of Sukinda mining area. *Our Nature* 8 (1): 63–71.

109 Essahale, A., Malki, M., Marín, I. et al. (2012). Hexavalent chromium reduction and accumulation by *Acinetobacter* AB1 isolated from Fez Tanneries in Morocco. *Indian Journal of Microbiology* 52 (13): 48–53.

110 Pei, Q.H., Shahir, S., Santhana Raj, A.S. et al. (2009). Chromium(VI) resistance and removal by *Acinetobacter haemolyticus*. *World Journal of Microbiology and Biotechnology* 25 (6): 1085–1093.

111 Maqbool, Z., Hussain, S., Ahmad, T. et al. (2016). Use of RSM modeling for optimizing decolorization of simulated textile wastewater by *Pseudomonas aeruginosa* strain ZM130 capable of simultaneous removal of reactive dyes and hexavalent chromium. *Environmental Science and Pollution Research* 23 (11): 11224–11239.

112 Andreazza, R., Pieniz, S., Wolf, L. et al. (2010). Characterization of copper bioreduction and biosorption by a highly copper resistant bacterium isolated from copper-contaminated vineyard soil. *Science of the Total Environment* 408 (7): 1501–1507.

113 Ghosh, A. and Saha, P.D. (2013). Optimization of copper bioremediation by *Stenotrophomonas maltophilia* PD2. *Journal of Environmental Chemical Engineering* 1 (3): 159–163.

114 Achal, V., Pan, X., and Zhang, D. (2011). Remediation of copper-contaminated soil by *Kocuria flava* CR1, based on microbially induced calcite precipitation. *Ecological Engineering* 37 (10): 1601–1605.

115 Nag, S., Mondal, A., Roy, D.N. et al. (2018). Sustainable bioremediation of Cd (II) from aqueous solution using natural waste materials: kinetics, equilibrium, thermodynamics, toxicity studies and GA-ANN hybrid modelling. *Environmental Technology and Innovation* 11: 83–104.

116 Alburquerque, J.A., de la Fuente, C., and Bernal, M.P. (2011). Improvement of soil quality after "alperujo" compost application to two contaminated soils characterised by differing heavy metal solubility. *Journal of Environmental Management* 92 (3): 733–741.

117 Chen, M., Xu, P., Zeng, G. et al. (2015). Bioremediation of soils contaminated with polycyclic aromatic hydrocarbons, petroleum, pesticides, chlorophenols and heavy metals by composting: applications, microbes and future research needs. *Biotechnology Advances* 33 (6): 745–755.

118 Ok, Y.S., Uchimiya, S.M., Chang, S.X. et al. (2015). *Biochar Production, Characterization, and Applications*. New York;: CRC Press.

119 Mohan, D., Sarswat, A., Ok, Y.S. et al. (2014). Organic and inorganic contaminants removal from water with biochar, a renewable, low cost and sustainable adsorbent – a critical review. *Bioresource Technology* 160: 191–202.

120 Ahmed, M.B., Zhou, J.L., Ngo, H.H. et al. (2016). Progress in the preparation and application of modified biochar for improved contaminant removal from water and wastewater. *Bioresource Technology* 214: 836–851.

121 Rizwan, M., Ali, S., Qayyum, M.F. et al. (2016). Mechanisms of biochar-mediated alleviation of toxicity of trace elements in plants: a critical review. *Environmental Science and Pollution Research* 23 (3): 2230–2248.

122 Klüpfel, L., Keiluweit, M., Kleber, M. et al. (2014). Redox properties of plant biomass-derived black carbon (biochar). *Environmental Science and Technology* 48 (10): 5608–5611.

123 Saquing, J.M., Yu, Y.H., and Chiu, P.C. (2016). Wood-derived black carbon (biochar) as a microbial electron donor and acceptor. *Environmental Science and Technology Letters* 3 (2): 2230–2248.

124 Yuan, Y., Bolan, N., Prévoteau, A. et al. (2017). Applications of biochar in redox-mediated reactions. *Bioresource Technology* 246: 271–281.

125 Violante, A., Cozzolino, V., Perelomov, L. et al. (2010). Mobility and bioavailability of heavy metals and metalloids in soil environments. *Journal of Soil Science and Plant Nutrition* 10 (3): 268–292.

126 Senthil Kumar, P. and Gunasundari, E. (2017). *Bioremediation of Heavy Metals. Bioremediation: Applications for Environmental Protection and Management*. Singapore: Springer.

5

Bioremediation of Heavy Metal Ions Contaminated Soil

Agnieszka Saeid[1], Liliana Cepoi[2], Magdalena Jastrzębska[3], and Philiswa N. Nomngongo[4]

[1] *Wroclaw University of Science and Technology, Wroclaw, Poland*
[2] *Institute of Microbiology and Biotechnology, Chisinau, Republic of Moldova*
[3] *University of Warmia and Mazury, Olsztyn, Poland*
[4] *Department of Chemical Sciences, University of Johannesburg Doornfontein Campus, Johannesburg, South Africa*

5.1 Introduction

In the view of the current situation of the environment, the disturbance of its equilibrium, and the incorrect balance of unwanted elements in parts of the natural ecosystem such as in the soil, water, or air enforces steps to restore the natural environment. The presence of unwanted elements is observed on different trophic levels, which have a direct impact on human health. Unsolved issues with the utilization of waste originating in industry, agriculture, and waste processing sectors concerning its processing and management have led to increased accumulation of toxic elements in agricultural soils over the past few decades.

Soil pollution is a huge worldwide problem; there are over 2.5 million potentially contaminated sites in Europe [1]. Extreme accumulation of pollutants in agricultural soils not only causes environmental pollution but also results in the appearance of toxic substances in crops, which can directly affect food quality. Soil plays a central role in food safety as it determines the potential composition of food at the root of the food chain. The quality of soil resources, as defined by their potential impact on human health by the propagation of harmful elements through the food chain [2], should be controlled worldwide.

Variations of pollutant concentrations in the soil are dependent on natural and anthropogenic processes (leaching, mechanical transport, plowing,

Soil Bioremediation: An Approach Towards Sustainable Technology, First Edition.
Edited by Javid A. Parray, Abeer Hashem Abd Elkhalek Mahmoud, and Riyaz Sayyed.

landscaping). Toxic substances tend to accumulate in the topsoil, appearing in the soil biocoenosis. The main group is heavy metals, which are not biodegradable and can easily accumulate in living organisms. Heavy metal(loid)s commonly present in soils include nickel (Ni), lead (Pb), cadmium (Cd), arsenic (As), chromium (Cr), copper (Cu), cobalt (Co), zinc (Zn), manganese (Mn), aluminum (Al), and mercury (Hg). Among these heavy metal(loid)s, As, Pb, Cd, and Hg are included in the top 20 Hazardous Substances of the Agency for Toxic Substances and Disease Registry [3] and the United States Environmental Protection Agency [4, 5]. Polluted soils significantly affect soil microbial communities by decreasing the biomass of soil microbes and the number of active bacteria colonies [6]. More importantly, some of them tend to accumulate in the food product that has significant importance on the human diet [7]. Although some trace elements are essential for living organisms (i.e., Cu, Mn, Zn), they may also be toxic in excess. In contrast, other elements (i.e., Cd, Pb, Hg) may be toxic, even present in very small quantities. Toxic contaminants may move into the aquatic ecosystem, influencing its performance, affecting aquatic living organisms and other organisms in the trophic chain, including humans [8].

Heavy metal pollution in the soil, and the bioconcentration and transfer features of the soil–vegetation (root, stem, and leaf) system are not well known. Currently progressing works aim to discover the characteristics of the transfer and bioconcentration of unwanted elements in the environment. Investigations on the status of heavy metal pollution within the ecosystem, trophodynamics, bioconcentration factor (BCF), and the translocation factor (TF) of trace metals in the food chain are imperative in order to implement habitat protection and appropriate conservation measures [9]. To estimate the transferability of heavy metals from soil to crop, two indices, the BCF and the TF can be used [10]. $BCF = C_{root}/C_{soil}$, where C_{root} and C_{soil} is the concentration of each heavy metal in the crop root and the corresponding value in the soil, respectively, but also BCF can be expressed as the general ratio of concentration of toxic elements in the biomonitor (plant/animal) and concentration of toxic elements in water or soil. While the $TF = C_{grain}/C_{root}$, where C_{grain} is the amount of heavy metal in the crop grain [10] and expresses the transfer within the plant.

Biomagnification of heavy metals in organisms from different trophic levels (predator–prey relationship) can be quantified and evaluated by calculating the Biomagnification Factor (BF) [11]. $BF = C_{prey}/C_{predator}$, where: C_{prey} is the concentration of toxic elements in the prey and $C_{predator}$ is the concentration of toxic elements in the predator.

Bioremediation is a technique that applies some of the organisms, which express a tendency to accumulate a significant quantity of toxic elements when present in the environment, for use as a medium to uptake unwanted elements

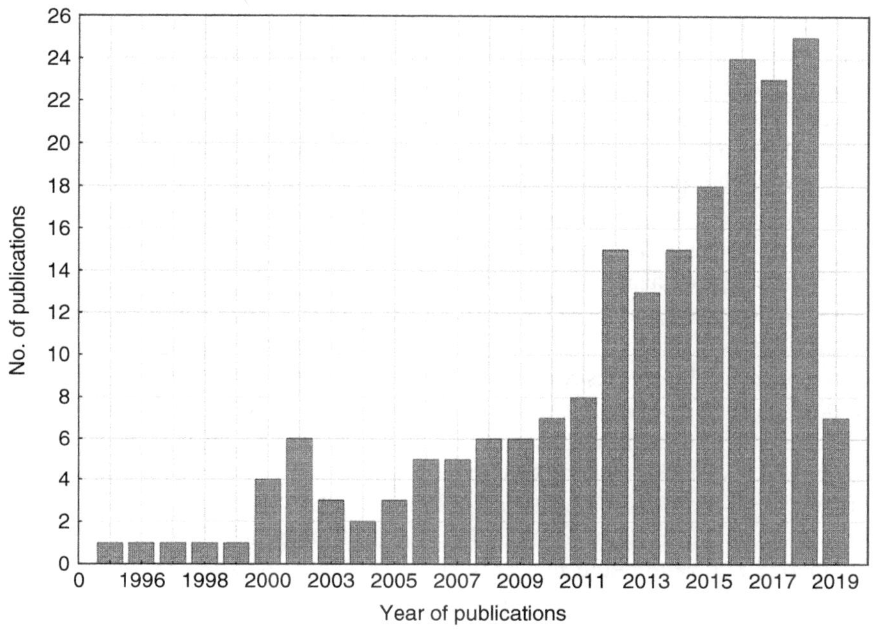

Figure 5.1 The numbers of articles published since 1995 (according to the Web of Science, April 11, 2019).

from the environment without side effects for itself. It is often used for heavy metal removal from the environment utilizing inherent biological mechanisms to eliminate or reduce amounts of toxic contaminants using microorganisms, plants, or their products to restore contaminated environments to their original state [12]. Many examples of hyperaccumulators and heavy-metal tolerant plants/animals provide a broad prospect for the bioremediation of environmental pollution. This is reflected in the increased number of articles concerning bioremediation of heavy metal ions contaminated soil published since 1995 (Figure 5.1).

5.2 Bioremediation

The role of metals in life processes is enormous. The confirmation of this is the development of a new research field, directed toward the studies or the role that metal ions have in the biological world, including their trafficking, uptake, transport, and storage. Since 2004 this research direction was defined by a special term – "metallomics" [13, 14]. Metal ions such as Zn^{2+}, Fe^{3+}, Cu^{2+}, Ni^{2+}, Mo^{2+}, and Mn^{2+} are part of numerous metal-enzymes, without which vital basic processes like the synthesis of nucleic acids and proteins, respiration, and nitrogen fixation are impossible. The biological effect of any metal is determined by its content of its cells and the environment. High concentrations of metals are mostly toxic. The toxic effect is determined by the accelerated formation of free radicals, which attack the biomolecules, altering their structure and functions [15].

Due to anthropogenic activities, the exposure of the environment to heavy metals and their accumulation in water and soil is significantly increased. Some natural sites accumulate such quantities of metals that all living organisms have to endure major hazards. At the same time, many microorganisms (bacteria, fungi, and microalgae) have the exceptional ability to resist and populate soil environments that have a high content of heavy metals. The comparison between different microbial species reveals that heavy metal resistance was provided by a multistage system. Some elements of this system are widespread and ensure a basic defense of the cells against heavy metals, while others are very specific [16]. The protective mechanisms against metals toxicity employed by microbial species are: metal elimination, extracellular and intracellular sequestration, biosorption, precipitation, production of chelating agents, intracellular bioaccumulation, compartmentalization, and biotransformation by specific enzymes and cellular metabolites[16–18]. From our point of view, all these mechanisms serve to achieve three major strategies of resistance of microbial cells to heavy metals: metals transport, metal sequestration, and metal biotransformation.

The strategies and mechanisms of microbial resistance to high concentrations of heavy metals can serve as effective tools for their use in soil bioremediation. Bioremediation is the microorganisms-mediated process for clearance or immobilization of the contaminants. If some organic toxic compounds can be completely cleansed from the polluted medium by the microorganisms, these are unable to convert heavy metals into harmless compounds. In fact, microorganisms should be used in the bioremediation process according to their specialization for the type of contaminants and specific detoxification mechanisms [19].

5.2.1 Metals Transport

Metal influx/efflux are essential both to ensure the needs of cells in metals, and for the protection of microorganisms from the toxic action of these xenobiotics, as well as for the bioremediation of a contaminated environment. The influx of metals across bacterial membranes include siderophores, ion pumps, ion channels, carrier-mediated transport, endocytosis, complex permeation, and lipid bilayer permeation. This mechanism has been reported to be associated with the transport of iron, mercury, lead, silver, cadmium, nickel, and other heavy metals [20, 21]. Efflux of the metals from microbial cells is a mechanism of metal homeostasis maintenance inside cells. The microorganisms use energy-dependent metal efflux, which effectively pumps toxic ions that have entered the cell back out of the cell [22]. The process is carried out by transmembrane transporters. In gram-negative bacteria the efflux of metals occurs in two stages: first – the transport of metals from the cytoplasm to periplasm; and second – the transport of metals from the periplasm to the extracellular space. Both steps are carried out by several families of transmembrane transporters such as P-type ATPase, cation/proton antiporter, cation diffusion facilitator transporters, resistance-nodulation-cell division (RND) proteins (which is also called the CBA efflux system), membrane fusion proteins, and outer membrane factors [16, 23, 24]. Pumping systems are involved in Cu, Cd, Zn, Co, and Ni efflux from bacterial cells. Efflux of Cd, Zn, and Co from gram-negative bacteria cells is provided by a cation-proton antiporter (CzcABC), encoded by specific *czc* genes; Cd efflux from gram-positive bacteria cells is ensured by Cd-efflux ATPase; Cu elimination is associated with an ion-dependent Cu antiporter [25–27].

5.2.2 Extracellular Sequestration

Extracellular sequestration of heavy metals is carried out by the microbial capsule, which consists predominantly of exopolysaccharides (EPS), and of a minor fraction of proteins, nucleic acids, humic substances, lipids, and some low molecular weight compounds. Extracellular sequestration of metals can also be performed by the cell wall. This process is known as biosorption. Capsules and cell walls provide heavy metal sequestration to restrict their penetration inside the microbial cells. The process is possible due to the presence of negatively charged groups: carboxyl, hydroxyl, sulfonate, amine, sulfhydryl, and phosphonate [19, 24, 28]. In gram-positive bacteria the peptidoglycan layers that contain alanine, glutamic acid, meso-di-aminopimelic acid, and teichoic acid play the main role in biosorption. In gram-negative bacteria the enzymes, glycoproteins, lipopolysaccharides, lipoproteins, and phospholipids of the cell walls ensure metal sequestration [29]. The EPS of the cellular capsule and the components of the bacterial cell wall are the active sites that act as ligands for binding metal ions, resulting in their ultimate remediation from contaminated environments [30]. The major polymers implicated in metal sequestration in fungi are glomalin, chitin, and melanin, which can take part in biosorption, especially for Cu [31].

Extracellular metal chelation by siderophores, small molecular mass metabolites, peptides, and proteins eliminated by cells, is another way to achieve the sequestration of metals. Bacterial and fungal cells can secret glutathione, oxalates, and extracellular mucilaginous materials with excellent toxic metal binding capabilities [32, 33].

5.2.3 Intracellular Sequestration

Intracellular sequestration of heavy metals is carried out by metal-binding metallothioneins, phytochelatins, and siderophores, which play a significant role in immobilization of metal ions [17, 22]. Metallothioneins are low molecular weight proteins rich in cysteine with a high affinity for cadmium, zinc, copper, silver, and mercury. Their primary function is metal detoxification. Metallothionein synthesis in microbial cells is induced by the presence of a toxic quantity of metals. Metal binding by metallothioneins can result in cellular accumulations [22]. Phytochelatins – oligomers of glutathione are typical for eukaryotic microorganisms (such as fungi, microalgae), and cyanobacteria. They are small molecules consisting of 2–11 repeats of γ-Glu-Cys dipeptide with a terminal Gly residue. They are capable of binding to various metals including Cd, Cu, Zn, or As via the sulfhydryl and carboxyl residues [34]. Siderophores are small organic molecules produced by microorganisms under iron-limiting conditions, which enhance microbial iron uptake. Although the principal role of siderophores is to chelate ferric iron, they can also detoxify heavy-metal-contaminated samples by binding to a wide array of toxic metals [35, 36].

Another strategy used by microorganisms, which accumulate metals against the concentration gradient, is compartmentalization. Metal compartmentalization occurs in many species of bacteria, yeasts, fungi, and microalgae. As a result of this process, metals become less toxic to cells, and microorganisms that possess compartmentalization mechanisms can accumulate metals at high intracellular concentrations. It is therefore important to know that metal compartmentalization is a short-term strategy, and subsequent metals should be a subject to efflux or other processes, including biotransformation. In eukaryotic microorganisms, e.g., algae, yeasts, and fungi, compartmentalization of metals (Co^{2+}, Mn^{2+}, Zn^{2+}) occurs in vacuoles, where they can remain in ionic form or binding with polyphosphates [37].

5.2.4 Biotransformation

Biotransformation of metals is the most important detoxification mechanism performed by microorganisms. In this process, the toxic metal is converted to less toxic or easily recoverable forms. Biotransformation can occur in two ways: by oxidation–reduction or by mineralization–demineralization reactions [38].

Some metals such as arsenic, selenium, chromium, and uranium are utilized by microorganisms as final electron acceptors for anaerobic respiration. In this process, the metals undergo dissimilatory reduction [39]. Certain microorganisms may possess reduction mechanisms that are not coupled to respiration but instead are thought to impart metal resistance [22]. For example, the microbial consortia can perform aerobic and anaerobic reduction of Cr(VI) to Cr(III), having less mobility and toxicity [40]. Dissimilatory reduction of selenate is known to support the anaerobic respiration in several microorganisms, and the dissimilatory reduction of soluble selenate and selenite to nanoparticulate elemental selenium greatly reduces the toxicity [41]. Indigenous microbes in varied sediments reduced U(VI) to insoluble U(IV) [42]. Intracellular reduction of As^{5+} to As^{3+} prior to extrusion or cytoplasmic sequestration is crucial in prokaryotic arsenic detoxification. Toxic As^{3+} resulting from arsenate reduction could be further inactivated by volatilization [43]. Generally, volatilization mechanisms involve turning metal ions into a volatile state. This is only possible with Se, As, and Hg, which have volatile states.

Many microbial transformations of metals occur due to their energy substrate use in which the metal is oxidized. A number of metals and metalloids are subject to oxidation including iron, manganese, selenium, and arsenic. Oxidation products may be less toxic. For example, As(III) oxidation can serve as a detoxification reaction since As(III) is up to 50 times more toxic to bacterial cells than As(V) in most biological systems [22].

Reactions between metals and various minerals are essential in biogeochemical cycles. A significant role in these reactions is played by microorganisms, which have mechanisms to solubilize metals from minerals and mechanisms of metal mineralization. These reactions change the degree of metal mobility. Biomineralization of metals can lead to removing metals from solutions providing a means of detoxification and metal biorecovery. The most common metal precipitates within microbial transformation include oxides, phosphates, sulfides, and oxalates [44].

An example of mineralization reaction is methylation–demethylation. For example, methylation is inherent to mercury, lead, selenium, arsenic, tin. The formed organometals are more lipophilic than the metal species and can be actively accumulated into the cells. The result of methylation (addition of methyl or ethyl groups to metal) can increase the metal volatility and metal toxicity, as in the case of mercury, which is transformed by methylation from Hg^{2+} to CH_3Hg^+. This reaction is carried out by anaerobic soil bacteria. Aerobic mercury-resistant bacteria perform the demethylation process – convert methyl mercury to Hg(II), which is 100-fold less toxic than methyl mercury. Arsenic is another example of a metal that is methylated as a resistance mechanism. It is methylated by some bacteria, such as *Rhodopseudomonas palustris*, and many fungi, such as *Scopulariopsis brevicaulis* to mono-, di-, and tri-methylarsine, volatile forms of arsenic. However, methylation of some metals, such as selenium, decreases their toxicity. By means of methylation, a considerable quantity of metals is removed from contaminated soils [22, 45]. Biotransformations are important components of metal biogeochemical cycles and may also be exploited in bioremediation of metal contaminated soils.

5.2.5 Limits and Prospects

Bioremediation technology is a good option as compared to other conventional technologies in terms of cost, eco-friendship, socially acceptance, and sustainability. The process of bioremediation is widely explored. Many bacteria, fungi, cyanobacteria, and microalgae with bioremediation potential have been isolated and characterized [46].

The process of bioremediation of soil contaminated with heavy metals can complete with two different results. The first result of soil bioremediation process is metal removal. Total metal removal can be achieved in *ex-situ* technologies by removing the biosorbent or metal precipitate, but it is also possible *in situ* (for example, by metal volatilization). For this purpose, it is necessary to increase metal solubility and enhance their recovery. Metal recovery is an expensive process, but it can be considered as ideal because after recovery-technology application the treated soil is available for reuse [22].

The second result consists in immobilization of the metals by their binding to different chelating agents, eliminated by soil microbiota, or by reduction reactions. As results of these reactions the metal solubility and toxicity decrease. *In-situ* metal sequestration technologies, due to their easy operation, low costs, and rapid remediation effects, are applied widely. However, the immobilized toxic metal remains in the soil, and its reuse may be limited because of the continued potential risk [18, 22, 47].

Among the limitations of the bioremediation approach are low adaptability, time-consuming, lack of microbial competitiveness, and low bioavailability to the target pollutants [46]. Despite the advances reached in bioremediation, most of the research concerning this issue is limited to laboratory scale studies. Moreover, a deeper understanding of the complexity of bioremediation process is necessary to transfer the methods and techniques from basic research into applied science and for optimization *in situ* remediation using microorganisms [48].

The clean-up of polluted environments using indigenous microorganisms has not yielded many positive results [17]. One of the causes of this phenomenon is the study of separated species with high resistance to pollutants. For bioremediation purposes, it is necessary to know not only the cellular processes and metabolic activity in reference to a particular species, but also, the microbial community composition and its metabolic activity in the presence of stress induced by toxic compounds [46]. From this point of view, *systems biology* and *metabolic engineering* approaches are attractive for bioremediation studies. Systems biology, which is used to study intricate biological systems at the molecular, cellular, population, community, and ecosystem levels, provides valuable detailed information about gene expression, enzymes, biosynthetic pathways, and secondary metabolites in microbes and can reveal a modification of already existing pathways under conditions of stress caused by various pollutants. The "omics" techniques (genomics, proteomics, transcriptomics, metabolomics, and metallomics) are extensively used to study microbial bioremediation [46, 49]. Metabolic engineering uses the information, offered by systems biology for alteration of microbial metabolic pathways to neutralize the particular single or multiple pollutants, including heavy metals [50].

An efficient way to remove toxic metal contaminants from the environment and stabilize the ecosystem is to make use of genetically engineered microorganisms to treat polluted environments by converting toxic heavy metals into nonhazardous forms. For example, indigenous bacteria cannot remove heavy metals such as Hg from the environment. However, recombinant DNA technology plays a major role in bioremediation of heavy metal contamination [17]. Genetic engineering technology has been focused on the development of genetically modified microorganisms that are used for the removal of metals from polluted sites [51]. The most promising targets in the genetic engineering of heavy metals are tolerant microorganisms that include: increased secretion of extracellular metal chelators; overexpression of metal transporters; overproduction of intracellular metal chelators; and genetic modification of the regulatory network of heavy metal stress defense [32]. For example, expression of phytochelatins in the bacterial host increased the accumulation of intracellular arsenic 50-fold by sequestering arsenic ions (likely arsenite) in a nontoxic form and competing with the arsenic efflux transporter [52]. *Escherichia coli* and *Moreaxella* sp., which express phytochelatin on the cell wall can accumulate 25 times more Cd or Hg compared to the wild-type strains [51].

Transgenic technologies have provided opportunities to remove metals from contaminated soil. In order to improve bioremediation performances, genetically modified microorganisms could be as efficient as inoculants. Although currently there is no evidence that the deliberate release of engineered bacteria has caused a measurable antagonistic influence on the natural microbiota, the risks associated with such microorganisms are still unclear. So, considering safety measures, the prospects of engineered bacteria need extensive research [53]. The successful application of soil bioremediation process in practice is possible but requires interdisciplinary involvement with engineering, nanotechnology, microbiology, geology, ecology, and chemistry [18]. The microbial community of polluted sites must be treated as a whole. The bioremediation potential of this community can be enhanced by genetically modified microorganisms. Systems biology and metabolic engineering approaches are best suited for this purpose.

5.3 Phytoremediation

Consistent with the origin of the term (*phyto* = plant in Greek, and *remedium* = restoring balance in Latin), phytoremediation comes down to the use of plants to remediate contaminated sites/medium (soil, water or sediments). Phytoremediation (also: phytotechnology/ies) is often described as a "new" technology, as it was initially used on a technological scale in the early 1980s [54]. However, its earliest beginnings date back to the discovery of metal tolerant plants in the sixteenth century [55]. A recent increase in the number of concepts in phytoremediation and new terms introduced for various processes, mechanisms, or techniques may lead to confusion, all the more so as the nuances between concepts are often narrow [56].

Phytoremediation includes the removal of pollutants (phytoextraction), their extraction from aqueous solution (e.g., rhizofiltration), transformation (e.g., phytodegradation, phytovolatilization), or immobilization (e.g., phytostabilization) [56].

5.3.1 Mechanism

Phytoremediation of toxic heavy metal and metalloid (hereafter termed "metal/s") contaminated soils is primarily based on the practical use of the plant's physiological response to the presence of these toxic elements in the environment [57]. Plants able to colonize metal-contaminated soils (metallophytes) evolved mechanisms to deal with this stress factor and are classified as resistant [58]. Heavy metal resistance can be achieved by avoidance and/or tolerance. Avoiders are plants that are able to protect themselves by preventing metal ions from entering their cellular cytoplasm, while tolerant plants are able to detoxify metal ions that have crossed the plasma membrane or internal organelle biomembranes. Based on strategies used by metallophytes they are classified into three categories: metal excluders, indicators, and accumulators [59]. Metal excluders limit the translocation of heavy metals and maintain low levels of contaminants in their aerial tissues over an extensive range of soil concentrations. Metal indicators accumulate metals in their harvestable biomass and these levels are generally reflective of the metal concentration in the soil. Metal accumulators can concentrate metal in their harvestable parts to levels far exceeding those in the soil. Plants that can absorb high levels of contaminants concentrated either in their roots, shoots, and/or leaves are termed hyperaccumulators [60]. Plant accumulators can cope with the effects of high internal metal concentrations through the upregulation of the antioxidant defense system [61]. This system is activated in order to respond to the deleterious effects caused by reactive oxygen species, making a plant tolerant and enhancing its metal-accumulating ability [58, 61].

Biogeochemical behavior of heavy metals in the soil–plant system is of supreme importance in understanding phytoremediation mechanisms. In general, the same physiological and genetic mechanisms are involved in the uptake, translocation, and storage of both toxic and essential elements [62]. These processes have been extensively studied, however, there are still issues to be explained more fully [63]. In order for plants to take metal from the soil solution, it must present it in a biologically available form. In soil, heavy metals exist in colloidal, ionic, particulate, and dissolved states. They also have a high affinity for humic acids, organo clays, and oxides coated with organic matter [60]. Only free metal ions and soluble metal complexes in the soil solution, and possibly metal ions occupying ion exchangeable sites and specifically adsorbed on inorganic soil constituents are readily available to plants [64]. The solubility of metals in soil is mainly controlled by soil pH, amount of metal, other elements presence and concentration, cation exchange capacity, organic carbon content, the oxidation state of the mineral components, and the redox potential of the system [60]. Soil pH seems to be the most important: a greater retention and lower solubility of metals occur at a high soil pH.

Plants possess highly specialized mechanisms both to stimulate and reduce metal bioavailability in the soil [65]. Plant root exudates, consisting of a complex mixture of organic acid anions, phytosiderophores, sugars, vitamins, amino acids, purines, nucleosides, inorganic ions (e.g., HCO_3^-, OH^-, H^+), gaseous molecules (CO_2, H_2), enzymes, and root border cells, have major direct or indirect effects on the metal ions' mobilization [66]. They can change soil properties determining metal solubility/mobility or bind metals into complex compounds, both soluble and insoluble. There can be a parallel mobilization and immobilization of metals by the same procedure, depending on the conditions [67].

After being mobilized, the metal may be caught by the root cells. Root cells can regulate to a limited extent the quantity of heavy metals taken from the environment. The root cell wall is the first barrier for metals. The functional groups (–OH, –COOH, –SH) present in the plant cell wall are responsible for metal binding via hydrogen bonding [68]. It was proved that metals associated with cell wall pectins are not completely immobilized – they can be taken by the plant by endocytosis of the metal-pectin complex. Under stress conditions induced by the presence of metals, the physical properties of the cell wall can be changed [69]. Primarily, the number of transverse bonds between wall constituents (protein molecules, saccharides, and phenols) is increased. The stress can also lead to lignification or suberization of the cell wall, which causes the wall to become more compact and stiffer, and thus less permeable [58, 61].

The plasma membrane is the first living selective barrier determining metal entry into the cytoplasm. The range of known transport mechanisms or specialized proteins embedded in the plant cell plasma membrane involved in ion uptake and translocation include: (i) proton pumps (H^+-ATPases that consume energy and generate electrochemical gradients), (ii) co- and anti-transporters (proteins that use the electrochemical gradients generated by H^+-ATPases to drive the active uptake of ions), and (iii) channels (proteins that facilitate the transport of ions into the cell) [62]. Special plant membrane proteins recognize the chemical structure of essential metals and bind them, and they are then ready for uptake and transport [70]. As these transporters are not characterized by high specificity, they may also take nonessential metals the chemical characteristics or size of which are close to the essential ones. Proteins taking metals into the cell often show specificity to metal ions at a particular degree of oxidation. Low substrate specificity of some metal transporters may lead to the accumulation of metals in plants [58]. Plants may decrease the uptake of metal ions through modified ion channels, and/or by removing metals from cells with active efflux pumps and/or with root exudates.

After heavy metals have entered the roots they are either stored in the root (a majority of plants, including excluders) or translocated to the shoots and stored there (hyperaccumulators) [60]. In hyperaccumulators, metals are efficiently loaded to xylem and transported to the shoots. It should also be noted that metals may be distributed within the plant via phloem [71]. Metal tolerant plants can sequester the surplus of metals in cell walls and extracellular spaces, and/or intracellular compartments, i.e., organelles where the metals cannot interfere with the physiological activities of the cells [72]. The vacuoles are important components of metal ions storage. Heavy metals may also be deposited inside idioblasts, trichomes, cuticle, or in some plant organs, the latter often being shed (e.g., older leaves). Metals may be excluded via stomatal cells, hydathodes, or ectodesms [61]. Metals are translocated within the plant both in the form of free hydrated ions or in complex with ligands [73]. Chelation of metal ions by specific high-affinity ligands reduces the solution concentration of free metal ions, thereby reducing their phytotoxicity. The chelation of metals allows for the restriction of metal uptake, the uptake of metal ions, their transport within the plant and compartmentation [58]. Ligands for a number of metals have been identified and proteins involved in the transport of these ligands and of metal–ligand complexes [73]. Metal chelation can be performed by compounds of both thiol origin (glutathione, phytochelatins, metallothioneins) and nonthiol origin (histidine, nicotianamine, organic acids) [74]. Transporters belonging to several protein families (e.g., P-ATPase – subfamily P_{1B}, CDF, NRAMP, ZIP, YSL, ABC) participate in the uptake and transport of metal ions through membranes (plasma membrane, intracellular membranes, e.g., tonoplast) [75]. Plant heavy metal-transport proteins can be classified as metal-uptake proteins (e.g., ZIP, NRAMP, and YSL) and metal-efflux proteins (e.g., CDF, P_{1B}-ATPases, ABC) [76]. The former can transport metals into cytoplasm, while the latter are a group of detoxification proteins, which can efflux excess and toxic heavy metals from the cytoplasm, or move these metals into the vacuole. The energy necessary for transporting metals through membranes comes from two sources: from the decomposition of ATP or from the primary proton motive force produced by proton pumps [75].

The phytoremediation of heavy metals from soil is based on three techniques: phytoextraction, phytostabilization, and phytovolatilization [77]. Phytoextraction refers to the removal of metal contaminants from the soil through their uptake by plant roots and translocation into the aboveground parts of the plants (shoots, leaves, etc.) [77]. According to current knowledge phytoextraction mechanism has five major steps: (i) metal mobilization by secretion of chelates and by acidification of the rhizosphere, (ii) metal ion or metal-chelate complex uptake by plant roots, (iii) translocation toward above ground plant parts, (iv) metal sequestration in plant tissues, and (v) heavy metal tolerance [78]. Flowingly, the plant biomass rich in metals is removed from the polluted place. Cropping of plants in contaminated soil is repeated until the metal concentration drops to an acceptable level.

Huge quantities of harvested contaminated (hazardous) biomass need proper disposal and management [60]). Direct disposal (deposition at landfill sites) is the least complicated approach, but it would cause environmental problems. Recycling, i.e., further utilization as a secondary raw material for some industries (e.g., timber, packaging, furniture), as a renewable energy source (biogas, biofuels, and combustion for energy production and heating), and/or for metal recovering (phytomining, "bio-ore"), is claimed to be more reasonable than deposition on dumps [60, 79, 80]. Treatment options of contaminated biomass (e.g. anaerobic digestion, composting, compaction, incineration, pyrolysis, gasification, liquid extraction, thermal-oxidation) were critically reviewed [81, 82]. Medicinal purposes are also being studied [83].

Phytostabilization refers to immobilization of a metal in soil through absorption and accumulation by roots, adsorption the root surface, or precipitation within the root zone of plants [84]. Unlike other technologies of phytoremediation, phytostabilization does not mean the physical removal of the pollutant from the soil. Heavy metals are retained in the place of contamination, but their mobility and bioavailability are limited, and consequently, toxic effects on other ecosystem components are reduced. Phytostabilization is used when soils are heavily contaminated (e.g., mine tailings) and the removal of metals using plants is certainly difficult or would take an unrealistic amount of time [85]. This strategy focuses on creating a vegetation cover, the biomass of which is not removed and processed further. The role of vegetation in stabilizing heavy metals is multidimensional. According to Bolan et al. [84] it is based on the following processes: (i) establishment of vegetation barrier that reduces the likelihood of physical contact with the soil by animals and humans, (ii) mechanical stabilization of the site to minimize erosion by wind and water, (iii) enhancement of evapotranspiration, thereby reducing the leaching of contaminants, (iv) alteration of soil factors that influence the speciation and immobilization of contaminants (pH, organic matter, redox levels), (v) root exudates that regulate the precipitation and immobilization of the contaminants, and (vi) uptake and sequestration of contaminants in the root system.

Phytovolatilization is an approach that involves the conversion of the absorbed metal into a volatile form and its release into the atmosphere [60]. This strategy is unique to a select group of heavy metals including selenium (Se), mercury (Hg), and arsenic (As). It was proved that volatile forms of Se and Hg, which can be lost to the atmosphere, are dimethylselenide [86] and Hg^0 [87], respectively. In turn, Sakakibara et al. [88] determined that vapor released from the frond of *Pteris vittata* included arsenic compounds, arsenite and arsenate.

According to many authors (i.e., [78, 84, 89]), there are several parameters that predestine plants to soil remediation through phytoextraction (*pe*), phytostabilization (*ps*), and/or phytovolatization (*pv*): (i) tolerance to the toxic effects of the target heavy metals and other stress factors (drought, compaction, extreme acidity/alkalinity, excess salinity, low/no nutrients), along with good adaptation to prevailing environmental conditions (*pe*, *ps*, *pv*); (ii) high biological absorption coefficient (BAC) of target heavy metals (>1) (*pe, ps, pv*) and high (>1; *pe*, *pv*) or low (<1, *ps*) TF; (iii) well-developed root system, high growth rate, and large above-ground biomass (*pe*, *ps*, *pv*); (iv) resistance to pathogens and pests (*pe*, *ps*, *pv*); (v) no attraction to herbivores (*pe*, *ps*, *pv*); and (vi) easy cultivation and harvest (*pe*, *pv*?). Plant hyperaccumulators are preferred for phytoextraction, while excluders are preferred for phytostabilization.

5.3.2 Obstacles and Prospects

Phytoremediation is commonly considered as an eco-friendly (nondestructive, solar driven) and relatively inexpensive strategy for remediation of heavy metal contaminated soil [90]. Being a promising approach, phytoremediation that involves natural accumulators/hyperaccumulators and/or excluders faces some limitations. Primarily, the efficacy of most metal natural phytoremediators is generally restricted by their low biomass and slow growth rate [78]. What is more, as the plant's heavy metal resistance is limited, even in the case of hyperaccumulators, phytoremediation is usually applicable for sites that have low to moderate levels of metal pollution; in highly contaminated soils plant growth is unsustainable. Abiotic (e.g., climate, weather) and biotic (e.g., diseases, pests) factors may also reduce a plant's usefulness for phytotechniques. There is also a risk of further interferences in ecosystems: metals accumulated in plant biomass (both below and above ground) may enter into food chains and the introduction of invasive plant species as phytoremediators may affect indigenous floral diversity. Lengthy period (several years) needed for soil phytoremediation is often indicated as well. For phytoextraction, a significant restriction is limited by the bioavailability of heavy metals in the soil. Correct disposal of the polluted biomass is also a challenge [80]. Due to the fact that volatile metal may return to the ecosystem by precipitation, phytovolatilization seems a questionable method [88].

Considering the limitations of the traditional phytoremediation, modern tools and approaches have been involved to enhance its efficiency. The improvement was reached through application of amendments, agricultural practices, microbes involving/introduction, and genetically engineered plants ("designer" plants) [90]. However, this new approaches for phytoremediation may pose new constraints and concerns.

Using of synthetic chelates (e.g., EDTA, HEDTA, DTPA, EGTA, NTA, EDDS, EDDHA) to stimulate metal uptake by plant roots (chelate-assisted phytoextraction) and the application of various organic and inorganic soil amendments (e.g., zeolites, phosphogypsum, biosolids, sewage sludge, composts) to promote metal fixation by plants (aided phytostabilization) were reported (i.e. [77, 84, 89, 91]). When it comes to agricultural practices, fertilization can promote plant growth and enhance the resistance of plants, tillage is beneficial – turning heavy metals into the rooting zone, which promotes the contact between plants and heavy metals, crop rotation, and intercropping or interplanting are beneficial to reduce the influence of weeds, diseases, and insect pests [92]. In addition, the application of more than two hyperaccumulators can improve remediation efficiency. Indigenous plant-associated (rhizosphere microbes and endophytes) and introduced plant growth promoting microorganisms (microbe-assisted phytoremediation) may affect metal mobility/phytoavailability, plant growth, tolerance to metal and other stress factors, and metal accumulation through increasing the absorptive surface area of plant roots (mycorrhizal fungi hyphal network) and production of various metabolites (e.g., phytohormones, siderophores, antibiotics, enzymes, organic acids, biosurfactants) [93]. On the other hand, some of the above factors that increase the mobility of metals may result in their increased release into groundwater.

Genetic engineering approaches to enhance plant tolerance and the accumulation of toxic metals holds great potential for phytoremediation [94]. Recently, several studies employing "omics" technologies (genomics, transcriptomics, proteomics, metabolomics) have been carried out to elucidate the genetic determinants and pathways involved in heavy metal tolerance in plants. Three main biotechnological approaches are being used to engineer plants for phytoremediation of heavy metals: (i) manipulating metal transporter genes and uptake systems, (ii) enhancing metal ligand production, and (iii) conversion of metals to less toxic and volatile forms [95]. Despite recent progress in biotechnology, its potential for phytoremediation has still not been fully exploited [95].

Transgenic plant-mediated phytoremediation poses potentially negative ecological consequences (gene flow, potentially invasive species, the impact of increased toxicity of heavy metals-enriched biomass on wildlife *in situ*), and the legal and societal restrictions limit their potential for land applications [94, 96]. Conventional breeding and similar traditional methodologies (e.g., chemical mutagenesis, somatic hybridization) have been seen by some researchers as promising alternatives to genetic transformation [97, 98].

Developing successful phytoremediation techniques still requires interdisciplinary scientific research and accompanying activities, with focus on: (i) exploring existing floristic diversity to screen new native effective hyperaccumulators/excluders; (ii) better understanding of interactions in metal-soil-microbes-plant system, biological mechanisms of metal resistance/tolerance, and fate of metal ions in plant tissues; (iii) conventional breeding and/or genetic engineering to further improve phytoremediators (ideal prototype: effective, productive, inedible, sterile, capable of multiple metal accumulation, eurybiont, noninvasive, nonexpansive species); (iv) selecting cost effective, locally available, and environmentally safe amendments and practices to enhance phytoremediation; (v) intensifying long-term field trials; (vi) intensifying practical approaches for clean-up of heavily contaminated industrial, abandoned, and urban areas; (vii) expanding ecotoxicological studies on the fate of accumulated pollutants in biomass during the various stages of its utilization; (viii) combining phytoremediation approach with bioenergy purposes; (ix) coupling phytoremediation with biofortification; (x) expanding risk assessment studies and economic evaluations; (xi) developing of an international forum/platform to discuss and search for solutions to the challenges faced by phytoremediation; (xii) integrating environmental, social, economic, and normative aspects of phytomediation; and (xiii) educating industrial and government sectors (i.e. [78, 79, 90, 94, 95, 99–102]).

According to Gerhardt et al. [100], knowledge of the limitations of phytoremediation and the conditions under which success is highly probable will favor greater implementation of this remedial strategy, and result in taking phytoremediation from proven technology to accepted practice in the near future.

5.4 Analytical Methods in Bioremediation of Metals

In bioremediation, the biomass abilities to bind heavy metal ions, as well as the binding mechanism, are examined using several advanced analytical techniques. In addition, these analytical techniques are used to determine the composition of heavy metal enriched biomass [103]. These include Fourier Transform infrared spectroscopy (FTIR), inductively coupled plasma-optical emission spectrometry (ICP-OES), inductively coupled plasma-mass spectrometry (ICP-MS) flame atomic absorption spectrometry (FAAS), electrothermal atomic absorption spectrometry (ETAAS), X-ray fluorescence (XRF), transmission electron micrographs (TEM), and scanning electron microscopy with energy dispersive X-ray spectroscopy (SEM–EDX) [104–110]. These analytical techniques can be categorized into two groups, that is direct and indirect methods [104, 111].

Indirect measurement methods such as spectrometric techniques ICP-OES, ICP-MS; ETAAS, and FAAS are destructive analytical methods because they require sample mineralization before quantification. In addition, the mineralization process enables the determination of the total metal ion content in the whole volume rather than on the surface [103, 104, 108, 109, 112]. Among the spectrometric techniques, ICP-OES and ICP-MS have gained more attention because of their attractive features such as simultaneous detection, short analysis time, and less sample consumption. In addition, the use of ICP-OES or ICP-MS in the bioremediation process proved to be highly efficiency [103]. Furthermore ICP-OES and ICP-MS have attracted a lot of attention for the determination of toxic heavy metal and micronutrient ions concentration in different biomatrices [106–109, 111, 113–115]. In some instances, ICP-OES were used to determine the bioremediation mechanism. For instance, in a study reported by Mota et al. [114] numerous culture fractions of the unicellular cyanobacterium Cyanothece sp. CCY 0110 were used in bioremediation of common heavy metals found in the water system, these included Cu, Cd, and Pb. The bioremediation was carried out separately or in combined systems [114]. The summary of their findings revealed that the released polysaccharides (RPS) were effective in bioremoval of selected heavy metals. An ICP-OES was used to investigate the binding mechanism of each metal on the RPS. The possibility of an occurrence of an ion exchange mechanism was carried out by analyzing release of the light metals Ca^{2+}, Mg^{2+}, and Na^{+} in the presence of Cu^{2+}, Cd^{2+}, or Pb^{2+} in solution was investigated [114]. The results revealed that the removal mechanism was driven by complexation due to the presence of the organic functional groups rather than to ion exchange.

Direct nondestructive analytical methods such as SEM–EDX and TEM enables the determination of the density of heavy metals bound on the surface of biomass [104]. Furthermore, analysis achieved by TEM and FTIR, among others, assists in the identification of contributing forces in the bioremediation process. The process can be driven by electrostatic interaction, complex formation of ions on the cell and intracellular uptake [104, 108]. In addition, these techniques facilitate the identification of the organic functional groups (such as carboxyl groups, hydroxyl groups, and amine) that play a role in the bioremediation process [104, 108–110, 114].

5.4.1 Combination of Analytical Methods for the Determination of Heavy Metal Concentrations

The bioremediation of heavy metals involves both the total content of metal ions in the whole mass of the material and density of heavy metals bound on the surface [104, 105, 108, 109, 114–116]. Therefore, in order to obtain the information for both total content and the analyte on the surface of the material, a combination of a direct and an indirect analytical method is crucial. In addition, if the direct method is used on its own, the results obtained should be validated by correlating them with those obtained using ICP-MS/OES [104, 108, 109, 113, 115].

Amorim et al. [116] investigated the possibility of using *Meyerozyma guilliermondii* and *Meyerozyma caribbica* strains for bioremediation of manganese (Mn^{2+}). The bioremoval of Mn^{2+} by yeast was confirmed by the SEM/EDX. The results obtained revealed that the Mn was adsorbed by the cell walls of the isolated strains. These findings were validated using ICP-OES. In another study, chemically modified biomass of a coastal seaweed community was used for bioremediation of heavy metal ions [113]. SEM and FTIR were used to investigate the chemical and physical properties of composite biosorbent before and after heavy metal ion biosorption and the remaining heavy metal ion concentrations were quantified using ICP-MS [113]. Chojnacka et al. [103] investigated the use of XRF and ICP-OES in bioremediation studies. Their findings revealed that there were high positive correlations estimated for Cu, Mn, and Zn [103]. These results demonstrated that the XRF technique can be used together with traditional techniques such as ICP-OES for the production of quantitative bioremediation data. Determination of the elemental composition of enriched in Green Macroalga *Vaucheria sessilis* using bioaccumulation process was investigated using SEM–EDX and ICP-OES [109]. According to the authors, the combination of ICP-OES and SEM–EDX allowed a better understanding of bioaccumulation of metal ions. This is because SEM revealed the interaction between metal ions and *V. sessilis's* surface, which was signified by significant changes in the morphology of the algae [109]. ICP-OES analysis before and after the bioaccumulation process indicated that metal ions from aqueous solution were bound on the surface of *V. sessilis* [109]. Other applications of a combination of analytical techniques in the bioremediation process are reported in the literature [105, 107, 111, 113, 117–121]. As can be seen all the studies discussed above confirmed that, in bioremediation, it is important to use more than one analytical technique to obtain information about heavy metals.

5.5 Conclusion

The chapter discussed bioremediation techniques divided into microbial and plant utilization. Also interesting is the concluding review of possible analytical techniques that can be applied in procedures that aim to evaluate the effectivity of the whole process. The biggest challenge is to scale-up the process, as in all techniques that apply the natural ability of living organisms (biomass). The transfer of the bioremediation techniques from basic research in practice requires interdisciplinary involvement and system biology and metabolic engineering approaches application. Bioremediation technology is a good option when compared to other conventional technologies in terms of cost, ecofriendliness, social acceptance, and sustainability.

References

1 Shi, Z., Tang, Z., and Wang, C. (2017). A brief review and evaluation of earthworm biomarkers in soil pollution assessment. *Environmental Science and Pollution Research* 24 (15): 13284–13294.

2 Tóth, G., Hermann, T., Da Silva, M.R. et al. (2016). Heavy metals in agricultural soils of the European Union with implications for food safety. *Environment International* 88: 299–309.

3 ATSDR (2012). Case studies in environmental medicine. http://www.atsdr.cdc.gov/csem/csem.html ().

4 Khalid, S., Shahid, M., Niazi, N.K. et al. (2017). A comparison of technologies for remediation of heavy metal contaminated soils. *Journal of Geochemical Exploration* 182: 247–268.

5 Penga, H., Chena, Y., Wenga, L. et al. (2019). Comparisons of heavy metal input inventory in agricultural soils in North and South China: a review. *Science of the Total Environment* 660: 776–786.

6 Brookes, P.C. and McGrath, S.P. (1984). Effects of metal toxicity on the size of the soil microbial biomass. *European Journal of Soil Science* 35 (2): 341–346.

7 Xie, L.H., Tang, S.Q., Wei, X.J. et al. (2017). The cadmium and lead content of the grain produced by leading Chinese rice cultivars. *Food Chemistry* 217: 217–224.

8 Alves, R.I.S., Sampaio, C.F., Nadal, M. et al. (2014). Metal concentrations in surface water and sediments from Pardo River, Brazil: human health risks. *Environmental Research* 133: 149–155.

9 Wang, Z., Liu, X., and Qin, H. (2019). Bioconcentration and translocation of heavy metals in the soil-plants system in Machangqing copper mine, Yunnan Province, China. *Journal of Geochemical Exploration* 200: 159–166.

10 Rezapoura, S., Atashpaza, B., Moghaddam, S.S. et al. (2019). Heavy metal bioavailability and accumulation in winter wheat (*Triticum aestivum* L.) irrigated with treated wastewater in calcareous soils. *Science of the Total Environment* 656: 261–269.

11 Shilla, D., Pajalb, G., Routh, J. et al. (2019). Trophodynamics and biomagnification of trace metals in aquatic food webs: the case of Rufiji estuary in Tanzania. *Applied Geochemistry* 100: 160–168.

12 Timková, I., Sedláková-Kaduková, J., and Pristaš, P. (2018). Biosorption and bioaccumulation abilities of *Actinomycetes/Streptomycetes* isolated from metal contaminated sites. *Separations* 5 (4): 54.

13 Grasso, G. (2017). Metals are main actors in the biological world. *Metals* 7 (10): 422–425. https://doi.org/10.3390/met7100422.

14 Haraguchi, H. (2004). Metallomics as integrated biometal science. *Journal of Analytical Atomic Spectrometry* 19 (1): 5–14. https://doi.org/10.1039/B308213J.

15 Jomova, K., Baros, S., and Valko, M. (2012). Redox active metal-induced oxidative stress in biological systems. *Transition Metal Chemistry* 37 (2): 127–134. https://doi.org/10.1007/s11243-012-9583-6.

16 Nies, D.H. (2003). Efflux-mediated heavy metal resistance in prokaryotes. *FEMS Microbiology Reviews* 27 (2–3): 313–339. https://doi.org/10.1016/S0168-6445(03)00048-2.

17 Ojuederie, O. and Babalola, O. (2017). Microbial and plant-assisted bioremediation of heavy metal polluted environments: a review. *International Journal of Environmental Research and Public Health* 14 (12): 1504. https://doi.org/10.3390/ijerph14121504.

18 Yadav, K.K., Gupta, N., Kumar, V. et al. (2017). Bioremediation of heavy metals from contaminated sites using potential species: a review. *Indian Journal of Environmental Protection* 37 (1): 65–84.

19 Dixit, R., Malaviya, D., Pandiyan, K. et al. (2015). Bioremediation of heavy metals from soil and aquatic environment: an overview of principles and criteria of fundamental processes. *Sustainability* 7 (2): 2189–2212. https://doi.org/10.3390/su7022189.

20 Issazadeh, K., Jahanpour, N., Pourghorbanali, F. et al. (2013). Heavy metals resistance by bacterial strains. *Annals of Biological Research* 4 (2): 60–63.

21 Schalk, I.J., Hannauer, M., and Braud, A. (2011). New roles for bacterial siderophores in metal transport and tolerance. *Environmental Microbiology* 13 (11): 2844–2854. https://doi.org/10.1111/j.1462-2920.2011.02556.x.

22 Roane, T.M., Pepper, I.L., and Gentry, T.J. (2015). Microorganisms and metal pollutants. In: *Environmental Microbiology*, 3e (eds. I.L. Pepper, C.P. Gerba and T.J. Gentry), 415–439. Academic Press https://doi.org/10.1016/B978-0-12-394626-3.00018-1.

23 Hall, J.L. (2002). Cellular mechanisms for heavy metal detoxification and tolerance. *Journal of Experimental Botany* 53 (366): 1–11. https://doi.org/10.1093/jexbot/53.366.1.

24 Kushwaha, A., Hans, N., Kumar, S. et al. (2018). A critical review on speciation, mobilization and toxicity of lead in soil-microbe-plant system and bioremediation strategies. *Ecotoxicology and Environmental Safety* 147: 1035–1045. https://doi.org/10.1016/j.ecoenv.2017.09.049.

25 Gube, M. (2016). 4 Fungal molecular response to heavy metal stress. In: *Biochemistry and Molecular Biology. The Mycota*, A Comprehensive Treatise on Fungi as Experimental Systems for Basic and Applied Research, vol. III (ed. D. Hoffmeister), 47–68. Cham: Springer https://doi.org/10.1007/978-3-319-27790-5_4.

26 Kamal, S., Prasad, R., and Varma, A. (2010). Soil microbial diversity in relation to heavy metals. In: *Soil Heavy Metals* (eds. I. Sherameti and A. Varma), 31–63. Berlin: Springer https://doi.org/10.1007/978-3-642-02436-8_3.

27 Karelová, E., Harichová, J., Stojnev, T. et al. (2011). The isolation of heavy-metal resistant culturable bacteria and resistance determinants from a heavy-metal-contaminated site. *Biologia* 66 (1): 18–26.

28 Coelho, L.M., Rezende, H.C., Coelho, L.M. et al. (2015). Bioremediation of polluted waters using microorganisms. In: *Bioremediation of polluted waters using microorganisms* (ed. N. Shiomi). Shanghai: InTech https://doi.org/10.5772/60770.

29 Ayangbenro, A.S. and Babalola, O.O. (2017). A new strategy for heavy metal polluted environments: a review of microbial biosorbents. *International Journal of Environmental Research and Public Health* 14 (1): 94. https://doi.org/10.3390/ijerph14010094.

30 Gupta, V.K., Nayak, A., and Agarwal, S. (2015). Bioadsorbents for remediation of heavy metals: current status and their future prospects. *Environmental Engineering Research* 20 (1): 1–18. https://doi.org/10.4491/eer.2015.018.

31 Ferrol, N., González-Guerrero, M., Valderas, A. et al. (2009). Survival strategies of arbuscular mycorrhizal fungi in Cu-polluted environments. *Phytochemistry Reviews* 8: 551–559. https://doi.org/10.1007/s11101-009-9133-9.

32 Pócsi, I. (2011). Toxic metal/metalloid tolerance in fungi – a biotechnology – oriented approach. In: *Cellular Effects of Heavy Metals* (ed. G. Banfalvi), 31–58. Dordrecht: Springer https://doi.org/10.1007/978-94-007-0428-2_2.

33 Wysocki, R. and Tamás, M.J. (2010). How Saccharomyces cerevisiae copes with toxic metals and metalloids. *FEMS Microbiology Reviews* 34 (6): 925–951. https://doi.org/10.1111/j.1574-6976.2010.00217.x.

34 Inouhe, M. (2005). Phytochelatins. *Brazilian Journal of Plant Physiology* 17 (1): 65–78. https://doi.org/10.1590/S1677-04202005000100006.

35 O'Brien, S., Hodgson, D.J., and Buckling, A. (2014). Social evolution of toxic metal bioremediation in *Pseudomonas aeruginosa*. *Proceedings of the Royal Society B: Biological Sciences* 281 (1787): 20140858. https://doi.org/10.1098/rspb.2014.0858.

36 Saha, M., Sarkar, S., Sarkar, B. et al. (2016). Microbial siderophores and their potential applications: a review. *Environmental Science and Pollution Research* 23 (5): 3984–3999. https://doi.org/10.1007/s11356-015-4294-0.

37 Sathendra, R.E., Praveen Kumar, R., and Baskar, G. (2017). Microbial transformation of heavy metals. In: *Waste Bioremediation. Energy, Environment, and Sustainability* (eds. S. Varjani, E. Gnansounou, B. Gurunathan, et al.), 249–263. Singapore: Springer https://doi.org/10.1007/978-981-10-7413-4_13.

38 Juwarkar, A.A. and Yadav, S.K. (2010). Bioaccumulation and biotransformation of heavy metals. In: *Bioremediation Technology* (ed. M.H. Fulekar), 266–284. Dordrecht: Springer https://doi.org/10.1007/978-90-481-3678-0_9.

39 Lovley, D. (2013). Dissimilatory Fe(III)- and Mn(IV)-reducing prokaryotes. In: *The Prokaryotes: Prokaryotic Physiology and Biochemistry* (eds. E. Rosenberg, E.F. DeLong, S. Lory, et al.), 287–308. Berlin, Heidelberg: Springer https://doi.org/10.1007/978-3-642-30141-4_69.

40 Nancharaiah, Y.V., Dodge, C., Venugopalan, V.P. et al. (2010). Immobilization of Cr(VI) and its reduction to Cr(III) phosphate by granular biofilms comprising a mixture of microbes. *Applied and Environmental Microbiology* 76 (8): 2433–2438. https://doi.org/10.1128/AEM.02792-09.

41 Eswayah, A.S., Smith, T.J., and Gardiner, P.H.E. (2016). Microbial transformations of selenium species of relevance to bioremediation. *Applied and Environmental Microbiology* 82 (16): 4848–4859. https://doi.org/10.1128/AEM.00877-16.

42 Newsome, L., Morris, K., Trivedi, D. et al. (2014). Microbial reduction of uranium(VI) in sediments of different lithologies collected from Sellafield. *Applied Geochemistry* 51: 55–64. https://doi.org/10.1016/j.apgeochem.2014.09.008.

43 Voica, D.M., Bartha, L., and Banciu, H.L. (2016). Heavy metal resistance in halophilic Bacteria and Archaea. *FEMS Microbiology Letters* 363 (14): fnw146. https://doi.org/10.1093/femsle/fnw146.

44 Gadd, G.M. and Pan, X. (2016). Biomineralization, bioremediation and biorecovery of toxic metals and radionuclides. *Geomicrobiology Journal* 33 (3–4): 175–178. https://doi.org/10.1080/01490451.2015.1087603.

45 Lu, X., Yurong, L., Alexander, J. et al. (2016). Anaerobic mercury methylation and demethylation by geobacter bemidjiensis Bem. *Environmental Science and Technology* 50 (8): 4366–4373. https://doi.org/10.1021/acs.est.6b00401.

46 Dangi, A.K., Sharma, B., Hill, R.T. et al. (2019). Bioremediation through microbes: systems biology and metabolic engineering approach. *Critical Reviews in Biotechnology* 39 (1): 79–98. https://doi.org/10.1080/07388551.2018.1500997.

47 Peng, J., Song, Y., Yuan, P. et al. (2009). The remediation of heavy metals contaminated sediment. *Journal of Hazardous Materials* 161 (2–3): 633–640. https://doi.org/10.1016/j.jhazmat.2008.04.061.

48 Alvarez, A., Saez, J.M., Davila Costa, J.S. et al. (2017). Actinobacteria: current research and perspectives for bioremediation of pesticides and heavy metals. *Chemosphere* 166: 41–62. https://doi.org/10.1016/j.chemosphere.2016.09.070.

49 de Lorenzo, V. (2011). Systems biology approaches to bioremediation. *Comprehensive Biotechnology* 6 (6): 15–24. https://doi.org/10.1016/B978-0-08-088504-9.00460-8.

50 Sharma, B., Dangi, A.K., and Shukla, P. (2018). Contemporary enzyme based technologies for bioremediation: a review. *Journal of Environmental Management* 210: 10–20. https://doi.org/10.1016/j.jenvman.2017.12.075.

51 Senthil Kumar, P. and Gunasundari, E. (2017). Bioremediation of heavy metals. In: *Bioremediation: Applications for Environmental Protection and Management* (eds. S. Varjani, A.K. Agarwal, E. Gnansounou, et al.), 165–195. Singapore: Springer https://doi.org/10.1007/978-981-10-7485-1_9.

52 Páez-Espino, D., Tamames, J., De Lorenzo, V. et al. (2009). Microbial responses to environmental arsenic. *BioMetals* 22 (1): 117–130. https://doi.org/10.1007/s10534-008-9195-y.

53 Sharma, H. and Prasad, M.N.V. (2019). Metabolic engineering of rhizobacteria associated with plants for remediation of toxic metals and metalloids. In: *Transgenic Plant Technology for Remediation of Toxic Metals and Metalloids* (ed. M.N.V. Prasad), 299–318. London: Academic Press https://doi.org/10.1016/b978-0-12-814389-6.00014.

54 Oh, K., Cao, T., Li, T. et al. (2014). Study on application of phytoremediation technology in management and remediation of contaminated soils. *Journal of Clean Energy Technologies* 2 (3): 216–220.

55 Stephenson, C. and Black, C.R. (2014). One step forward, two steps back: the evolution of phytoremediation into commercial technologies. *Bioscience Horizons: The International Journal of Student Research* 7 https://doi.org/10.1093/biohorizons/hzu009.

56 Conesa, H.M., Evangelou, M.W.H., Robinson, B.H. et al. (2012). A critical view of current state of phytotechnologies to remediate soils: still a promising tool? *The Scientific World Journal* 2012: 173829. https://doi.org/10.1100/2012/173829.

57 Jabeen, R., Ahmad, A., and Iqbal, M. (2009). Phytoremediation of heavy metals: physiological and molecular mechanisms. *Botanical Review* 75 (4): 339–364.

58 Mehes-Smith, M., Nkongolo, K., and Cholewa, E. (2013). Coping mechanisms of plants to metal contaminated soil. In: *Environmental Change and Sustainability* (eds. S. Silvern and S. Young), 53–89. IntechOpen https://doi.org/10.5772/55124.

59 Baker, A.J.M. and Walker, P.L. (1990). Ecophysiology of metal uptake by tolerant plants. In: *Heavy Metal Uptake by Tolerant Plants: Evolutionary Aspects* (ed. A.J. Shaw), 155–177. Boca Raton: CRC Press.

60 Ghosh, M. and Singh, S.P. (2005). A review on phytoremediation of heavy metals and utilization of its byproducts. *Applied Ecology and Environmental Research* 3 (1): 1–18.

61 Małachowska-Jutsz, A. and Gnida, A. (2015). Mechanisms of stress avoidance and tolerance by plants used in phytoremediation of heavy metals. *Archives of Environmental Protection* 41 (4): 104–114.

62 Jan, S. and Parray, J.A. (2016). Heavy metal uptake in plants. In: *Approaches to Heavy Metal Tolerance in Plants* (eds. S. Jan and J.A. Parray), 1–18. Singapore: Springer.

63 Singh, S., Parihar, P., Singh, R. et al. (2016). Heavy metal tolerance in plants: role of transcriptomics, proteomics, metabolomics, and ionomics. *Frontiers in Plant Science* 6: 1143. https://doi.org/10.3389/fpls.2015.01143.

64 Anju, M. (2017). Biotechnological strategies for remediation of toxic metal(loid)s from environment. In: *Plant Biotechnology: Recent Advancements and Developments* (eds. S.K. Gahlawat, R.K. Salar, P. Siwach, et al.), 315–359. Singapore: Springer.

65 Lasat, M.M. (2002). Phytoextraction of toxic metals: a review of biological mechanisms. *Journal of Environmental Quality* 31: 109–120.

66 Dakora, F.D. and Phillips, D.A. (2002). Root exudates as mediators of mineral acquisition in low-nutrient environments. *Plant and Soil* 245: 35–47.

67 Luptakova, A., Macingova, E., Slesarova, A. et al. (2007). Solubilization and immobilization of toxic metals by bacteria (Slovak Republic and Italy). *Proceedings of IMWA Symposium*, Cagliari, Italy (27–31 May 2007). Cagliari: Mako Edizioni.

68 Krzesłowska, M. (2011). The cell wall in plant cell response to trace metals: polysaccharide remodeling and its role in defense strategy. *Acta Physiologiae Plantarum* 33 (1): 35–51.

69 Le Gall, H., Philippe, F., Domon, J.M. et al. (2015). Cell wall metabolism in response to abiotic stress. *Plants* 4 (1): 112–166.

70 Jadia, C.D. and Fulekar, M.H. (2009). Phytoremediation of heavy metals: recent techniques. *African Journal of Biotechnology* 8 (6): 921–928.

71 Yoneyama, T., Gosho, T., Kato, M. et al. (2010). Xylem and phloem transport of Cd, Zn and Fe into the grains of rice plants (Oryza sativa L.) grown in continuously flooded Cd-contaminated soil. *Soil Science and Plant Nutrition* 56 (3): 445–453.

72 Sunitha, M., Prashant, S., Kumar, S.A. et al. (2013). Cellular and molecular mechanisms of heavy metal tolerance in plants: a brief overview of transgenic plants over-expressing phytochelatin synthase and metallothionein genes. *Plant Cell Biotechnology and Molecular Biology* 14 (1–2): 33–48.

73 Kutrowska, A. and Szelag, M. (2014). Low-molecular weight organic acids and peptides involved in the long-distance transport of trace metals. *Acta Physiologiae Plantarum* 36 (8): 1957–1968.

74 Anjum, N.A., Hasanuzzaman, M., Hossain, M.A. et al. (2015). Jacks of metal/metalloid chelation trade in plants – an overview. *Frontiers in Plant Science* 6: 192. https://doi.org/10.3389/fpls.2015.00192.

75 Hall, J.L. and Williams, L.E. (2003). Transition metal transporters in plants. *Journal of Experimental Botany* 54 (393): 2601–2613.

76 Colangelo, E.P. and Guerinot, M.L. (2006). Put the metal to the petal: metal uptake and transport throughout plants. *Current Opinion in Plant Biology* 9 (3): 322–330.

77 Sarwar, N., Imran, M., Shaheen, M.R. et al. (2017). Phytoremediation strategies for soils contaminated with heavy metals: modifications and future perspectives. *Chemosphere* 171: 710–721.

78 Ali, H., Khan, E., and Sajad, M.A. (2013). Phytoremediation of heavy metals – concepts and applications. *Chemosphere* 91: 869–881.

79 Burges, A., Alkorta, I., and Epelde, L. (2018). From phytoremediation of soil contaminants to phytomanagement of ecosystem services in metal contaminated sites. *International Journal of Phytoremediation* 20 (4): 384–397.

80 Mohanty, M. (2016). Post-harvest management of phytoremediation technology. *Journal of Environmental & Analytical Toxicology* 6: 398. https://doi.org/10.4172/2161-0525.1000398.

81 Kovacs, H. and Szemmelveisz, K. (2017). Disposal options for polluted plants grown on heavy metal contaminated brownfield lands – a review. *Chemosphere* 166: 8–20.

82 Sas-Nowosielska, A., Kucharski, R., Malkowski, E. et al. (2004). Phytoextraction crop disposal – an unsolved problem. *Environmental Pollution* 128 (3): 373–379.

83 Pandey, V.C., Bajpai, O., and Singh, N. (2016). Energy crops in sustainable phytoremediation. *Renewable and Sustainable Energy Reviews* 54: 58–73.

84 Bolan, N.S., Park, J.H., Robinson, B. et al. (2011). Phytostabilization: a green approach to contaminant containment. *Advances in Agronomy* 112: 145–204.

85 Yang, S., Liang, S., Yi, L. et al. (2014). Heavy metal accumulation and phytostabilization potential of dominant plant species growing on manganese mine tailings. *Frontiers of Environmental Science & Engineering* 8 (3): 394–404.

86 Terry, N., Zayed, A.M., de Souza, M.P. et al. (2000). Selenium in higher plants. *Annual Review of Plant Biology* 51: 401–432.

87 Leonard, T.L., Taylor, G.E. Jr., Gustin, M.S. et al. (1998). Mercury and plants in contaminated soils: 1. Uptake, partitioning, and emission to the atmosphere. *Environmental Toxicology and Chemistry* 17 (10): 2063–2071.

88 Sakakibara, M., Watanabe, A., Inoue, M. et al. (2010). Phytoextraction and phytovolatilization of arsenic from as-contaminated soils by *Pteris vittata*. *Proceedings of the Annual International Conference on Soils, Sediments, Water and Energy*. https://scholarworks.umass.edu/soilsproceedings/vol12/iss1/26

89 Radziemska, M., Koda, E., Bilgin, A. et al. (2018). Concept of aided phytostabilization of contaminated soils in postindustrial areas. *International Journal of Environmental Research and Public Health* 15 (1): 24. https://doi.org/10.3390/ijerph15010024.

90 Muthusaravanan, S., Sivarajasekar, N., Vivek, J.S. et al. (2018). Phytoremediation of heavy metals: mechanisms, methods and enhancements. *Environmental Chemistry Letters* 16: 1339–1359.

91 Bian, X., Cui, J., Tang, B. et al. (2018). Chelant-induced phytoextraction of heavy metals from contaminated soils: a review. *Polish Journal of Environmental Studies* 27 (6): 2417–2424.

92 Wu, C., Zhang, X., and Deng, Y. (2017). Review in strengthening technology for phytoremediation of soil contaminated by heavy metals. *Proceedings of the IOP Conference Series: Earth and Environmental Science* 78 (1): 012015.

93 Jiang, Y., Huang, R., Jiang, S. et al. (2018). Adsorption of Cd(II) by rhizosphere and non-rhizosphere soil originating from mulberry field under laboratory condition. *International Journal of Phytoremediation* 20 (4): 378–383.

94 Vázquez-Núñez, E., Peña-Castro, J.M., Fernández-Luqueño, F. et al. (2018). A review on genetically modified plants designed to phytoremediate polluted soils: biochemical responses and international regulation. *Pedosphere* 28 (5): 697–712.

95 Mosa, K.A., Saadoun, I., Kumar, K. et al. (2016). Potential biotechnological strategies for the cleanup of heavy metals and metalloids. *Frontiers in Plant Science* 7: 303. https://doi.org/10.3389/fpls.2016.00303.

96 Suman, J., Uhlik, O., Viktorova, J. et al. (2018). Phytoextraction of heavy metals: a promising tool for clean-up of polluted environment? *Frontiers in Plant Science* 871: 1476. https://doi.org/10.3389/fpls.2018.01476.

97 Brewer, P.E., Saunders, A.J., Angle, S.J. et al. (1999). Somatic hybridization between the zinc accumulator Thlaspi caerulescens and Brassica napus. *Theoretical and Applied Genetics* 99 (5): 761–771.

98 Nehnevajova, E., Herzig, R., Federer, G. et al. (2007). Chemical mutagenesis – a promising technique to increase metal concentration and extraction in sunflowers. *International Journal of Phytoremediation* 9 (2): 149–165.

99 Beans, C. (2017). Phytoremediation advances in the lab but lags in the field. *Proceedings of the National Academy of Sciences of the United States of America* 114 (29): 747–7477.

100 Gerhardt, K.E., Gerwing, P.D., and Greenberg, B.M. (2017). Opinion: taking phytoremediation from proven technology to accepted practice. *Plant Science* 256: 170–185.

101 Mahar, A., Wang, P., Ali, A. et al. (2016). Challenges and opportunities in the phytoremediation of heavy metals contaminated soils: a review. *Ecotoxicology and Environmental Safety* 126: 111–121.

102 Song, U. and Park, H. (2017). Importance of biomass management acts and policies after phytoremediation. *Journal of Ecology and Environment* 41 (1): 13. https://doi.org/10.1186/s41610-017-0033-4.

103 Chojnacka, K., Samoraj, M., Tuhy, Ł. et al. (2018). Using XRF and ICP-OES in biosorption studies. *Molecules* 23: 2076.

104 Chojnacka, K. and Mikulewicz, M. (2019). Green analytical methods of metals determination in biosorption studies. *TrAC Trends in Analytical Chemistry* 116: 254–265. https://doi.org/10.1016/j.trac.2019.02.013.

105 Ding, H., Luo, X., Zhang, X. et al. (2019). Alginate-immobilized Aspergillus niger: characterization and biosorption removal of thorium ions from radioactive wastewater. *Colloids and Surfaces A: Physicochemical and Engineering Aspects* 562: 186–195. https://doi.org/10.1016/j.colsurfa.2018.11.032.

106 Dutta, S. and Das, A.K. (2010). Analytical perspective on waste management for environmental remediation. *TrAC Trends in Analytical Chemistry* 29: 636–644.

107 Krishna, K.R. and Philip, L. (2005). Bioremediation of Cr (VI) in contaminated soils. *Journal of Hazardous Materials* 121: 109–117.

108 Michalak, I., Chojnacka, K., and Marycz, K. (2011). Using ICP-OES and SEM-EDX in biosorption studies. *Microchimica Acta* 172: 65–74.

109 Michalak, I., Marycz, K., Basińska, K. et al. (2014). Using SEM-EDX and ICP-OES to investigate the elemental composition of green macroalga *Vaucheria sessilis*. *The Scientific World Journal* 2014: 891928.

110 Mota, R., Pereira, S.B., Meazzini, M. et al. (2015). Effects of heavy metals on Cyanothece sp. CCY 0110 growth, extracellular polymeric substances (EPS) production, ultrastructure and protein profiles. *Journal of Proteomics* 120: 75–94.

111 Cocârţă, D.M., Neamţu, S., and Reşetar Deac, A.M. (2016). Carcinogenic risk evaluation for human health risk assessment from soils contaminated with heavy metals. *International journal of Environmental Science and Technology* 13: 2025–2036. https://doi.org/10.1007/s13762-016-1031-2.

112 Agnello, A.C., Bagard, M., Van Hullebusch, E.D. et al. (2016). Comparative bioremediation of heavy metals and petroleum hydrocarbons co-contaminated soil by natural attenuation, phytoremediation, bioaugmentation and bioaugmentation-assisted phytoremediation. *Science of the Total Environment* 563: 693–703.

113 Deniz, F. and Karabulut, A. (2017). Biosorption of heavy metal ions by chemically modified biomass of coastal seaweed community: studies on phycoremediation system modeling and design. *Ecological Engineering* 106: 101–108. https://doi.org/10.1016/j.ecoleng.2017.05.024.

114 Mota, R., Rossi, F., Andrenelli, L. et al. (2016). Released polysaccharides (RPS) from Cyanothece sp. CCY 0110 as biosorbent for heavy metals bioremediation: interactions between metals and RPS binding sites. *Applied Microbiology and Biotechnology* 100: 7765–7775.

115 Wang, F., Tan, L., Liu, Q. et al. (2015). Biosorption characteristics of Uranium (VI) from aqueous solution by pollen pini. *Journal of Environmental Radioactivity* 150: 93–98.

116 Amorim, S.S., Ruas, F.A.D., Barboza, N.R. et al. (2018). Manganese (Mn2+) tolerance and biosorption by Meyerozyma guilliermondii and Meyerozyma caribbica strains. *Journal of Environmental Chemical Engineering* 6: 4538–4545.

117 Alipanahpour Dil, E., Ghaedi, M., Ghezelbash, G.R. et al. (2017). Highly efficient simultaneous biosorption of Hg^{2+}, Pb^{2+} and Cu^{2+} by live yeast Yarrowia lipolytica 70562 following response surface methodology optimization: kinetic and isotherm study. *Journal of Industrial and Engineering Chemistry* 48: 162–172. https://doi.org/10.1016/j.jiec.2016.12.035.

118 Beolchini, F., Dell'Anno, A., De Propris, L. et al. (2009). Auto-and heterotrophic acidophilic bacteria enhance the bioremediation efficiency of sediments contaminated by heavy metals. *Chemosphere* 74: 1321–1326.

119 Chardin, B., Dolla, A., Chaspoul, F. et al. (2002). Bioremediation of chromate: thermodynamic analysis of the effects of Cr (VI) on sulfate-reducing bacteria. *Applied Microbiology and Biotechnology* 60: 352–360.

120 Cid, H.A., Flores, M.I., Pizarro, J.F. et al. (2018). Mechanisms of Cu^{2+} biosorption on *Lessonia nigrescens* dead biomass: functional groups interactions and morphological characterization. *Journal of Environmental Chemical Engineering* 6: 2696–2704. https://doi.org/10.1016/j.jece.2018.03.034.

121 Kumari, S., Mahapatra, S., and Das, S. (2017). Ca-alginate as a support matrix for Pb(II) biosorption with immobilized biofilm associated extracellular polymeric substances of *Pseudomonas aeruginosa* N6P6. *Chemical Engineering Journal* 328: 556–566. https://doi.org/10.1016/j.cej.2017.07.102.

6

Bioremediation of Dye Contaminated Soil

Manikant Tripathi[1], Shailendra Kumar[1], Durgesh Narain Singh[2], Rajeev Pandey[3], Neelam Pathak[4], and Hera Fatima[1]

[1] *Centre of Excellence, DST-FIST Supported Department of Microbiology, Dr. Rammanohar Lohia Avadh University, Ayodhya, Uttar Pradesh, India*
[2] *Department of Zoology, University of Delhi, Delhi, India*
[3] *Department of Environmental Sciences, Dr. Rammanohar Lohia Avadh University, Ayodhya, Uttar Pradesh, India*
[4] *Department of Biochemistry, Dr. Rammanohar Lohia Avadh University, Ayodhya, Uttar Pradesh, India*

6.1 Introduction

Industrialization has given rise to many types of hazardous chemicals that accumulate in the environment and cause adverse effects to the ecosystem. Most of the industries do not have appropriate waste treatment facilities and discharge huge quantities of waste. The majority of toxic xenobiotics discharged from industries are mixed in with the natural water systems and to the soil of the biosphere [1]. Among the various contaminants, textile dyes are identified as the most predominant pollutants of our environment.

Dyes are organic ingredients used in textiles and other materials for colorization. They are classified as: (i) chromophore groups in their chemical structures as azo dyes, anthraquinone dyes, and phthalocyanine dyes etc.; and (ii) their usage or application method as disperse dyes for polyester and reactive dyes for cotton [2]. About 100 000 commercial dyes are manufactured including several types of dyes such as acidic, basic, reactive, azo, diazo, and anthraquinone-based dyes with an annual production of over 7×10^5 metric tons. Approximately 50% of the dyes are discharged as industrial effluents [3].

Dyes used in the textile industry are classified into three classes: (i) anionic (direct, acid, and reactive dyes); (ii) cationic (all basic dyes); and (iii) nonionic

Soil Bioremediation: An Approach Towards Sustainable Technology, First Edition.
Edited by Javid A. Parray, Abeer Hashem Abd Elkhalek Mahmoud, and Riyaz Sayyed.

(dispersed dyes). Basic and reactive dyes are extensively used in the textile industry because of their favorable characteristics of bright color, being easily water soluble, cheaper to produce, and easier to apply to fabric [2]. Dye production in India is estimated to be around 60 000 t/year, or about 6.6% of the world production [4]. It is estimated that about 40–65 l of textile waste is generated per kilogram of cloth produced [2]. About 15–30% of these dyes remain in the effluents for the coloring of the final products. The presence of color or color-causing compounds in waste has always been undesirable. The unaesthetic colored dye waste pollutes waterway as well as agricultural land and causes ecosystem toxicity. Colored effluents are toxic, mutagenic, and carcinogenic [2]. Due to their complexity, azo dyes are difficult to remediate in nature.

Moreover, environmental aspects are also gaining importance around the world in protection of the global environment. Many issues related to these aspects arise during the processing of textiles. Therefore, it becomes necessary for the textile processor to either curtail their chemical usage or to design ecofriendly approaches for performing various processes such as bleaching, dyeing, printing, and finishing so as to minimize the effluent load. The wastewater generated by such dyeing contains huge amounts of chemicals as well as unfixed dyestuffs, which make the effluent liquor deeply colored and polluted. In order to diminish the contamination of such color effluent, an inexpensive and ecofriendly method is necessary.

Microbial decolorization of dyes using bacteria has recently gained attention. It began under static/anaerobic/microaerophilic situations by an enzymatic transformation process [5] in which a reductive cleavage of azo bond leads to the formation of colorless toxic aromatic amines. These amines are further changed to simpler nontoxic forms under aerobic conditions [5]. Every microorganism requires specific cultural and nutritional conditions for the optimum growth and extent of dyes discoloration. These factors are cultural (pH, temperature, anaerobic/aerobic/microaerophilic conditions, type and concentration of dye, inoculum size, etc.) and nutritional (carbon and nitrogen sources). Many researchers reported dye decolorization by microorganisms [5–8].

Since microbial dye decolorization varies with changes in cultural and nutritional factors their optimization should always be considered a necessary step for bioremediation of such toxicants from environment. The conditions optimized for synthetic pure dye decolorization may not be applicable to real effluent decolorization/detoxification, as the composition of textile waste is variable, and may also contain many other toxic chemicals. Different methods (physical, chemical, and biological) and factors (oxygen, hydrogen ion concentration, carbon and nitrogen supplements, temperature, dye structure, concentration, and toxicity) affecting the decolorization and detoxification of azo dyes are discussed under specific headings. This chapter will focus on acquainting readers with bioremediation methods for azo dyes in contaminated sites in the environment.

6.2 History and Usage of Dyes

The use of natural dyes for painting and dyeing has been known since ancient times. From primitive people humans have been endeavoring to add color to the world around them. They used natural material to stain hides, decorate shells and feathers, and paint stories on the walls of ancient caves. The dyes used were derived from natural sources. The dyes like tyrian purple were particularly prized as an important part of the local economy and world trade of the time [9]. Scientists have been able to date the black, white, yellow, and reddish pigments made from ochre, and used by the primitive men in cave paintings, to over 15 000 BCE. The earliest written record of the use of dyestuffs in China is 2600 BCE. In 327 BCE, Alexander the Great mentions "beautiful printed cottons" in India. By 715 BCE, wool dyeing was established as a craft in Rome [10]. The recent discovery, in the Chauvet-Pont-d'Arc caves in France, of 30 000-year-old Paleolithic rock paintings provides the most ancient testimony of the millenary use of inorganic pigments such as hematite, manganese oxide, soot, and ochre [9].

Organic natural colorants also have a timeless history of application, especially as textile dyes. Most dyeing techniques in use until the nineteenth century were established by the ancient Egyptians, who developed methods using plant extracts, sometime in conjunction with a mordant [11]. Until the sixteenth century, the dyeing processes were kept secret; however, with the growth of commercial trips and an expansion of knowledge, they experienced a rapid increment and diffusion. In 1548, Giovanventura Rossetti published the "Plichto dei tintori," which described not only some dyeing and active constituent extraction methods, but also some chemical preparations [10]. Until the middle of the nineteenth century, all colorants applied were of natural origin. The use of natural dyes in textile processing operations was very limited [12, 13]. The first synthetic dye was picric acid, which was produced in 1770s from the interaction of indigo and nitric acid. The first dye was named "mauve," synthesized accidentally by William Henry Perkin in 1856. He patented and commercialized his accidental innovation, and then a new process was developed on a large scale for the synthesis of same dye, which led to the beginning of a new chemical industry [3]. By the early 1880s, diazotization was a known reaction, which was employed to synthesize commercially useful dyes. Researchers have focused mainly on dyes with improved stability, thus more resistant to treatment process [14]. The word natural dye covers all the dyes derived from natural resources: those derived from plants and animals [15].

6.3 Classification of Dyes

The various classes of dyes are acid dyes, cationic dyes, direct dyes, reactive dyes, and disperse dyes. These dyes are generally classified according to their utilization method and for the convenience of the dye user. The best classification method available is that used in Colour Index (https://colour-index.com/) a publication sponsored by the Society of Dyers and Colourists (UK) and the American Association of Textile Chemists and Colorists. This classification of dyes also gives the various different methods of dyeing of various fabrics with different dyes.

6.3.1 Acid Dyes

Acid dyes are usually in the form of sodium salts of the color acids that contain sulfonic acid or phenolic group. These dyes are mostly used to dye fibers in the basic group, such as wool, silk, and polyamides. Application is generally made under acidic conditions, which causes protonation of the basic groups.

6.3.2 Basic or Cationic Dyes

Basic or cationic dyes have a basic amino group, which is protonated under the acidic conditions of the fibers by formation of salt linkages with anionic or acidic groups in the fibers. They are generally known for their intense and brilliant shades but give poor light fastness.

6.3.3 Direct Dyes

Direct dyes are usually bearing sulphonic acid groups, which are strongly adsorbed on cellulose. However, these dyes are not considered as acid dyes because the sulfonic groups are not used as a means of attachment to the fiber. These dyes are generally used for dyeing wool and silk fibers using a neutral bath. A typical example of a direct dye is Congo red.

6.3.4 Reactive Dyes

Reactive dyes are a new class of dye that form covalent bonds with fibers possessing hydroxyl or amino groups. One important type of reactive dye has a chlorine atom, which reacts with the hydroxyl groups in cellulose in the presence of an alkali. Another important type of reactive dye involves an activated vinyl group, which can react with a cellulose hydroxyl group in the presence of base. Reactive dyes offer excellent fastness to washing since the dye becomes part of the fiber.

6.3.5 Disperse Dyes

Disperse dyes are a class of water insoluble dyes originally introduced for dyeing cellulose acetate and usually applied by fine aqueous dispersion. Disperse dye molecules are generally small and have some hydroxyl or amino groups to give finite water solubility at dyeing temperatures. The main uses of disperse dyes includes dyeing of cellulose acetate, nylon, polyester, and polyacrylonitrile fibers. Most polyester fibers must be dyed under pressure or with the use of organic swelling agents. The washing fastness of disperse dyes on these fibers is excellent. Other applications of disperse dyes include dyeing woolen sheepskins and for surface dyeing of plastics.

6.4 Problems Due to Colored Textile Effluent

The organic and inorganic toxic pollutants in untreated wastewater need to be removed for harmless discharge into the aquatic system [16]. The uncontrolled industrial and urban wastewater discharge has posed a significant adverse effect on the environment [17]. The textile industry employs various types of synthetic dyes and discharges huge quantities of colored wastewater as the uptake of these dyes by fabrics is very poor. The discharge of untreated colored wastewater adversely affects the flora and fauna found in soil and water. It may also be lethal to some marine species due to the presence of metals and chlorine in synthetic dyes. So, the colored wastewater must be treated before being discharged into the environment. The dye color and other organic and inorganic toxicants present in contaminated soil and water adversely affect environmental health.

6.5 Physico-Chemical Quality of Textile Effluents

In the dying process 2% of the dyes is directly discharged as aqueous effluent and 10% is lost during textile processing. The textile industry uses huge amounts of water mainly as process water (90–94%), and cooling water (6–10%), which become loaded with different pollutants, dyes, surfactants, acids or bases, salts, heavy metals, and suspended solids [18]. Textile manufacturing process uses mixtures of dyes with various chemical additives such as solvents, antifoaming agents, whitening agents, and pH conditioners.

Textile wastewater chemistry and behavior are different due to the combination of various dyes, chemicals, suspended particles, nutrients, and bacteria in it. Textile wastewater contains high levels of pollutants: suspended solids, chemical oxygen demand (COD), heat, color, acidity, and other substances. The sediments, suspended and dissolved solids, are important repositories for toxic heavy metals and dyes that cause rapid depletion of dissolved oxygen content in water [19]. The metals and dyes are recalcitrant in nature and accumulate throughout the food chain. These contaminants have direct and indirect toxic impacts on human health [20]. It was observed that most of the physico-chemical parameters studied exceeded the BIS (Bureau of India Standards) limits.

The wastewater discharged from industries is either used for irrigation or it runs off into natural water courses [21]. Research has shown that the inefficiency in dyeing has resulted in 10–15% [22] and nearly 15% [23] of unused dyestuff entering the processing and wastewater. However, the color of the textile effluent depends on the dyes used. Both synthetic and natural colors easily disperse in water [24, 25]. The color of effluent is pH dependent and it increases as effluent pH is raised or lowered. Different colors of the effluent may be the combined results of pH, temperature, and acidic conditions that do not allow the chromophore group of the dye to degrade thus making the effluent highly colored. The bad odor of textile effluent could be due to presence of volatile compounds [26].

The Environmental Protection Agency (EPA) recommends pH value of range 6–9 for effluent to be discharged into a stream, as either high or low pH value will be harmful to man, aquatic animals, and will disturb biological activity of the stream if discharged untreated. Total alkalinity has been reported as a major factor, which influences pH [27, 28]. Alkalinity is the capacity of water to neutralize acid and is characterized by the presence of hydroxyl ions. The hardness of water is not a chemical parameter but indicates the water quality mainly in terms of Ca^{2+} and Mg^{2+} and expressed as $CaCO_3$. The hardness has no known adverse effect. Their low concentration in the textile wastewater may be due to the use of soft water in the dyeing and printing process to avoid coagulation of dyes. The electric conductivity (EC) is often used for indicating the total concentration of the ionized constituents of water and an indirect measure of ions or charge-carrying species in the wastewater [29]. High total suspended solids (TSS) and total dissolved solids (TDS) have been detected in textile effluent.

6.6 Dye Decolorization/Degradation Techniques

Figure 6.1 shows the different physical, chemical, and biological treatment techniques such as adsorption, coagulation, flocculation, membrane filtration, ozonation, electrochemical, radiolysis, bacterial, fungal, algal, and advanced oxidation processes that can be employed to remove color from dye contaminated waste [2, 30, 31]. In general, each technique has advantages and limitations. Dye

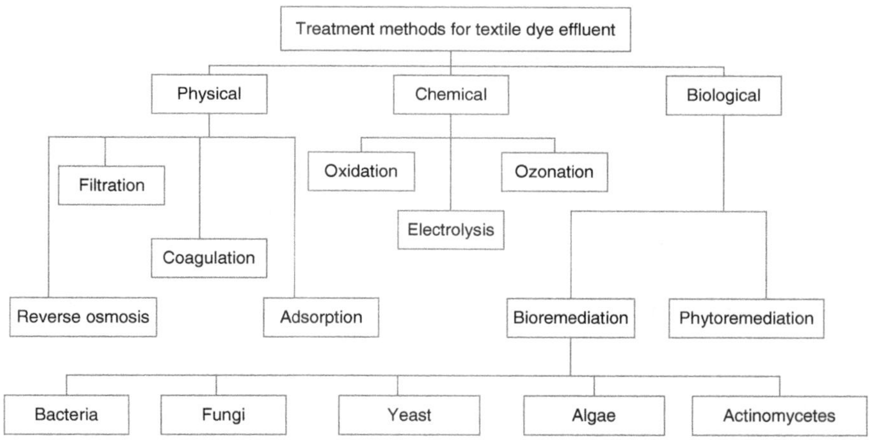

Figure 6.1 Strategies for treatment of textile dyes from contaminated sites.

removals strategies therefore mostly have a combination of different techniques [14]. In the following sections provide an overview of the most important techniques.

6.6.1 Physical Methods

6.6.1.1 Membrane Filtration

This method has the potential to remove, clarify, and concentrate the color from dye effluent continuously [23]. There are certain special features, which are not present in other physical methods: resistance to temperature and microbial attack. Nanofiltration, reverse osmosis, ultrafiltration, and microfiltration can be employed for separation, purification, and the reuse of salts and larger molecules including dyes from dye bath effluents and bulk textile-processing wastewaters [32]. Nanofiltration and reverse osmosis can be applied to main or post treatment of dye effluent. In nanofiltration, the membrane effectively acts as a molecular filter retaining material with a relative molecular mass higher than ~200. Whereas, in reverse osmosis, the effluent is forced to pass under moderate pressure across the semipermeable membrane, removing ~98% of the impurities of relative molecular mass greater than ~100 [33]. Despite higher level of efficiency, nanofiltration and reverse osmosis suffer from certain disadvantages. The membranes are required to be cleared regularly, and their surface characteristics may be altered due to the action of dye and/or other constituents of the effluent. Additionally, these techniques have high capital requirement as well as running costs [34]. Ultrafiltration and microfiltration (bigger pores) can be successful only as a pretreatment because the pore size is too large to retain the dye molecules from passing through the membrane. While filtration is potentially applicable for the production of pure water, low molecular mass dyes can pass through the filters.

6.6.1.2 Electrokinetic Coagulation or Flocculation

This is a low-cost effective method for dye removal from the textile effluents. The inorganic (lime, aluminum, magnesium and iron salts) or organic (anionic, cationic, or nonionic polymers) coagulants have been used for coagulation in the treatment of textile wastewaters to partly remove TSS, biological oxygen demand (BOD), chemical oxygen demand (COD), and color over several years [35, 36].

6.6.1.3 Sorption

Among various physical methods, sorption techniques are very popular due to their efficiency and economic viability in treatment of wastewaters having very stable pollutants generally not removed by conventional treatment methods. Dye effluents are multicomponent mixtures with different concentrations of individual species and their varied extent of adsorption. In some cases, weaker bonds are formed with the adsorbent and some material can be released back into the stream [33]. The use of any sorbent for removing dye species from a liquid stream depends on the equilibrium between the adsorbed and the free species.

6.6.1.4 Activated Carbon

It is reasonably effective at removing many different dyes from aqueous streams [37]. Along with color, other organic compounds present in the effluent are also removed by activated carbon/charcoal. This is very effective for adsorbing cationic, mordant, and acidic dyes, while it has slightly poor potential for removing disperse, direct, vat, pigment, and reactive dyes. Furthermore, the effective cost of high temperature regeneration process, including the replacement cost and waste sludge yield, make their regeneration unattractive to small establishments [38]. The carbon loaded with the adsorbed dye need to be reactivated otherwise disposal of the concentrate poses another problem, and reactivation leads to 10–15% loss of the sorbent. Therefore, various low-cost adsorbents have been investigated as an alternative to activated carbon.

6.6.1.5 Peat

Peat is an exceptionally good choice due to its cellular structure. The peat exhibits high potential to adsorb transition metals as well as polar organic compounds from dye-containing wastewaters. Unlike activated carbon, peat does not require activation, and also it is very cheap. The spent peat may be burned for steam generation or as a substrate in solid-state fermentation (SSF) for single-cell protein production.

6.6.1.6 Wood Chips

They are effective only in the removal of acid dyes. However, due to their hardness they are not very effective as compared to other available adsorbents [39]. Furthermore, an extended contact time would be required when wood chips are used as biosorbents. Generally, the biosorbed chips are burned for power generation, but they can also be used as substrate for solid-state fermentation.

6.6.1.7 Fly Ash and Coal Mixture

This mixture may be substituted for an activated carbon with 1 : 1 ratio of fly ash to carbon. The increase in concentration of fly ash enhances the rate of adsorption owing to increased surface area available for adsorption [40].

6.6.1.8 Silica Gel

This is a good adsorbent for removal of basic dyes only. However, its commercial use is restricted due to side reactions (air binding and air fouling) with the particulate matter.

6.6.1.9 Other Materials

Biosorbents are naturally occurring cheap biodegradable materials that have high dye-binding potential and can act as ion exchangers. A wide range of biomaterials can be used as biosorbents: corn cobs, wheat straw, rice husk, sawdust, bark, bagasse pith, cotton waste, bacterial, fungal, yeast, algal biomass, chitin, chitosan, hen feathers, etc.

6.6.1.10 Ion Exchange

Standard ion-exchange systems have not been widely used on a commercial scale for the treatment of dye effluents, because of the high cost of organic solvents (a major disadvantage) employed to regenerate the ion exchanger, and also due to huge inorganic load of the effluent [33]. The ion exchanger can't be used for a wide range of dyes due to the restricted type of charge on different dyes present in the textile effluent. The ion-exchange method is effective only for cationic and anionic dyes, and is not very effective for the dispersed type of dyes [23]. The advantages of this method include: (i) no loss of ion-exchange resin during regeneration, (ii) reclamation of solvent after use, and (iii) removal of soluble dyes.

6.6.2 Chemical Methods

6.6.2.1 Electrolysis

The electrochemical destruction technique, developed in the mid-1990s, is very efficient for dye removal from colored wastewater. In this process of treatment, BOD, COD reduction, and coagulation of TSS occur, and there is neither use of chemicals nor build-up of any sludge. The electrolytically treated wastewater can be safely released to the nearby water bodies, as all the breakdown metabolites are generally not hazardous to living entities [41]. The process is simply based on applying an electric current through wastewater immersed with sacrificial iron electrodes. This generates Fe(II) and –OH ions. The $Fe(OH)_2$ is formed and soluble/insoluble dyes are removed from the effluent. Furthermore, Fe(II) can reduce azo dyes to arylamine, and water may also be oxidized resulting in the formation of O_2 and O_3. The process exhibits high efficiency (~90%) for color removal and degradation of recalcitrant pollutants. However, this process is cost intensive because of a high energy requirement, the limited life of electrodes, and uncontrolled radical reactions [14].

6.6.2.2 H_2O_2-Fe(II) Salts (Fenton's Reagent)

Fenton's reagent is highly suitable for the treatment of wastewaters containing pollutants, which are toxic to biomass or resistant to biological treatment process [42]. The process is based on the formation of reactive hydroxyl radical species, which efficiently degrade the pollutants present in wastewater streams. The Fenton oxidation process can decolorize a wide range of acid, direct, vat, mordant, and reactive dyes, but the resultant floc is of poor quality and does not settle well. This process is relatively cheap compared to ozonation, and also results in larger COD reduction. However, this process suffers from certain disadvantages: (i) the textile wastewaters usually have high pH, while the Fenton process requires low pH; and (ii) the sludge containing a concentrated floc of chemicals and dye molecules is formed, which poses a disposal problem. The sludge is conventionally incinerated to generate power. The Fenton system is recommended to be combined with other post treatment such as biological process.

6.6.2.3 Photocatalytic Process

Advanced oxidation processes can oxidize a broad range of organic dyes quickly and nonselectively. Photocatalytic or photochemical dye degradation has emerged as an important technique for dye degradation, since it results in complete mineralization with mild operational conditions of pH and temperature. The photoactivated chemical reactions are characterized by a free radical mechanism initiated by the interaction of photons of a proper energy level with the molecules of pollutant species present in the effluent, with or without the presence of a catalyst. The free radicals can be easily produced using UV radiations. UV light has been tested in combination with catalysts such as H_2O_2, TiO_2, Fenton's reagent, and O_3 for the decolorization of dye solutions [34]. The process was found more promising with UV/TiO_2 combination as compared to UV/H_2O_2 process, which appeared too slow and cost intensive, and less effective for potential full-scale application.

6.6.2.4 Ozonation

Ozone is a very powerful and efficient oxidizing agent that can react with most species containing multiple bonds (e.g., —C=C—, —C=N—, —N=N—). The high instability of O_3 as compared to other chemicals (chlorine, H_2O_2, etc.) makes O_3 a more powerful oxidizing agent. Oxidation by ozone ensues degradation of phenols, pesticides, and aromatic and chlorinated hydrocarbons. The decolorization of water-soluble dyes by O_3 is faster than insoluble dyes (vat and disperse dyes). Textile wastewater usually contains other refractory constituents (in addition to dyes) that react with ozone, and thus increase its requirement [43, 44]. The colored groups (chromophores) in the dyes are organic compounds with conjugated double bonds that can be broken down by ozone, thereby forming relatively small molecules, which are colorless. However, these resultant colorless molecules are toxic and carcinogenic; therefore, ozonation may be employed along with other physical and/or biological methods of treatment for total mineralization of dyes. The gaseous state of ozone offers one major advantage, which does not allow the increase in volume of wastewater and sludge. However, the ozone treatment also suffers from certain drawbacks: (i) the half-life of ozone is only 20 minutes, which is further shortened if dyes and other pollutants are present in the effluent, (ii) ozone decomposition is accelerated under alkaline conditions (pH >10) and therefore, proper monitoring of pH is essential, and (iii) the relatively high cost of ozone generation is required owing to its short half-life. The advantage of this technique is no requirement of sludge disposal.

6.6.2.5 Sodium Hypochloride (NaOCl)

This method involves destruction of the amino group of the dye molecule by the chloride of the hypochloride, which in turn is responsible for azo bond cleavage [45]. There is a direct correlation between increases in chloride concentration and dye dechlorination. However, the residual chlorine in wastewater imparts a negative effect on the aquatic environment. The formation and release of aromatic amines exhibit carcinogenic and toxic effect on life forms [45]. However, this method is unsuitable for the decolorization of disperse dyes.

6.6.3 Microbial Degradation of Dyes

Microbial remediation is a process that employs bacteria, yeast, or fungi to degrade or biotransform toxic substances to nontoxic or less toxic substances. The benefits of biological methods in addition to being ecofriendly include: complete mineralization of pollutants is attained in economically viable way; and biological treatment techniques of synthetic dyes results in breakdown of dye molecules into several by-products. Microorganisms produce enzymes required for the reduction of dye molecules. Different microorganisms, which include bacteria, fungi, yeasts, and algae, have been utilized for the biodegradation of synthetic dyes. The efficacy of microbial remediation relies on the structure of dye molecule, and the survival and activity of microorganisms [46]. Cultivation of bacteria is easy and less time consuming due to their fast growth rate and thus, bacterial degradation of dye molecules is an efficient and faster process in comparison to fungi and yeast. Biodegradation of azo dyes takes place under aerobic or anaerobic conditions. The reductive cleavage of the azo bond (—N=N—) is the first step in the bacterial degradation of azo dyes. In comparison to aerobic degradation the anoxic environment is more favorable for the reduction process; however, the intermediate (aromatic amines) of anaerobic degradation are recalcitrant, which can be degraded through an aerobic process before its discharge into the environment [38, 47, 48]. A two-step biodegradation process has been adopted to overcome the problem associated with the degradation of recalcitrant intermediates produced during anaerobic degradation. In the first step, the reductive cleavage of the azo bond (—N=N—) under anaerobic conditions generates colorless aromatic amines (Figure 6.2). In the second step, aromatic amines are treated by aerobic cultures.

Several researchers studied and discussed dye detoxification using microorganisms such as bacteria, fungi, algae, and mixed cultures (Table 6.1). Moreover,

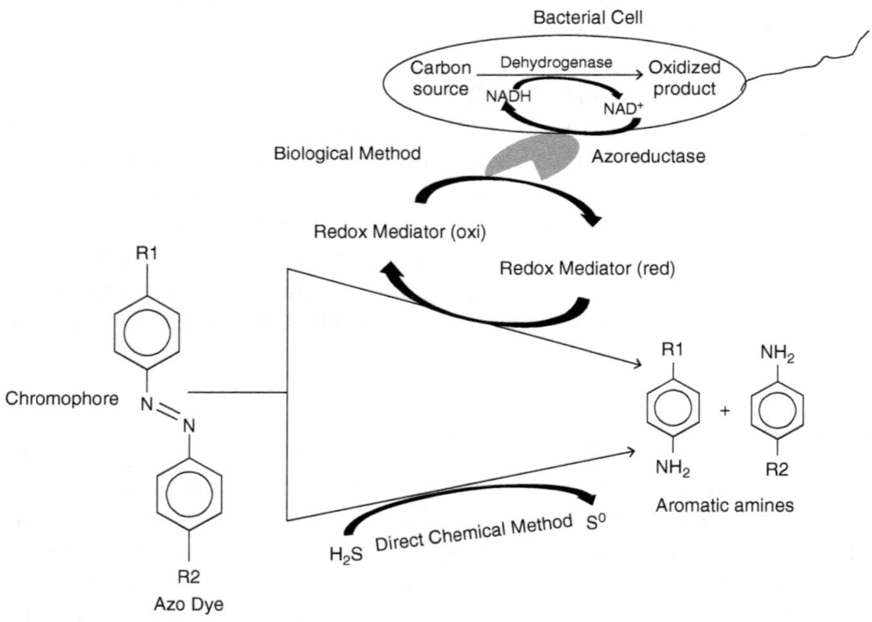

Figure 6.2 Mechanism of dye decolorization and degradation.

Table 6.1 Microbial dye decolorization mechanisms under varied culture conditions.

Bacterial culture	Name of dye and concentration	Culture conditions (pH, temperature (°C), agitation)	Time (h)	Decolorization (%)	Reaction mechanism	References
Bacillus subtilis	Congo Red	–	–	–	Biosorption	[49]
Aspergillus niger	Nigrosin	25 °C	168	93.33	–	[50]
	Basic fuchsin			92.85		
	Malachite green			90.05		
	Dye mixture			32.33		
Bacillus cereus MAM-B11	Congo Red; 25 mg/l	7.0; 37; 150 rpm	72	89.28	–	[51]
Escherichia coli and *Pseudomonas sp.*	Alizarin Red S; 500 mg/l	7.0; 37	12	69.17 and 78.04	Reductive	[52]
Escherichia coli and *Pseudomonas sp.*	Alizarin Red S; 100 mg/l	7.0; 37	12	68 and 72	–	[53]
Bacillus cereus	Acid Orange 7; 100 mg/l	8.0; 33; static	96	68.5	Reductive	[7]
Pseudomonas putida SKG-1	Orange II; 100 mg/l	8.0; 30; static	96	92.8	Reductive	[54]
Bacillus sp.	Reactive Red 5B; 20 mg/l	30, 130 rpm	96	60	–	[6]
Micrococcus glutamicus NCIM 2168	Reactive Green 19A	6.8; 37; static	42	100	Oxidative and reductive	[55]
Rhizobium radiobacter MTCC8161	Reactive red 141; 50 mg/l	7.0; 30; static	48	90	Oxidative and reductive	[56]
*Comamonas sp.*UVS	Direct red 5B; 1100 mg/l	6.5; 40; static	13	100	LiP and Laccase	[48]
*Pseudomonas sp.*SUK1	Reactive red 2; 5 g/l	6.2–7.5; 30; static	6	96	LiP and azoreductase	[57]
Pseudomonas desmolyticum NCIM 2112	Red HE7B; 100 mg/l	6.8–8.0; 30; static	72	95	LiP and azoreductase	[58]

(*Continued*)

Table 6.1 (Continued)

Bacterial culture	Name of dye and concentration	Culture conditions (pH, temperature (°C), agitation)	Time (h)	Decolorization (%)	Reaction mechanism	References
Unidentified bacterium KMK 48	Reactive red 2, reactive red 141, reactive orange 4, reactive orange 7, reactive violet 5; 200–1000 mg/l each	Neutral pH, room temp, aerobic	30	100	Reduction	[59]
Pseudomonas aeruginosa NBAR12	Reactive blue 172; 500 mg/l	7.0–8.0; 40; static	42	83	Azoreductases, flavin reductases, or peroxidases	[60]
Four bacterial isolates consortium *Bacillus*	Acid red 88, acid red 119, acid red 97, acid red 113, reactive red 120; 60 mg/l	7.0; 35; 100 rpm	24	78	Azoreduction	[61]
cereus (BN-7),			24	99		
Pseudomonas putida						
(BN-4), *Pseudomonas fluorescence*			24	94		
(BN-5) and *Stenotrophomonas acidaminiphila* (BN-3)			24	99		
			24	82		
Bacterial consortium SV5	Ranocid fast blue; 100 ppm	7.0; 37; static	24	100	–	[62]
Aeromonas hydrophila	Red RBN; 30 000 mg/l	5.5–10.0; 20–35	8	90	–	[46]
Desulfovibrio desulfuricans	C.I. Reactive orange 96	28; anaerobic	2	95	Reduction	[63]
Pseudomonas luteola	C.I. Reactive red 22	Agitation, without aeration and with a constant dye loading rate of 200 mg/h		113.7 mg dye/g cell/h		[64]

Proteus mirabilis	Red RBN; (1 g/l)	6.5–7.5; 30–35; static	20	95	Reductive followed by biosorption	[65]
Sphingomonas sp. BN6	Amaranath (sulfonated azo dye)				Azoreduction	[66]
Klebsilella pneumoniae R5–13	Methyl red; 100 mg/L	6.0–8.0; 30; 200 rpm	168	100	Reduction	[67]
Mixed bacterial cultures	Mixture of azo- and diazo-reactive dyes	–	96	80	–	[39]
Pseudomonas cepacia 13NA	C.I. acid orange 12 C.I. acid orange 20 C.I. acid red 88		68	90		[68]
Aeromonas hydrophila var. 24 B	Various azo dyes; 10–100 mg/l		24	50–90	Azoreductase (cell free extract)	[69]

phytoremediation techniques have also been used for the removal of dye in wastewater. It attempts to use plants and microbes associated with the plant root system to protect the environment by removal of pollutants in the form of organic and inorganic wastes [70]. Different microorganisms involved in dye decolorization are discussed below.

6.6.3.1 Bacteria

Bacterial degradation and decolorization of dyes can be an environmentally friendly and cost-competitive alternative to physico-chemical processes. Many bacteria carry out decolorization processes such as *Bacillus cereus, Pseudomonas putida* [7]. The bacterial isolate *Psedomonas aeruginosa* was capable of decolorizing 12 different dyes with varying decolorization efficiency. Dye degradation using isolated *Pseudomonas* species as biocatalysts has shown that biological and combination treatment can offer a low-cost alternative system that involves the elimination of the reactive dyes. *P. aeruginosa* and *Brevibacillus choshinensis* isolated from textile effluent, was observed to decolorize the effluent in the presence of 10% glucose very efficiently within seven days [71]. *Bacillus subtilus, B. cereus, Bacillus mycoides, Pseudomonas sp.*, and *Micrococcus sp.* have shown to be among the best microbes for bioremediation of textile effluents [72]. The complete degradation of Mordant Yellow 3 (azo dye) was achieved using *Sphingomonas* sp. BN6 and 5-aminosalicylate degrading bacterium in same reaction. Bacteria, which possess azoreductase, are able to degrade azo dyes aerobically [73]. Some of the bacteria require organic carbon as source of energy as they cannot utilize dye molecules as a carbon source [74]. *P. aeruginosa* degrades Navitan Fast Blue S5R only in a glucose supplemented medium [10].

A major step involved in the bacterial degradation of azo dyes is the reduction of the —N=N— bond. The reduction step catalyzed by enzymes facilitates the transfer of reducing equivalent generated by the oxidation of substrate to azo dye (Figure 6.2). In the 1970s, degradation of azo dye was reported by *Bacillus* sp. [47], and *Aeromonas hydrophila* [69]. Later, other dye degrading *Pseudomonas* spp. and *Aeromonas* species were reported [75]. Moreover, further dye degrading bacteria for example, *B. cereus*, *Pseudomonas* sp., *Achromobacter* spp., *Rhodococcus* spp., and *Plesiomonas* spp., were identified [2]. Garg et al. [3] reported that *B. cereus* isolate was capable of decolorizing reactive orange 4 dye effectively under broad cultural and nutritional conditions. They optimized various process parameters using conventional and statistical methods suggesting that dye concentration is the most important parameter for efficient dye decolorization. They also reported that at the bioreactor level, 100% dye decolorization was achieved within 60 hours under optimum cultural (pH 8.0, 35 °C, 40 mg dye/l, 4.0% *B. cereus* inoculum) and nutritional (1.0% glucose and 0.2% ammonium nitrate) conditions. Therefore, under a specific range of cultural and nutritional conditions is required for biodecolorization of textile dye effluent.

6.6.3.2 Actinomycetes

Several actinomycetes are known for decolorization and degradation of highly recalcitrant compounds. The efficiency of a soil actinomycetes *Streptomyces chromofuscus* was compared with a fungus *Phanerochaete chrysosporium* by Paszczynski et al. [76]. They found that the extent of dye decolorization by actinomycetes was less compared to white-rot fungus. Other researchers reported dye decolorization by actinomycetes [77, 78]. Complete decolorization and degradation of azo dye reactive blue-59 (reactive navy-blue RX) by *Streptomyces krainskii* SUK-5 isolate at pH 8.0 of dye containing nutrient medium occurs within 24 hours under shaking/aerobic condition [78]. This suggests that the precondition for azo dye reduction is the presence and availability of a carbon/energy substrate that acts as an electron donor for the reduction of azo dye [39]. The induction in the activities of lignin peroxidase, NADH-DCIP reductase, and methyl red reductase indicates the role of these enzymes in dye degradation [78].

6.6.3.3 Fungi

Fungal cell walls are made up of macromolecule that contains many functional groups such as carboxyl group, amino, phosphates, lipids, sulphates, and hydroxides. These groups act as metal sorption sites. Fungi remove the dyes essentially by absorption and oxido-reduction reaction. Most of the research concerning bioremediation with these fungi has centered on a single species *Ph. chrysosporium*, which is known to metabolize a wide range of xenobiotic compounds. Nonlignolytic fungi from basidiomycetes such as *Aspergillus* species have been reported to decolorize many dyes. The strain of *Aspergillus sojae* has shown decolorization of the azo dyes Amaranth, Congo red, and Sudan III in nitrogen-poor media. Several other wood rotting fungi like *Aspergillus fumigatus* and *Aspergillus oryza* are capable of decolorizing a wide range of structurally different dyes and have been found to be more effective than *Ph. chrysosporium* [2, 31].

6.6.3.4 Algae

The use of microalgae or macro-algae for the removal of pollutants from wastewater is known as phycoremediation. Number of microalgae like *Chlorella marina*, *Isochrysis galbana*, *Tetraselmis sp*, *Nanochloropsis sp*, and *Dunaliella salina* are concluded as the best method for the removal of textile dye from effluent. Algae have been showed to be capable of removing color from various dyes through mechanisms such as biosorption, bioconversion, and bioagulation. The use of microalgae in bioremediation of the colored wastewater has attracted a great interest due to their central role in carbon dioxide fixation. Additionally, the algal biomass generated has a great potential as feedstock for biofuel synthesis. Moreover, it was reported that azo compounds were biodegraded and decolorized by *Chlorella* sp. where azo dyes were decomposed into simpler aromatic amines [79].

6.6.3.5 Dye Decolorization Enzymes

Enzymes represent an attractive option for wastewater treatment for numerous reasons, which include their biocompatibility as well as the ease and simplicity of process control. Most of the work related to lignolytic enzymes and decolorization of dyes has been concentrated on selected white-rot basidiomycetes. Laccase enzyme offers an attractive solution because of its potential to degrade wide spectrum dyes having diverse chemical structure. The use of lignin degrading white-rot fungi has attracted increasing scientific attention as these organisms are able to degrade a wide range of recalcitrant organic compounds such as polycyclic aromatic hydrocarbons, chlorophenol, and various azo, heterocyclic, and polymeric dyes. The major enzymes associated with lignin degradation are laccase, lignin peroxidase, and manganese peroxidase. The lacasses are the multicopper enzymes, which catalyze the oxidation of phenolic and non-phenolic compounds. Fungi such as *Trametes modesta, Trametes versicolor*, *Trametes hirsuta*, and *Sclerotium rolfsii* have been used for laccase production and for the decolorization of synthetic dyes.

6.6.3.6 Use of Dead Biomass in Decolorization

The use of biomass has its advantages, especially if the dye containing effluent is very toxic. Dead biomass is preferable for wastewater treatment since it is not affected by toxic wastes and chemicals and does not pollute the environment by releasing toxins and or propagates. Dead and dried biomass can be stored for long periods at room temperature with little risk of putrefaction making it easier to use and transport. Dead biomass is also generated as a waste product from established industrial processes. Fungal biomass is not pathogenic to humans and animals and it can be produced cheaply using simple fermentation techniques or obtained as a waste from various industrial fermentation processes.

6.7 Factors that Control the Discoloration of Microbial Dye

Microorganisms are labile in the presence of chemical substances such as dyes, high salinity, high content of organic compounds, change in pH, and temperature. For bioremediation processes, the most useful microorganisms are those that are isolated from dye-contaminated environments including soil, effluents, and sludge from wastewater treatment plants, because they are adapted to grow under extreme situations. The biodiscoloration process is dependent on the following factors: the azo dye structure/concentration, carbon/nitrogen sources, salinity, pH, temperature, type of microorganism, inoculum size, and the presence/absence of oxygen [2]. Many researchers studied the effect of various cultural and nutritional factors on decolorization of extensively used dyes by a variety of bacterial cultures: *Aeromonas hydrophia* for red RBN and remazol black B [46]; *Bacillus* sp. for methyl orange and terasil black [80, 81]; *Enterobacter faecalis* for orange II [82]; *Pseudomonas* sp. for amaranth [83]; *Klebsiella* sp. for reactive yellow 107, reactive red 198, reactive black 5, and direct blue 71 [6]. The behavior of different bacterial strains varied for dye decolorization depending upon the dye used for the treatment process. The conditions optimized for synthetic pure dye decolorization may not be applicable to real effluent decolorization/detoxification, as the composition of textile wastewater is varied, and may also have organics, nutrients, salts, heavy metals, and other toxicants, in addition to dyes. Discussion of the effects of various factors follows.

6.7.1 Effects of the Azo Dye Structure

It has been observed that the enzymatic degradation of dye is highly influenced by its structure. The azo dyes with electron-withdrawing groups, such as SO_3-, are easier to degrade than those with electron-releasing groups, such as—NH—triazine. Therefore, azo dyes that contain more electron withdrawing groups show faster discoloration. If these groups are in para or ortho positions to the azo bond, the degradation proceeds faster than when they are in a meta position, because they provide more effective resonance effect, making azo dyes highly electrophilic resulting in faster reductive discoloration [84]. Optimal decolorization has been reported with an azo compound that has less stearic hindrance near the azo bond [84]. Microorganisms adapted to extreme conditions that have the ability to decolorize dyes can be acclimatized by gradually raising the concentration of the dye to induce enzymatic activity; this induction is dye specific. Therefore, the general approach of bioremediation is to improve the natural degradation capacity of the isolated microorganisms [85].

6.7.2 Influence of Carbon and Nitrogen Sources

Carbon and nitrogen sources have an important influence on the extent of dye decolorization using microorganisms. Different microbial metabolic characteristics lead to differences in the uptake of sources, thus affecting azo dye discoloration. Carbon sources have two purposes: as sources of carbon and energy for the growth and survival of the microorganisms and as electron donors, which are necessary for the breakage of the azo bond [73]. Carbon sources are accepted differently by different microorganisms, and have an important effect on the extent of dye decolorization. In several cases, the microbial discoloration of azo dyes/textile effluents is increased in the presence of glucose.

An interesting goal for microorganisms is to use the dye as a carbon source or even as the sole source of carbon and nitrogen. This goal can be achieved in some cases by successively increasing the amount of dye and diminishing the carbon source until they can survive with the azo dye alone to acclimatize the microorganisms. Examples are the growth of *Saccharomyces cerevisiae* with methyl red; *Pseudomonas, Arthrobacter*, and *Rhizobium* consortium with acid orange 7, *A. fumigatus* with a mixture of reactive black RC, reactive yellow HF2-GL, reactive blue BGFN, reactive black B-150, and reactive red A-6BF. *Oscillatoria curviceps* was able to use acid black 1 as a nitrogen source in an oligotrophic environment.

Several researchers have reported the requirement of a carbon source (glucose, sucrose) supplementation for azo dye decolorization [81, 82, 84]. Ponraj et al. [85] reported sucrose (at 1.0%, w/v) as the best cosubstrate for maximum 87.8% decolorization of orange 3K by *Bacillus* sp. Whereas, Alalewi and Jiang [86] reported starch as the best carbon cosubstrate for acid orange 7 and direct blue 75 decolorization by *Comamonas acidovorans* and *Bacillus cepacia* isolates.

Nitrogen sources are also important for microbial dye decolorization. Different nitrogen sources have been studied to improve the dye decolorization rate. The metabolism of organic nitrogen sources is considered essential for the regeneration of NADH [55]. Discoloration of the azo dyes ruby red and flame orange occurred only in the presence of NADH or NADPH as a cosubstrate. Mendez-Paz et al. [87] observed that inorganic nitrogen supplement (NH_4Cl) was suitable for decolorization of azo dye orange 7 under batch-fed and continuous anaerobic culture conditions.

6.7.3 Influence of Salinity, Color Concentration, pH, Temperature, and Oxygen

Environmental conditions affect the efficiency of microorganisms to decolorize azo dyes, such as the presence of salts, concentration of the dyes, pH, temperature, and oxygen. Generally, azo dye removal decreases under saline conditions. However, there are examples of halotolerant microorganisms that are able to decolorize azo dyes in the presence of salts. *Bacillus* sp. decolorized navy blue in 48 hours, but this process took 18 hours in the presence of $CaCl_2$, because extracellular lipid peroxidase (LiP) and intracellular Lac activities were induced by $CaCl_2$. *Exiguobacterium acetylicum*, *Exiguobacterium indicum*, and *Staphylococcus gallinarum* were able to decolorize reactive black 5 even in cultures containing

60 000 ppm NaCl [88]. Adsorption and enzymatic activity are dependent on the pH. As the extent of dye discoloration is influenced by the medium pH it also affects the color of the solution and the solubility of the dye. Congo red was decolorized 13%, 56%, 70%, and 7% by the bacterium VT-II at pH 2, 5, 7, and 10, respectively [89].

Temperature affects microbial growth and enzyme production, and consequently, the percentage of dye decolorization. *P. aeruginosa* degraded 97% of 50 ppm remazol red at 40 °C, 72% at 10 °C, and 82% at 30 °C [90]. The presence of oxygen can either favor or inhibit the microbial degradation of azo dyes. Shaking increases mass and oxygen transfer between cells and the medium, and enzyme activity depends on the presence of oxygen if the mechanism is aerobic. The time to decolorize methyl red using *Micrococcus* strain was reduced from 24 to 60 hours under the supply of oxygen [91]. Orange II decolorization with *Aspergillus niger* improved with shaking (84%) compared to static growth (61%) [90].

It is evident that bacteria can be employed as a vital biological tool for developing decentralized wastewater treatment systems for decolorization of dye effluents through biodegradation. Such systems could be particularly useful for small-scale textile or dyeing units because most of the microbial strains can be supported on locally available and low-cost growth substrates. Nevertheless, there is a need to formulate efficient bacterial strains in a product that can be delivered to remote places.

6.8 Conclusions

A variety of synthetic dyestuffs released by the textile industry pose a threat to environmental safety. Azo dyes account for the majority of all dyestuffs produced because they are extensively used in the textile, paper, food, leather, cosmetics, and pharmaceutical industries. The recalcitrance and color of dyes is attributed to azo bond(s), and the presence of reactive groups (e.g., sulfonate) make such dyes xenobiotic in nature. Several physico-chemical methods have been used for the removal of dyes from textile effluent. However, implementation of physical/chemical methods have the inherent drawbacks of (i) being economically impractical (as they require more energy and chemicals), (ii) being unable to completely remove the recalcitrant azo dyes and/or their organic metabolites, (iii) generating significant quantities of sludge that may cause secondary pollution/disposal problems, and (iv) involving complicated procedures. However, microbial decolorization is an ecofriendly cost-competitive alternative to physico-chemical processes. The chapter covered the whole range of microbial treatment methods that are adopted for dye removal to create an ecofriendly system.

References

1 Singh, L. and Singh, V.P. (2015). Textile dye degradation: a microbial approach for biodegradation of pollutants. In: *Microbial Degradation of Synthetic Dyes in Wastewaters, Environmental Science and Engineering* (ed. S.N. Singh), 187–204. Springer International Publishing Switzerland https://doi.org/10.1007/978-3-319-10942-8_9.

2 Garg, S.K. and Tripathi, M. (2017). Microbial strategies for discoloration and detoxification of azo dyes from textile effluents. *Research Journal of Microbiology* 12: 1–19.

3 Zollinger, H. (1987). *Color Chemistry: Synthesis, Properties and Applications of Organic Dyes and Pigments*. New York;, USA: VCH Publishers.

4 Down to Earth (2020). United Colours of Industry. https://www.downtoearth.org.in/coverage/united--colours-of--industry-9113 (accessed October 9, 2020).

5 Garg, S.K., Tripathi, M., and Lal, N. (2015). Statistical design for optimization of process parameters for biodecolorization of reactive orange 4 azo dye by *Bacillus cereus* isolate. *Research Journal of Microbiology* 10 (11): 502–512.

6 Prasad, A.S.A. and Rao, K.V.B. (2010). Physicochemical characterization of textile effluent and screening of dye decolorizing bacteria. *Global Journal of Biotechnology and Biochemistry* 5 (2): 80–86.

7 Garg, S.K. and Tripathi, M. (2013). Process parameters for decolorization and biodegradation of orange II (acid orange 7) in dye-simulated minimal salt medium and subsequent textile effluent treatment by *Bacillus cereus* (MTCC 9777) RMLAU1. *Environmental Monitoring and Assessment* 185 (11): 8909–8923. https://doi.org/10.1007/s10661-013-3223-2.

8 Holkar, C.R., Jadhav, A.J., Pinjari, D.V. et al. (2016). A critical review on textile wastewater treatments: possible approaches. *Journal of Environmental Management* 182: 351–366.

9 Chippindale, C.P. and Tacon, S.C. (1998). *The Archeology of Rock-Art*. Cambridge: Cambridge University Press.

10 Welham, A. (2000). The theory of dyeing (and the secret of life). *Journal Society of Dyers and Colourists* 116: 140–143.

11 Carr, C.M. (1995). *Chemistry of the Textile Industry*, 276. Glasgow, UK: Blackie academic and professional.

12 Dyer, A. (1977). *Dyes from Natural Sources*. Newton, MA: Charles T Branford Company.

13 Severkow, U. (1988). Natural dyes an alternative to synthetic dyes. *Melliand Textilberichte* 69: 271–276.

14 Van der Zee, F.P. (2002). Anaerobic azo dye reduction. PhD thesis, Wageningen University, Wageningen, Netherlands.

15 Benkhaya, S., Harfi, S.E., and Harfi, A.E. (2017). Classifications, properties and applications of textile dyes: a review. *Applied Journal of Environmental Engineering Science* 3: 311–320.

16 Devi, R. and Dahiya, R.P. (2008). COD and BOD removal from domestic wastewater generated in decentralised sectors. *Bioresource Technology* 99: 344–349.

17 Qadir, M., Malik, N.R., and Husain, Z.S. (2007). Spatio-temporal variations in water quality of Nullah Aik-tributary of the river Chenab, Pakistan. *Environmental Monitoring and Assessment* 140: 43–59.

18 Zaharia, C. and Suteu, D. (2012). Textile organic dyes characteristics, polluting effects, and separation/elimination procedures from industrial effluents. A critical overview. In: *Organic Pollutants—Ten Years after the Stockholm Convention. Environmental and Analytical Update* (eds. T. Puzyn and A. Mostrag-Szlichtyng), 55–86. Rijeka: Intech Publisher Inc.

19 Alihameed, N. and Ahmed, A. (2008). Physicochemical characterization and bioremediation perspective of textile effluent dyes and metals by indigenous bacteria. *Journal of Hazardous Materials* 164: 322–328.

20 Kant, R. (2012). Textile dyeing industry and environmental hazard. *Natural Science* 4 (1): 12–26.

21 Ahlawat, K. and Kumar, A. (2009). Analysis of industrial effluents and its comparison with other effluents from residential and commercial areas in Solan H.P. *Journal of Theoretical and Applied Science* 1 (2): 42–46.

22 Spadarry, J.T., Isabella, L., and Ranganathan, V. (1994). Hydroxyl radical mediated degradation of azo dyes: evidence for benzene generation. *Environmental Science and Technology* 28: 1389–1393.

23 Mishra, G. and Tripathy, M. (1993). A critical review of the treatments for decolorization of textile effluent. *Colourage* 40: 35–38.

24 Tortman, S.R. and Tortman, E.R. (1964). *The Bleaching, Dyeing Chemical Technology of Textile Fibers*. London: Charles Griffin.

25 Cock, J.G. (1964). *Handbook of Textile Fibers*. Watford, UK: Merrow Publication Co. Ltd.

26 Arul, J.M., Revathi, M.N., and Saravanan, J. (2011). Decolorization and physicochemical analysis of textile azo dye by *Bacillus*. *International Journal of Applied Bioengineering* 5: 35–39.

27 Wetzel, R.G. (1972). Primary production. In: *Water Pollution Microbiology* (ed. R.A. Mitchel), 230. New York;: Wiley Inter science.

28 Verma, S.R. and Dalela, R.C. (1975). Studies on the pollution of the Kalinadi by industrial wastes near Mansurpur part-II: biological index of pollution and biological characteristics of the river. *Acta Hydrochimica et Hydrobiologica* 3 (3): 259–274.

29 Sultana, M., Islam, M., Sahaa, R. et al. (2009). Impact of the effluents of textile dyeing industries on the surface water quality inside D.N.D (Dhaka-Narayanganj-Demra) Embankment, Narayanganj. *Bangladesh Journal of Scientific and Industrial Research* 44: 65–80.

30 Chaudhari, K., Bhatt, V., Bhargava, A. et al. (2011). Combinational system for the treatment of textile wastewater: a future prospective. *Asian Journal of Water, Environment and Pollution* 8: 127–136.

31 Gupta, V.K., Khamparia, S., Tyagi, I. et al. (2015). Decolorization of mixture of dyes: a critical review. *Global Journal of Environmental Science and Management* 1: 71–94.

32 Kim, T.H., Park, C., and Kim, S. (2005). Water recycling from desalination and purification process of reactive dye manufacturing industry by combined membrane filtration. *Journal of Cleaner Production* 13: 779–786.

33 Southern, T.G. (1995). Technical solutions to the color problem: a critical review. In: *Color in Dye House Effluent* (ed. P. Cooper), 73–91. Bradford, UK: Society of Dyers and Colourists.

34 Hao, O.J., Kim, H., and Chiang, P.C. (2000). Decolorization of wastewater. *Critical Reviews in Environmental Science and Technology* 30: 449–505.

35 Allegre, C., Moulin, P., Maisseu, M. et al. (2006). Treatment and reuse of reactive dyeing effluents. *Journal of Membrane Science* 269: 15–34.

36 Golob, V., Vinder, A., and Simonia, M. (2005). Efficiency of the coagulation/flocculation method for the treatment of dyebath effluents. *Dyes and Pigments* 67: 93–97.

37 Robinson, T., McMullan, G., Marchant, R. et al. (2001). Remediation of dyes in textile effluent: a critical review on current treatment technologies with a proposed alternative. *Bioresource Technology* 77: 247–255.

38 Forgacs, E., Cserhatia, T., and Orosb, G. (2004). Removal of synthetic dyes from waste-waters: a review. *Environment International* 30: 953–971.

39 Nigam, P., McMullan, G., Banat, I.M. et al. (1996). Decolorization of effluent from textile industry by a microbial consortium. *Biotechnology Letters* 18: 117–120.

40 Gupta, A.K. and Yogita P (2000). Environmental pollution from textile industry. Indian Institute of Technology, Delhi, NCUTE – programme on environmental problems in chemical processing of textiles.

41 Daneshvar, N., Sorkhabi, H.A., and Tizpar, A. (2003). Decolorization of orange II by electrocoagulation method. *Separation and Purification Technology* 31: 153–162.

42 Slokar, Y.M. and Le Marechal, A.M. (1998). Methods of decoloration of textile wastewaters. *Dyes and Pigments* 37: 335–356.

43 Muthukumar, M., Sargunamani, D., and Selvakumar, N. (2005). Statistical analysis of the effect of aromatic, azo and sulphonic acid groups on decoloration

of acid dye effluents using advanced oxidation processes. *Dyes and Pigments* 65: 151–158.

44 Chen, T.Y., Kao, C.M., Hong, A. et al. (2009). Application of ozone on the decolorization of reactive dyes – orange-13 and blue-1. *Desalination* 249: 1238–1242.

45 Banat, I., Nigam, P., Singh, D. et al. (1996). Microbial decolorization of textile dye containing effluents: a review. *Bioresource Technology* 58: 217–227.

46 Chen, K.C., Wu, J.Y., Liou, D.J. et al. (2003). Decolorization of textile dyes by newly isolated bacterial strains. *Journal of Biotechnology* 10: 57–58.

47 Wuhrmann, K., Mechsner, K., and Kappeler, T. (1980). Investigation on rate-determining factors in the microbiol reduction of azo dyes. *European Journal of Applied Microbiology and Biotechnology* 9: 325–338.

48 Jadhav, S.U., Kalme, S.D., and Govindwar, S.P. (2008). Biodegradation of methyl red by *Galactomyces geotrichum* MTCC 1360. *International Biodeterioration and Biodegradation* 62: 135–142.

49 Khan, M.S., Kukreja, K., Shah, I. et al. (2019). Biosorption of direct textile dye Congo red by *Bacillus subtilis* HAU-KK01. *Bioremediation Journal* 23 (3): 185–195. https://doi.org/10.1080/10889868.2019.1641466.

50 Ranjitha, J., Shalini, P., Anand, M. et al. (2018). Detoxification of dyes by *Aspergillus niger* isolated from contaminated soil effluent from the sites of textile industry. *Research Journal of Chemical Environment* 22: 1–5.

51 Abo-State, M.A.M., Saleh, Y.E., and Hazaa, H.A. (2017). Decolorization of congo red by bacterial isolates. *Journal of Ecological Health and Environment* 2: 41–48.

52 Illakkiam, D., Subha, D., Ahila, V. et al. (2016). Decolorization of alizarin red S dye bacterial strains isolated from industrial effluents. *International Journal of Plant, Animal, Environmental Science* 6: 268.

53 Shinkafi, M.S., Mohammed, I.U., and Audu, A.A. (2015). Degradation and decolorization of textile dyes effluents. *European Journal of Biotechnology and Bioscience* 3: 268–275.

54 Garg, S.K., Tripathi, M., Singh, S.K. et al. (2012). Biodecolorization of textile dye effluent by *Pseudomonas putida* SKG-1 (MTCC 10510) under the conditions optimized for monoazo dye orange II color removal in simulated minimal salt medium. *International Biodeterioration and Biodegradation* 74: 24–35.

55 Saratale, R.G., Saratale, G.D., Kalayani, D.C. et al. (2009). Enhanced decolorization and biodegradation of textile azo dye Scarlet R by using developed microbial consortium-GR. *Bioresource Technology* 100: 2493–2500.

56 Telke, A.A., Kalyani, D.C., Dawkar, V.V. et al. (2009). Influence of organic and inorganic compounds on oxidoreductive decolorization of sulfonated azo dye C.I. reactive orange 16. *Journal of Hazardous Materials* 172: 298–309.

57 Kalyani, D.C., Telke, A.A., Dhanve, R.S. et al. (2008). Ecofriendly biodegradation and detoxification of reactive red 2 textile dye by newly isolated *Pseudomonas* sp. SUK1. *Journal of Hazardous Materials* 163: 735–742.

58 Kalme, S., Ghodake, G., and Gowindwar, S. (2007). Red HE7B degradation using desulfonation by *Pseudomonas desmolyticum* NCIM 2112. *International Biodeterioration and Biodegradation* 60: 327–333.

59 Kodam, K.M., Soojhawon, I., Lokhande, P.D. et al. (2005). Microbial decolorization of reactive azo dyes under aerobic conditions. *World Journal of Microbiology and Biotechnology* 21: 367–370.

60 Bhatt, N., Patel, K., Keharia, H. et al. (2005). Decolorization of diazo-dye reactive blue 172 by *Pseudomonas aeruginosa* NBAR12. *Journal of Basic Microbiology* 45: 407–418.

61 Khehra, D.C., Saini, H.S., Sharma, D.K. et al. (2005). Comparative studies on potential of consortium and constituent pure bacterial isolates to decolorize azo dyes. *Water Research* 39: 5135–5141.

62 Mathew, S. and Madamwar, D. (2004). Decolorization of ranoacid fact blue dye by bacterial consortium SV5. *Applied Biochemistry and Biotechnology* 118: 371–381.

63 Yoo, E.S., Libra, J., and Adrian, L. (2000). Mechanism of decolorization of azo dyes in anaerobic mixed culture. *Journal of Environmental Engineering* 127: 844.

64 Chang, J.S., Chou, C., Lin, Y. et al. (2001). Kinetic characteristics of bacterial azo dye decolorization by *Pseudomonas luteola*. *Water Research* 35: 2041–2050.

65 Chen, K.C., Huang, W.T., Wu, J.Y. et al. (1999). Microbial decolorization of azo dyes by *Proteus mirabilis*. *Journal of Biotechnology* 23: 686–690.

66 Keck, A., Rau, J., Reemtsma, T. et al. (2002). Identification of quinoide redox mediators that are formed during the degradation of naphthalene-2-sulfonate by *Sphingomonas xenophaga* BN6. *Applied and Environmental Microbiology* 68: 4341–4349.

67 Wong, P.K. and Yuen, P.Y. (1996). Decolorization and biodegradation of methyl red by *Klebsiella pneumonia* RS-13. *Water Research* 30: 1736–1744.

68 Ogawa, T., Yatome, C., Idaka, E. et al. (1986). Biodegradation of azo acid dyes by continuous cultivation of *Pseudomonas cepacia* 13NA. *Journal of the Society of Dyers and Colourists* 102: 12–14.

69 Idaka, E. and Ogawa, Y. (1978). Degradation of azo compounds by *Aeromonas luteola*. *Bioresource Technology* 49: 91–94.

70 Bharathiraja, B., Jayamuthunagai, J., Praveenkumar, R. et al. (2018). Phytoremediation techniques for the removal of dye in wastewater. In: *Bioremediations: Application for Environmental Protection and Management, Energy, Environment, and Sustainability* (eds. S.J. Varjani, A.K. Agarwal, E. Gnansounou, et al.). Singapore: Springer Nature https://doi.org/10.1007/978-981-10-7485-1_12.

71 Annika, A.D., Arvind, R.G., and Sayali, R.N. (2012). Decolourisation of textile dyes and biological stains by bacterial strains isolated from industrial effluents. *Advances in Applied Science Research* 3 (5): 2660–2671.

72 Mahmood, R., Shariff, R., Ali, S. et al. (2013). Bioremediation of textile effluents by indigenous bacterial consortia and its effects on *Zea mays* L. CVC 1415. *Journal of Animal and Plant Sciences* 23 (4): 1193–1199.

73 Asgher, M., Bhatti, H.N., Shah, S.A. et al. (2007). Decolorization potential of mixed microbial consortia for reactive and disperse textile dyestuffs. *Biodegradation* 18: 311–316.

74 Stolz, A. (2001). Basic and applied aspects in the microbial degradation of azo dyes. *Applied Microbiology and Biotechnology* 56: 69–80.

75 Ogawa, T. and Yatome, C. (1990). Biodegradation of azo dyes in multistage rotating biological contactor immobilized by immobilized by assimilating bacteria. *Bulletin of Environmental Contamination and Toxicology* 44: 561–566.

76 Paszczynski, A., Pasti, M.B., Goszczyngki, S. et al. (1991). New approach to improve degradation of recalcitrant dyes by *Streptomyces* spp. and *Phanerochaete chrysosporium*. *Enzyme Microbial Technology* 13: 378–384.

77 Zhau, W. and Zimmerman, W. (1993). Decolorization of industrial effluents containing reactive dyes by actinomycetes. *FEMS Microbiology Letters* 107: 157–162.

78 Mane, U.V., Gurav, P.N., Deshmukh, A.M. et al. (2008). Degradation of textile dye reactive navy-blue Rx (reactive blue-59) by an isolated actinomycete *Streptomyces krainskii* SUK-5. *Malaysian Journal of Microbiology* 4 (2): 1–5.

79 Subashini, P.S. and Rajiv, P. (2018). An investigation of textile wastewater treatment using *Chlorella vulgaris*. *Oriental Journal of Chemistry* 34 (5): 2517–2524.

80 Pourbabaee, A.A., Malekzadeh, F., Sarbolouki, M.N. et al. (2005). Decolorization of methyl orange (as a model azo dye) by the newly discovered *Bacillus* sp. *Iranian Journal of Chemical Engineering* 24: 41–45.

81 Pourbabaee, A.A., Malakzadeh, F., Sarbolouki, M.N. et al. (2006). Aerobic decolorization and detoxification of disperse dye in textile effluent by a new isolate of *Bacillus* sp. *Biotechnology and Bioengineering* 93: 63–635.

82 Meitiniarti, V.I., Soetarto, E.S., Timotius, K.H. et al. (2007). Products of orange II biodegradation by *Enterococcus faecalis* ID6017 and *Chryseobacterium indologenes* ID6016. *Microbiology Indonesia* 1: 51–54.

83 Grekova-Vasileva, M., Popov, I., Vassilev, D. et al. (2009). Isolation and characterization of microbial strain AZ029 capable of azo dye decolourization. *Biotechnology and Biotechnology EQ* 23: 318–322.

84 Chen, B.Y. and Chang, J.S. (2007). Assessment upon species evolution of mixed consortia for azo dye decolorization. *Journal of Chinese Institute of Chemical Engineering* 38: 259–266.

85 Ponraj, M., Gokila, K., and Zambare, V. (2011). Bacterial decolorization of textile dye-orange 3R. *International Journal of Advances in Biotechnology Research* 2: 168–177.

86 Alalewi, A. and Jiang, C. (2012). Bacterial influence on textile wastewater decolorization. *Journal of Environmental Protection* 3: 1–12.

87 Mendez-Paz, D., Omil, F., and Lema, J.M. (2005). Anaerobic treatment of azo dye acid orange 7 under batch conditions. *Enzyme and Microbial Technology* 36: 264–272.

88 Ali, N., Hameed, A., Ahmed, S. et al. (2009). Decolorization of structurally different textile dyes by *Aspergillus niger* SA1. *World Journal of Microbiology and Biotechnology* 24: 1067–1072.

89 Sawhney, R. and Kumar, A. (2011). Congo red (azo dye) decolorization of local isolate VT-II inhabiting dye-effluent exposed soil. *International Journal of Environmental Sciences* 1: 1261–1267.

90 Jadhav, S.B., Phugare, S.S., Patil, P.S. et al. (2011). Biochemical degradation pathway of textile dye emazol red and subsequent toxicological evaluation by cytotoxicity, genotoxicity and oxidative stress studies. *International Biodeterioration and Biodegradation* 65: 733–743.

91 Olukanni, O.D., Osuntoki, A., and Gbenle, G.O. (2009). Decolorization of azo dyes by strain of *Micrococcus* isolated from a reuse dump soil. *Journal of Biotechnology* 8: 442–448.

7

Composting and Bioremediation Potential of Thermophiles

Mohammad Yaseen Mir[1], Saima Hamid[1], Gulab Khan Rohela[2], Javid A. Parray[3], and Azra N. Kamili[1]

[1] *Centre of Research for Development, Department of Environmental Sciences, University of Kashmir, Srinagar, Jammu and Kashmir, India*
[2] *Biotechnology Section, Moriculture Division, Central Sericultural Research & Training Institute, Central Silk Board, Ministry of Textiles, Government of India, Pulwama, Jammu and Kashmir, India*
[3] *Department of Environmental Sciences, Govt Degree College, Eidgah, Srinagar, Jammu and Kashmir, India*

7.1 Introduction

Composting as an age-old practice of the decomposition of plant and animal matter, resulting into nutritionally rich compost, dates back to the early Roman era; Pliny the Elder (23–79 BCE) refers to compost in his writings. It is, however, different from the natural process of decomposition of large amounts of organic material on a regular basis by microorganisms, which results in humus and is a means of nutrient turnover in the natural ecosystem. The latter takes place at a slow pace, which can be accelerated by creating conditions ideal for the growth and activity of the decomposers. This accelerated process of decomposition of organic matter by a mixed population of microorganisms in a warm, moist, and aerobic environment has been termed composting [1]. Composting can be either "passive" or "active," involving complex physico-chemical interactions between the organic matter and decomposers, developing into the end products of compost, carbon dioxide, water, and heat. The high organic matter content and biological activity of compost makes it effective in a wide range of applications [2]. Composting has also become an increasingly important strategy for the treatment of municipal organic waste, which is a far better and an ideal ecofriendly approach for the disposal and decomposition of wastes when compared to land fill and burning as the latter poses environmental

Soil Bioremediation: An Approach Towards Sustainable Technology, First Edition.
Edited by Javid A. Parray, Abeer Hashem Abd Elkhalek Mahmoud, and Riyaz Sayyed.

problems [3]. Regardless of the products formed, the active component mediating biodegradation and conversion process during composting is the resident microbial community [4]. Knowledge about the microorganisms present in composts, their coexistence, and the ways they replace each other during the different stages of biological degradation process should help to ensure a resulting high-quality compost product [3]. Thermophilic microflora constitutes the predominant component, which provides selectivity to compost. Thus, the optimization of compost quality is directly linked to composition and succession of microbial communities in the composting process. This means that tools are required to monitor and characterize the composition, patterns, and dynamics of species diversity during composting and to relate them to compost quality.

It is now becoming possible to define the causes of time-dependent changes in the health of the microbial community on the basis of observed genetic diversity [5]. These include assessment of microbial diversity based on *in-situ* extracted phospholipids ([6]), measurement of carbon source utilization by substrate-extracted microbial cell consortia [7], and nucleic acid-based techniques [4, 8]. The combination of signature lipid and nucleic acid-based analyses has greatly expanded the specificity and the scope for assessing the microbial community composition in composts [3]. These approaches also provided basic knowledge of mechanisms that control species coexistence, which have fundamental applications since they offer the framework that serves biologists interested in maintaining and restoring diversity in natural ecosystems.

This chapter covers various types of composting, physico-chemical aspects of composting, and the structure of thermophilic microbial community and their functionality in various compost ecosystems. Readers may find that available information on several aspects of compost ecosystem is scant, which is due to horizontal advancements in some areas and also because compost represents a complex ecological system from the viewpoint of microbial distribution and activity. A special emphasis has been given to mushroom compost ecology since this complex man-made ecosystem harbors a complete spectrum of microbial diversity and is also unique with respect to conditions under which the mushroom crop is grown and the relatively short time required to complete successional cycles, which are not matched elsewhere.

7.2 Heavy Metal Resistance Genes

Thermophilic microorganisms occupy natural habitats often enriched with heavy metals and hence develop strategies to resist metal toxicity, either as part of their metabolic activities or as a specialized mechanism to reduce metal toxicity thereby providing competitive advantage to survive and flourish under such extreme conditions [9]. Recent studies on the diversity of metal-resistant microbes across a range of environments including the most extreme clearly imply that metal-resistance systems were evolved long before an anthropogenic release of metal contaminants [10].

Prevalence of high metal tolerance between chemolithoautotrophic acidophilic bacteria and archaea and the presence of metal resistance genes in their genomes are the testimony for such inferences [11]. In recent times, heavy metal resistant thermophilic bacilli (*Bacillus sphaericus*, *Bacillus pumilus*, *Paenibacillus larvae*, and *Geobacillus stearothermophilus)* were isolated from Ma'een hot springs in the Jordan area and were used as immobilized, dead, and live biomass for bioremediation of heavy metals present in the wastewater effluent of Jordan Rotogravure Establishment (JRE) [12]. The investigator showed that although the overall sorption efficiency does not vary significantly within the temperature range of 37–70 °C, thermophilic bacilli exhibited a maximum metal-binding capacity at temperatures 60 and 70 °C.

The affinity of *Bacillus* for heavy metals adsorption was found to be Cu > Cr > Ni for *B. sphaericus* and Cr > Cu > Ni for other *Bacillus* species. Apart from the isolation of various thermophilic metal resistant microorganisms followed by molecular characterization of metal-resistant determinants, several studies have identified genes encoding metal resistance in thermophiles [13]. Several metal-resistant genes (*chrA, merA, cadA, cadC, arsB, arsC, arsR, arr, acr* (*3*) *p,* and *czc*) that mostly confer resistance by encoding efflux pumps and reductases (for arsenate, chromate, and mercury) have been detected in thermophilic bacteria and archea (Table 7.1).

Heavy metal toxicity on pure cultures of thermophilic methanogenic *Methanobacterium thermoautotrophicum* (MT) and a thermophilic acetate-decarboxylating methanogenic bacterium (TAM) was studied by Ahring and Westermann [15]. They found that Cd and Cu toxicity depends upon the presence of Ni in the medium. Llanos et al. [10] tested Cd, Zn, Co, and Ni sensitivity of 30 thermophilic strains (*Bacillus* spp., members of *Thermotogales* and *Thermococcales*) isolated from heavy metal-rich hydrothermal vent sites at Lau Basin together with their Cd accumulation potential. Compared to the members of *Thermotogales* that showed resistance to only Co, members of the *Thermococcales* were highly resistant to both Cd and Zn. Watkin et al. [9] studied Cu, Zn, Ni, and Co tolerance by a number of moderately thermophilic bacteria closely related to *Acidithiobacillus caldus*, *Acidimicrobium ferrooxidans*, and *Sulfobacillus thermosul fi dooxidans*. Among other metals, isolation, and characterization of Hg-resistant microbes from several geothermal springs got distinctive consideration [16–18]. Along with Pb and Cd, Hg belongs to class B heavy metals and is extremely poisonous at low concentration [18]. Among bacteria, mercury resistance is well studied and is

Table 7.1 Metal resistance genes detected in thermophilic bacteria and archaea.

Gene	Bacteria
ChrA	*Geobacillus kaustophilus*
ChrA	*Geobacillus* sp.
ChrA	*G. thermoglucosidasius*
ChrA	*Thermosipho melanesiensis* BI429
ChrA	*Fervidobacterium nodosum* Rt17-B1
ChrA	*Geobacillus thermoleovorans CCB US3*
merA	*Brevibacillus* sp.
merA	*Anoxybacillus* sp.
MerA	*G. kaustophilus*
MerA	*Hydrogenobacter thermophilus* TK-6
CadA	*Streptococcus thermophilus*
ArsB	*A. flavithermus*
ArsB	*Syntrophothermus lipocalidus*
ArsB	*Spirochaeta thermophila*
ArsB	*Fervidobacterium nodosum*
ArsB	*S. thermophila*
ArsB	*S. thermosul fi dooxidans*
ArsB	*Deferribacter desulfuricans* SSM1
ArsC	*A. flavithermus*
ArsC	*G. kaustophilus*
ArsC	*S. thermophila*
ArsC	*Campylobacter lari*
ArsC	*Synechococcus sp.* JA-2-3B'a(2–13)
ArsC	*Exiguobacterium sp.* AT1b
ArsR	*P. thermopropionicum*
ArsR	*S. lipocalidus*
ArsR	*Caldicellulosiruptor bescii*
ArsR	*Thermotoga lettingae*
ArsR	*F. nodosum*
ArsR	*C. saccharolyticus*
ArsR	*T. lettingae*
ArsR	*T. petrophila*

(*Continued*)

Table 7.1 (Continued)

Gene	Bacteria
ArsR	*H. thermophilus*
ArsR	*A. flavithermus*
ArsR	*G. kaustophilus*
ArsR	*D. desulfuricans*
ArsR	*Thermosynechococcus elongatus*
ArsR	*S. thermosul fi dooxidans*
Acr3	*P. thermopropionicum*
Acr3	*D. desulfuricans*
Acr3	*S. lipocalidus*
Arr	*D. desulfuricans*
CzcD	*A. flavithermus*

Source: Adapted from Sar, Kazy, and Paul [14].

usually plasmid borne [19]. Active resistance to Hg (II) is regulated by *merR*, a unique representative of the winged helix-turn-helix (WHTH) family of bacterial transcription factors [20]. MerR controls production of mercuric reductase (MerA) along with other related components [19]. Schelert et al. [17, 18] have studied the interaction between Hg and hyperthermophilic *Sulfolobus solfataricus* and revealed the occurrence of mercury resistance in members of the archaeal domain. Members of the phylum *Crenarchaeota*, including the *S. solfataricus*, inhabit metal-rich geothermal environments. Proliferation under these conditions is accompanied by induction of active metal detoxification pathways encoded by homologs of mercuric reductases (MerA) that operates under the control of MerR, a unique class of transcription factors that exert both positive and negative regulation on gene expression [17, 18]. In *S. solfataricus*, the level of resistance afforded by these proteins was found to be considerably lower than that of other mercury-resistant bacteria [21]. It has been found that *S. solfataricus* though sensitive to mercuric chloride, a low-level adaptive response, could be induced as a result of upregulation of *merA* transcription by mercury challenge. Mercuric ion (Hg 2^+) inactivates generalized transcription but simultaneously derepresses transcription of mercuric reductase (MerA) by interacting with the archaeal transcription factor MerR [18].

On either side of *merA*, two additional Hg (II)-inducible genes, *merH* and *merI*, were also detected. Further experiments by creating mutant strains that lack both *merA* and *merRA* Hg^{2+} –mediated transient derepression of *mer* expression "implicated the existence of a secondary system or alternative cellular reductases for metal detoxification" [18]. Chatziefthimiou et al. [16] have investigated mercury resistance in several facultative thermophilic and chemolithoautotrophic, thiosulfate oxidizing bacteria isolated from mercury-rich hot sulfidic springs and indicated that Mer mediated Hg (II) reduction was the dominant resistance mechanism among these strains. Based on several similar studies, it was suggested that bacterial genera affiliated to the order *Bacillales* are common in mercury-rich high temperatures and geothermal sources of mercury in acidic to slightly alkaline environments [16, 21]. Phylogenetic analysis of *merA* genes by later investigators further showed genetic incongruence, suggesting evolution by horizontal gene transfer (HGT) of the *merA* gene.

Protein phylogenetic analysis of mercury reductase (MerA) in bacteria and thermophilic archaea revealed some interesting findings. In bacterial origin, the catalytic site of MerA is comprised of a conserved pyridine nucleotide disulfide oxidoreductase domain (PFAM 0007) with two active cysteines (C207 and C212) [22]. Unlike other members of this large family, MerA has two additional unique regions: (i) a short C-terminal extension having two additional active cysteines (C628 and C629), and (ii) an extended N-terminus that promotes metal recruitment critical for metal reduction. Additionally, two conserved tyrosines (Y48 and Y605) that facilitate catalysis are also present in bacteria [21]. In contrast to these, MerA among the archaea lacked the N-terminal metal recruitment domain and a C-terminal active tyrosine.

Phylogenetic analysis of amino acid sequences of thermophilic *merA* gene retrieved from NCBI (National Center for Biotechnology Information) GenBank revealed presence of discrete clades well separated from those of mesophilic origin. Three distinct clades represented the MerA sequences from thermophilic bacterial, mesophilic bacterial, and archaeal domains were observed. Interestingly, MerA from *Anoxybacillus* spp., *Geobacillus kaustophilus*, showed a close relationship, forming a single monophyletic group. MerA from archeal (*S. solfataricus)* origin also formed a separate clade leaving the MerA from mesophiles as distinct members. Apart from these studies, several other investigations have yielded sequences of metal-resistant determinants (e.g., *cad, czc chr, ars, arr,* and *acr*) from thermophilic microorganisms and deposited in the NCBI database. Sequences of such genes/proteins was retrieved and analyzed to obtain a preliminary idea about their phylogenetic lineage.

7.2.1 Cadmium-Resistant Genes (*cad* and *czc*)

Either ATP-dependent cadmium efflux transporters or metallothionein-based metal binding and detoxification system provide cadmium resistance in bacteria. Cadmium resistance systems found in gram-positive bacteria encodes Cd exporting P-type ATPase (CadA pump) along with CadC (regulator protein). Function of these two genes was originally studied in *Staphylococcus aureus* [23]. Although, homologs of these genes were later identified in many mesophilic bacteria (e.g., *Bacillus firmus, Listeria monocytogenes, Lactococcus lactis, Listeria innocua*, and *Stenotrophomonas maltophilia*), their detection in thermophilic microorganisms remains scant. So far, both *cadA* and *cadC* genes have been reported only from two genera, namely, *Streptococcus thermophilus* and *G. stearothermophilus* [24]. Phylogenetic analysis of thermophilic CadA showed its lineage with that of mesophilic *Escherichia coli* CadA, while the CadC from thermophilic *S. thermophilus* and *G. thermophilus* represented its distinctness from *E. coli* CadC. In addition to the *cad* system, a *czc* system that encodes proteins responsible for transporting cobalt, zinc, and cadmium to or from cell interior is also detected in thermophilic bacteria *Anoxybacillus fl avithermus* [25]. This protein is a member of the RND (resistance-nodulation-division) protein family, which is involved in bacterial transport of heavy metals (*Cupriavidus metallidurans*), nodulation (*Mesorhizobium lot*), and cell division (*E. coli*) [26]. In bacteria and archaea, members of this superfamily are involved in transport of heavy metals, hydrophobic compounds, amphiphiles, and nodulation factors.

7.2.2 Chromium Resistance Gene (*chr* A)

Several bacterial mechanisms of resistance to chromate have been reported. The best-characterized mechanisms comprise a reduction of Cr(VI) to Cr(III) and efflux of chromate ions from the cell cytoplasm [27]. Chromate efflux by the ChrA transporter has been established. ChrA functions as a chemiosmotic pump that effluxes chromate from the cytoplasm using the proton-motive force [28]. In thermophilic bacteria, the protein for chromate reductase (ChrA) was detected only in *G. kaustophilus* and *Geobacillus thermoglucosidasius* [29]. Phylogenetic analysis revealed that although ChrA sequences from thermophilic origin are closely related, they share a common evolutionary node with that of mesophilic bacteria (*E. coli*).

7.2.3 Arsenite Resistance (*ars*) Operon

One of the best characterized, and probably the most widespread, arsenic resistance system in microorganisms is the *ars* gene system [30]. This system consists of a series of genes coding for a transmembrane pump system (*ars* B), arsenate reductase (*arsC*), and a regulatory gene (*arsR*) [31]. In this mechanism, ArsC catalyzes the first step in arsenic detoxification reduction, reduction of arsenate [As(V)] to arsenite [As(III)] using reducing equivalents derived from glutathione (GSH) via glutaredoxin (GRX). This protein was well characterized in many mesophilic bacteria including *Pseudomonas* sp., *Bacillus* sp., *Shewanella* sp., *Mycobacterium* sp., *Klebsiella* sp., *Clostridium* sp., and *Staphylococcus* sp. [32] along with a few thermophilic bacteria like *Anoxybacillus flavithermus*, *G. kaustophilus*, and *Spirochaeta thermophila* [25, 29, 33]. Analysis of ArsC sequences from several mesophiles and thermophiles revealed a distinct evolutionary lineage. All the ArsC sequences from thermophilic origin were highly related, forming a separate clade from that of mesophiles.

The other important members of *ars* operon are *arsB* and *arsR*. The gene *arsB* encodes the efflux pump membrane protein, while the *arsR* is a trans-acting inducer responsive repressor codes for a repressor protein that regulates *ars* gene expression [34]. Following reduction of arsenate, the reduced arsenite is pumped out from the cell through this pump. The ArsB protein is well characterized in most of the arsenate-reducing and arsenate-resistant mesophilic bacteria like *Pseudomonas* sp., *Shewanella* sp., *Bacillus* sp., and *Staphylococcus* sp. ([35, 36]), while in the case of thermophilic bacteria this protein was found only within limited number of spices including *Hydrogenobacter thermophilus*, *Pelotomaculum thermopropionicum*, *S. thermophila*, *A. flavithermus*, and *Syntrophothermus lipocalidus* [25, 33]. The thermophilic bacteria include *P. Thermopropionicum*, *Deferribacter desulfuricans* and *S. lipocalidus*, a further arsenite transporter Acr3 was also detected [29]. Acr3p transporters perform a similar function to ArsB, but the two proteins have no significant sequence similarity.

The ArsR protein was characterized in both mesophilic (*Klebsiella* sp., *Halobacterium* sp., *Bacillus* sp., *Microcystis* sp., *Pantoea* sp., *Pseudomonas* sp., *Staphylococcus* sp.) as well as thermophilic (*P. thermopropionicum, S. lipocalidus, Caldicellulosiruptor bescii, Thermotoga lettingae, H. thermophilus, S. thermophila, G. kaustophilus, D. desulfuricans*, etc.) bacteria [25, 33].

7.2.4 Arsenite Respiratory Reductase

The arsenite respiratory reductase (Arr) system is well distributed within diverse dissimilatory arsenate-reducing prokaryotes capable of using arsenate as a respiratory electron acceptor while oxidizing various aromatic compounds. The Arr that has been characterized from the *Shewanella* sp. strain ANA-3 [37] is a periplasmic dimethyl sulfoxide (DMSO)-type reductase (2596 bp, 854 amino acids, or 95.2 kDa) that reduces arsenate to arsenite and is considered as reliable biomarker for arsenate respiration [38]. Arsenite reductase gene (arrA), encoding the Arr protein, has been studied from a number of mesophilic bacteria (*Bacillus selenitireducens*, *Desulfitobacterium hafniense*, *Wolinella succinogenes*, *Bacillus arseniciselenatis*, *Chrysiogenes arsenatis,* and *Sulfurospirillum barnesii*) while its presence in thermophiles is much less [39]. The only thermophilic bacteria reported to have Arr protein is *D. desulfuricans* [40].

7.3 Biotransformation-Based Bioremediation

7.3.1 Oxidation-Based Metal Transformation

This section deals with oxidation of metals like arsenite and iron. Thermophiles may either develop resistance due to their metabolic activities like the transfer of electrons to and from iron during growth or can reduce toxic effects of metal by expulsion of accumulated metal using an efflux pump. Thus, they provide an advantage as they are able to thrive and flourish in harsh environmental conditions [9, 11, 16]. Some metals are metabolically essential while others are possibly toxic so it is necessary that thermophiles should maintain a tender balance to maintain homeostasis. Toxicity of metals is of particular concern in cases of metal-metabolizing microbial communities, which evade harmful metals released into the environment as a part of their energy metabolism [41].

7.3.1.1 Arsenite [As(III)] Oxidation

In autotrophic bacteria, As(III) oxidation involves As(III) as the electron donor and oxygen as the electron acceptor while in heterotrophic bacteria the As(III) oxidation process is described as a mode of detoxification that is mediated by the enzyme arsenite oxidase (Aio) [42]. The Aio carries out oxidation of arsenite to arsenate (a less toxic form) in *Acidianus brierleyi* and *Sulfolobus metallicus* [43]. The toxicity of As(III) is an important concern when bioleaching of arsenic ores releases arsenite during mineral partition. An adequate amount of Fe(III) should be available to oxidize As(III) and counteract the harmful consequences [43]. The oxidation of As^{3+} can be accomplished by phylogenetically unrelated bacteria with the help of Aio, a heterodimeric enzyme that was formerly known as Aox, Aro, and Aso [44, 45]. The enzyme Aio consists of two subunits: AioA (molybdopterin centre and a 3Fe—4S cluster) and AioB (Rieske, 2Fe—2S cluster) [46]. Phylogenetically distinct microbes like Actinobacteria, Aquificae, Bacteroidetes, Chlorobi, Chloroflexi, Crenarchaeota, Deinococcus-Thermus, Firmicutes, and Nitrospira, and members of Alpha proteobacteria, Beta proteobacteria, and Gamma proteobacteria contain homologs of genes involved in AioA synthesis [47, 48]. The presence of Aio in archaea, viz. *Aeropyrum pernix*, *Pyrobaculum calidifontis*, and *Sulfolobus tokodaii*, indicates their involvement in As^{3+} oxidation in high-temperature ecosystem [45]. The presence of the aoxB gene in the phylogenetically distinct microbes indicates that HGT played an important role in the spread of the aox gene in prokaryotic communities [49].

7.3.1.2 Iron [Fe(II)] Oxidation

Members of family Sulfolobaecae like *A. brierleyi*, *Sulfolobus acidocaldarius*, *Sulfolobus brierleyi*, *S. solfataricus*, and *Sufurococcus* sp. grow at temperatures up to 80 °C and oxidize iron for energy in geothermal environments [50]. The genome sequence data for the Sulfolobales indicates the presence of the Fe(II) oxidation (fox) gene, which encodes a novel terminal oxidase [51]. The fox gene cluster is present in Fe(II) oxidizers like *Metallosphaera sedula*, *S. metallicus*, and *S. tokodaii* and is absent in non Fe(II)-oxidizing species except for *Metallosphaera cuprina*. In *Metallosphaera prunae*, a mutated fox gene cluster, foxA', is present. However, this mutated gene cluster exhibits a reduced ability for uranium and fox [52]. SoxABCD, DoxBCE, and SoxM supercomplexes are possibly terminal oxidase clusters involved in fox [53]. Rusticyanin, sulfocyanin, and novel multicopper oxidases that possibly act as electron shuttles to terminal oxidase may also play a role in fox in hyperthermoacidophiles. In *Metallosphaera* spp., the high level of expression of rus-like blue copper proteins on the exposure of cells to chalcopyrite and triuranium octoxide indicates the role of the rus operon in the electron transport chain involved in fox [52]. Fe^{2+} and HS^- are important electron donors in environments like hydrothermal systems [53]. The respiratory clusters of thermoacidophilic Sulfolobales associated with fox are distinct from mesophilic microorganisms [41]. The fox mechanisms in Sulfolobales are of prime interest especially in case of mixotroph like *M. sedula*, which grow as a heterotroph in the presence of organic substrates while exhibiting chemolithotrophy in the presence of metal sulfides like pyrite and sulfur.

The gene expression studies of **M. sedula* have shown increased expression of genes within the fox cluster as well as the cbsAB-soxLN cytochrome ba complex in cultures grown on Fe(II) [53]. *S. metallicus* and other phylogenetically diverse Fe(II)-oxidizing archaea possess an acid-stable, redoxactive biomolecule [54]. *A. brierleyi* and *M. sedula*, when grown in the presence of Fe(II), also express a redox-active, membrane-bound yellow cytochrome, which could be possibly involved in fox and represent an uncommon thermoacidophilic redox-active enzyme [55]. It has been speculated that CsbA, a membrane associated cytochrome b566/588 involved in electron transfer in the pseudo-periplasmic space of *S. acidocaldarius*, could be related to yellow redox-active enzyme engaged in fox and present in high concentration when cells encounter FeS2 [56].

7.3.2 Reduction-based Metal Transformations

Thermophiles are able to reduce metals and metal-containing compounds like iron oxides, manganese, cobalt, molybdenum, and chromium, which are frequently encountered in geothermal environments [57]. The existence of thermophiles on the hot subsurface of Earth since evolution time is an indication of their involvement in biogeochemical cycles and also attracted attention of astrobiologists and evolutionary biologists [57]. Reduction of metals helps microorganisms fulfill various functions like assimilation, dissimilation, and detoxification. Dissimilation is involved in energy generation like formation of intracellular inclusions of iron and manganese formed after reduction of Fe(III) and Mn(IV) [58, 59]. Assimilation has a biosynthetic function like magnetosomes formed in magnetotactic bacteria and chelated Fe(III) reduction mediated by enzyme iron reductases either before or after its transport into the cell [60]. Detoxification helps in the immobilization of toxic metals as reduced forms of chromium, cobalt, uranium, mercury, etc., are less soluble than oxidized forms [57].

7.3.2.1 Chromium Reduction [Cr(VI) → Cr(III)]

In natural environments, Cr occurs in an oxidation state ranging from +2 to +6 and is frequently found in rocks, plants, animals, soil, volcanic dusts, and gases. Cr(III) is the most common form which is less toxic when compared to Cr(VI). Cr(VI) is a potential contaminant of groundwater. It has mutagenic and carcinogenic effects, due to its high oxidation potential. It does not interact directly with DNA, but its genotoxicity can be due to intracellular reduction to Cr(III) through reactive intermediates [61]. Thus, the DNA damages that ensue are oxidative DNA damage and Cr(III)-DNA interactions [62]. So, bioremediation of Cr(VI) by chromium-resistant bacteria is an effective approach for cleaning of the environment [63]. Chromium resistance is attributed to efflux of chromate from the cytosol followed by the reduction of soluble and toxic Cr(VI) to insoluble and nontoxic Cr(III) by chromate reductase. Reduction of chromium from Cr(VI) to Cr(III) can be carried out using whole cell or cell-free extracts or via immobilized enzymes [64]. The immobilized chromium reducing enzymes can serve as an excellent approach for bioremediation of chromate-containing wastes [65]. In bacteria, the enzymes involved in the reduction of chromium are chromate reductase, aldehyde oxidase, cytochrome P450, and DT-diaphorase, and several oxidoreductases like nitro reductase, iron reductase, quinine reductase, and flavin reductase [66].

The chromate reductase genes are generally carried by plasmid-mediated transfer but can be present on both plasmid and chromosome [67]. The chromate reductase activity is exhibited by cytoplasmic fractions of bacteria during an aerobic reduction process [68] whereas Cr(VI) acts as a terminal electron acceptor in anaerobic reduction and is, therefore, reduced in the plasma membrane [69]. There are several reports on aerobic chromate reduction by mesophiles like *Pseudomonas* spp.; however, aerobic reduction of chromate by thermophilic bacteria is yet not much explored [70]. Thermophiles like *Anoxybacillus* spp., *Bacillus* spp., and *Thermus* spp. are reported to be promising candidates for chromium reduction in aerobic conditions. In *Thermus scotoductus* SA-01, a membrane-associated chromate reductase exhibited coupling of Cr(VI) reduction-NAD(P)H oxidation [71]. Seventy-five percent of reduction of chromium in 24 hours has been reported by a thermophile *Anoxybacillus rupiensis*, isolated from a hot spring in Dapoli, Maharashtra, India [70]. Other species of thermophilic archaea and bacteria that can reduce Cr(VI) in the concentration ranging from 0.1 to 0.5 mM with lactate, glucose, or H2 as electron donors are *Deinococcus geothermalis*, *Pyrobaculum islandicum*, and *Thermoanaerobacter* spp. [72].

7.3.2.2 Mercury Reduction [Hg(II) → Hg(0)]

Hg is a major pollutant in water bodies. It is bioaccumulative and toxic for animals and humans due to its high affinity for thiols [73]. Microorganisms play a crucial role in transforming mercury either by methylation or degradation of methyl mercury or by reduction and oxidation of an inorganic form of mercury [74]. Mercury-resistant hyperthermophiles and thermophiles have been reported from deep-sea hydrothermal vents and terrestrial hot springs [16]. The plausible reason for mercury resistance in thermophilic microbial population could be due to their exposure to mercury present in hydrothermal habitats, and this, in turn, had imparted these high-temperature microbes as an adaptation to mercury toxicity [75].

7.3.2.3 Selenium Reduction [Se(IV)/Se(VI) → Se(0)]

In natural environments, selenium (Se) occurs in various forms, viz. elemental selenium (S0), selenide (Se^2–), selenium oxyanions selenate [Se(VI)], and selenite [Se(IV)]. Selenium has a dual role in biological systems. It is an important metal in some cellular processes and enzymes. It is toxic at high concentrations. It is considered to be one of the essential toxins like zinc, copper, and iron [76]. Selenium oxyanions, i.e., Se(IV) and Se(VI), are the most toxic forms of selenium. Microbial reduction of Se(IV) and Se(VI) has attracted wide attention because of the nanoparticle synthesis as bacteria reduce Se(VI)/Se(IV) to S0, which is visible as red-colored nanoselenium [77]. The microbial synthesis of nanoparticles is a keystone in the field of nanotechnology [78].

The synthesis of selenium nanoparticle (SeNP) formation was first reported in anaerobic bacteria [79]. The anaerobic synthesis of Se nanoparticles is quite challenging, as it demands reducing conditions throughout the process, making optimization, and scale-up processes difficult. Nevertheless, these problems can be overcome by aerobic selenium-resistant bacteria like *Bacillus cereus*, *Bacillus* spp., and *Pseudomonas aeruginosa* [80]. Microbial synthesis of nanoparticles can be accomplished by reduction, which is followed by precipitation, nucleation, and crystal growth. To date, the mechanisms behind the reduction are well studied but concepts related to biosynthesis of nanoparticles are still untapped [78]. Although several authors have reported the involvement of proteins in the biosynthesis of nanoparticles [81], it is yet to be explored. The SeNPs are biologically active molecules and possess antioxidant properties. They also elevate the antioxidant ability of glutathione peroxidase and thioredoxin reductase [82, 83]. The antioxidant activity of SeNPs is determined by size as SeNPs ranging from 5 to 200 nm are able to scavenge free radicals *in vitro* [84]. The understanding of the mechanisms or elaborated science behind the synthesis of SeNPs will help to control and modulate molecular switches involved in determining various parameters like size, shape, and stability of SeNPs [85].

Correa-Llantén et al. [78] reported that pH and temperature have a profound effect on the size and shape of bacteriogenic SeNPs. The SeNPs synthesized by thermophilic bacteria *Geobacillus weigleii* via the NADPH/NADH dependent reductase were 120 nm in size at pH 7 while under acidic and alkaline conditions the size of majority of nanoparticles was less than 100 nm. Fifty percent of nanoparticles were of size greater than 100 nm at 80 °C while at 100 °C, SeNPs were aggregated to nanorod structures. Thus, physiological parameters affect the size and morphology of SeNPs. Ghalib et al. [86] reported aerobic reduction of selenite to elemental selenium at an initial selenite concentration of 200 μg/ml up to a maximum of 500 μg/ml at 45 °C by two strains (TA2 and TA4) of *B. cereus*. The aerobic species of *Anoxybacillus* and *Geobacillus*, isolated from the Paralana hot springs of South Australia, have been reported to reduce selenium along with other heavy metals like vanadium, cobalt, and uranium [87] and can serve as potential candidates for bioremediation.

7.3.2.4 Uranium Reduction [U(VI) → U(IV)]

Uranium is a radioactive element and wastes from nuclear reactors, mining, and milling operations introduce it into an ecosystem. Removal of uranium by microbial reduction, precipitation, and immobilization are areas of interests from a bioremediation perspective [88]. Microorganisms use toxic U(VI) as an electron acceptor and reduce it to the less toxic U(IV), which can be easily removed by precipitation [89]. U contamination in the concentration ranging from 0.1 to 2 mM can be treated by thermophiles like *D. geothermalis*, *P. islandicum*, *Thermoanaerobacter* sp., and *T. scotoductus* [72].

In natural uranium ore deposits and uranium-contaminated environments, microorganisms exhibit a different transformation of uranium phosphate [90]. The microbial-mediated U(VI) phosphate reduction generally results in the formation of uraninite (UO2); however, thermophilic bacteria like *Thermoterrabacterium ferrireducens* can also form ningyoite as an end product [89, 90]. Cason et al. [91] reported the involvement of a 70-kDa protein in U(VI) reduction to U(IV) in *T. scotoductus* SA-01. It is a periplasmic and membrane-bound, ABC transporter, peptide-binding protein. It was reported earlier to reduce Au(III). The Uranium reduction is mostly mediated by multifunctional enzymes [91]. Wright et al. [87] reported that *Clostridium thermosuccinogenes* can reduce uranium from the toxic U(VI) (up to a concentration of 6 mM) to a precipitated nontoxic, easily removable variant. Uranium-reducing thermophilic bacteria offer a cheaper, autonomous solution to the decontamination of depleted and still-active sites.

7.3.2.5 Iron Reduction [Fe(III) → Fe(II)]

Brock and Gustafson [92] reported thermophilic reduction of Fe(III) in *S. acidocaldarius*. Thermophile ferric iron reducers are common in terrestrial ecosystems, hot springs, hydrothermal vents, and deep subsurface water [93]. A number of phylogenetically diverse iron-reducing thermophilic bacterial species have been reported, viz. *Bacillus infernus*, *Thermosinus carboxydovorans*, *Thermovenabulum ferriorganovorum*, and *T. ferrireducens*, and species of the genera *Anaerobranca*, *Sulfobacillus*, and *Thermoanaerobacter* [93]. Some significant iron reducers are as follows: moderately thermophilic Actinobacteria like *A. ferrooxidans* [94], *Deferribacter abyssi*, and *Deferribacter thermophilus* [95], and *Deltaproteobacteria Geothermobacter ehrlichii* [96]. *Geothermobacterium ferrireducens* and *Thermodesulfobacterium commune* (class Thermodesulfobacteria), *T. scotoductus* (Deinococcus group), *Sulfurihydrogenibium azorense*, and *Sulfurihydrogenibium subterraneum* (class Aquificae), and *Thermotoga species* (class Thermotogae) constitute another phylogenetic high-level thermophilic group of iron reducers [97].

The process of iron reduction is possibly mediated by enzymeiron reductase. The enzymeiron reductase from the hyperthermophilic archaeon *Archaeoglobus fulgidus* is a 18-kDa subunit that employs NADH and NADPH for reduction of Fe(III) while FMN (flavin mononucleotide) or FAD (flavin adenine dinucleotide) is required for catalytic activity. The cell suspensions exhibit iron-reduction ability [98]. c-type cytochromes, characteristic of mesophiles, are responsible for dissimilatory iron reduction reported from thermophiles like *T. ferrireducens*. It is hypothesized that in dissimilatory iron reduction, enzymes take part only in the transfer of electrons from metal while mineral formation is accomplished with the aid of physico-chemical factors [99]. Childers and Lovley [100] reported that *P. islandicum*, on exposure to an Fe(III)-containing medium, exhibited the NADP-dependent iron reductase activity, which was not dependent on c-type cytochromes [101].

7.3.3 Biosorption-Based Bioremediation

It has become clear that treatment of metal-polluted environments by conventional methods like chemical oxidation, adsorption, chemical precipitation, electrochemical treatment, evaporate recovery, ion exchange, reverse osmosis, and sludge filtration would be too expensive even without the consideration of the generation of toxic waste products [102]. Therefore, low-cost technologies with high efficiency should be employed to remediate heavy metals from contaminated water [103]. Bacterial biomass can be employed for the removal of toxic metals from wastewater. Bioremediation of metals can be achieved by the following: (i) biosorption, a phenomenon of metal binding to the microbial cell wall without the involvement of energy consumption, and (ii) bioaccumulation, the metal uptake by cells, which is dependent on energy consumption [104]. Removal of metals by microorganisms is dependent on various factors like composition of the cell wall, cell physiology, and physico-chemical factors (pH, temperature, salinity, contact time, etc.; chemical composition of metal ion, its ionic strength and concentration of metal) [105]. Metals can be absorbed by ionic and physico-chemical interactions to the cell wall and external surfaces of dead and living microbial biomass by energy independent mechanism. The amine, carboxyl, hydroxyl, phosphate, and sulfhydryl groups present on the microbial cell wall play a major role in metal biosorption. The distinct bacterial groups express the same kind of reactive functional groups on their surface and also react with metals in a similar manner [105]. Mesophilic bacterial species of genus *Bacillus* are well studied for metal biosorption, but few such reports are available on thermophiles. Thermophiles encounter elevated levels of metals in their natural habitats. Their cell walls and external surface structures play an important role in determining metal-cell wall interactions [106].

Özdemir et al. [107] examined the biosorption ability of dried cells of two thermophilic bacteria (*Anoxybacillus amylolyticus* and *Geobacillus thermantarcticus*) for heavy metals like Cd2+, Cu2+, Co2+, and Mn2+ ions. The optimum pH range for biosorption of Cd2+, Cu2+, and Co2+ for both the organisms was 4.0 to 5.0 while that for Mn2+ was 6.0. Both organisms have similar growth temperature optima. It was observed that 50 mg of dried cells of *G. thermantarcticus* was more efficient than *A. amylolyticus* in terms of metal adsorption. The metal adsorption by thermophilic bacteria may differ qualitatively as well as quantitatively compared to mesophiles. Their unique cell wall structure plays a crucial role in adsorption and also influences the degree of metal interaction with the cell wall [107].

Surface complexation model (SCM) has been employed to study the Cd^{2+} biosorption in thermophilic bacteria *G. stearothermophilus* DSM-6790 and *Geobacillus thermo catenulatus* DSM-720. SCM describes adsorption of a single dissolved chemical species or metal ion to a particular type of functional group present on the cell wall of bacteria by using a stability constant, K. SCMs are designed on a set of molecular-scale thermodynamics and serve as an important tool for developing advanced bioremediation techniques. SCM analysis of two species of Geobacillus indicated the presence of a functional group with a deprotonation constant pK value of approximately 3.8, which is low when compared to other bacterial species, indicating that the genus Geobacillus probably uses other functional groups for complexation reactions and biosorption of cadmium [106]. *Bacillus licheniformis* (KG9) (a thermophilic and haloalkalitolerant strain) immobilized on Amberlite XAD-4 was observed to effectively bioabsorb Cd2+, Cu2+, and Ni2+ ions from wastewater [108]. Biosorption, as a part of biotransformation, i.e., conversion of toxic metal components to metal sulfides, also presents a method of detoxification in microorganisms. Metal sulfides possess low solubilization ability and thus less toxicity, as they are unavailable. *Cyanidioschyzon*, a thermophilic red alga, has been reported to be more capable of cadmium removal when compared to the mesophilic green alga *Chlamydomonas*. *Cyanidioschyzon* was three fold more active in converting Cd(II) into metal sulfide as compared to *Chlamydomonas*. The possible reason for this could be the adaptability of *Cyanidioschyzon* to sulfur rich geothermal springs [109]. Biosorption studies of Cu2+, Zn2+, and Pb2+ on *Geobacillus thermocatenulatus* and *Geobacillus thermodenitrificans* revealed the affinity of metals in the order Pb2+ > Cu2+ > Zn2+ *G. thermocatenulatus* exhibited higher biosorption activity than *G. thermodenitrificans* [110]. Another important microbial cell structure that plays a crucial role in biosorption is EPS (microbial exopolysaccharide). The composition of EPS is highly diverse, comprising of polysaccharides, heteropolysaccharides, mucopolysaccharides, proteins, and various other components, depending on the conditions of culture medium provided. Polysaccharides are an important constituent of bacterial EPS due to its high affinity to bind with cations. EPSs sequester metals via mechanisms like acetylation, adsorption, precipitation, ion exchange, and binding of metal ions to functional groups (amine, carboxyl, phosphoric, etc.) [111]. During biosorption, the level of acetylation is critical in determining selectivity for metal binding sites on EPS [112].

Biosorption of metals via EPS is a potential alternate to other metal eradication systems, which can be used for removal of metal wastes during mining, industrial processes, sewage treatment, etc. [113]. Zhao et al. [114] reported that EPS of thermophile *Anoxybacillus* sp. R4-33, a heteropolysaccharide (mannanglucan) in structure, acidic, and of high molecular weight, can efficiently biosorb Zn(II) and Cd(II). EPS can be accumulated as biofilms in thermophilic archaea like *A. fulgidus*, *Sulfolobus*, *Thermococcus* sp., *Thermococcus litoralis*, and thermophilic bacteria, viz. *B. licheniformis*, *Bacillus thermantarcticus*, and *G. thermodenitrificans*. Cocultures of *Thermotoga maritime* and methanogen *Methanococcus jannaschii* have also been reported to produce EPS [115]. Thermostable EPS is produced by *Geobacillus tepidamans* V264. High-pressure anion exchange with pulsed amperometric detector (HPAE-PAD) revealed glucose to be the major component of the biopolymer. In NMR (nuclear magnetic resonance), the exo polysaccharide exhibited a weight loss due to the presence of water, centered at 50–60 °C. The EPS was found to be stable from 50–60 to 250 °C and decompose at 280 °C, indicating that EPS is very rigid. The decomposition at unusually high temperatures provides handling and storage advantages, which can be used for bioremediation [116]. The thermophilic microbial strains recovered from deep-sea hydrothermal vents are resistant to metal duress, and via EPS, which can tolerate elevated levels of toxic metals [117]. Thus, the formation of EPS aids in better acclimatization of thermophilic microorganisms to the metal-stressed environment.

The interaction of thermophilic EPS with alloys like steel has been documented in industrial set-ups where EPSs of thermophiles grow on steel, leading to microbial-influenced corrosion. The thermophilic sulfate-reducing bacteria *Desulfotomaculum* spp. have been recovered from moderately high-temperature sites where corrosion has been detected for example, condensers in electric power units and oil wells [118]. Lata et al. [119] reported an increase in corrosivity with the increase in bacterial activity and EPS production. The major component of EPS was uronic acid with minimum amounts of carbohydrate and proteins. When steel of different grades (304L, 316L and 2205) was treated with *Desulfotomaculum* spp., it causes biodegradation of its own EPS. The presence of compounds like nucleic acids, uronic acids, and lipids and functional groups like phosphates and carboxylic acids has been reported to increase the metal binding ability of EPS. It plays an essential role in forming microbial aggregates, which aids in survival of microorganisms in harsh conditions like metal stress. The EPS production has been observed to increase with the increase in concentration of chromium in the growth medium due to the toxic effect of Cr(VI) on cells [120].

7.3.4 Metallothioneins, Metallochaperones, and Other Metal-Binding Proteins

Metallothioneins (MTs) are a family of metal-chelating proteins, which regulate intracellular metal-ion concentration. They have been widely reported from eukaryotes to counteract the noxious effects of metals [121]. In bacteria, MTs are functionally homologous to the short cysteine-rich eukaryotic metal-binding proteins [122]. Bacterial metallothionein polypeptides differ in sequence and are completely different from eukaryotic MTs from an evolutionary viewpoint [123]. In *Thermosynechococcus elongatus*, tmt operon has been reported, which consists of two genes encoding TmtB, a transcription regulator, and TmtA, which functions as a metallothionein consisting highly conserved Cys-His residues involved in metal binding [124]. The tmt operon of *T. elongatus* is similar to the smt operon present in *Synechococcus* PCC 7942, which helps in maintaining metal homeostasis via sequestration of excess intracellular zinc [123]. The presence of metal-controlling operon in thermophilic cyanobacteria suggests the presence of advance metal-ion sensing systems in extreme environments.

Some bacterial species have Zn^{2+}-translocating metallochaperones, which are periplasmic proteins [125]. In *E. coli*, when cells were exposed to elevated levels of Zn2+ or Pb2+, expression of ZraP proteins that form Zn2+–binding metallochaperones was increased, indicating the role of ZraP in metal sequestration [126]. However, the functioning of metallo chaperones is not clear. It is being speculated that these proteins either associate with other metal resistance proteins or carry out detoxification mechanisms.

The iron–sulfur (FeS) clusters involve an interaction of metallochaperones in biological systems. The molten globule form of [3Fe–4S] [4Fe4S] ferredoxin (AaFd) consists of a zinc site in the thermophilic archaea Acidianus ambivalens. AaFd was stable at acidic pH with intact structural folding and metal centers, but with the rise in temperature, the zinc site was dissociated, the iron-sulfur cluster disassembled and the secondary structure of protein was also modified while cooling retains the apo ferredoxin state and characteristics similar to that of the native form. Thus, knowledge of structural flexibility of FeS clusters is essential for understanding insertion of metal center as molten globule states in FeS from a physiological viewpoint [127].

7.3.5 Metal Efflux System-Based Bioremediation

Metal ions, which can easily enter microbial cells, collect to toxic levels before detoxification via reduction, oxidation, or complexation. Therefore, metal efflux systems play a crucial role in balancing intracellular metal ion concentrations [128]. Paulsen et al. [129] categorized metal efflux pumps in four families: P-type ATPases involved in transportation of Cd^{2+}, Cu^{2+}, H^+, K^+, Na^+, Mg^{2+} and Ca^{2+}, and these ATPases transfer metal ions both in and out of the cell; ABC-type transporters, which transport Ni^{2+}, Mn^{2+}, Fe^{2+} and Mo^{2+}; RND (metal resistance/nodulation of legumes/cell division) transporters which efflux Ni^{2+}, Co^{2+}, Cd^{2+}, and Zn^{2+}; and the cation diffusion facilitator (CDF) family that transports Zn^{2+}, Cd^{2+}, and Co^{2+}. CDF has CzcD as a prototype, which is the regulatory component of a cobalt–zinc–cadmium-resistant determinant.

P-type ATPases and CDF transporters remove metals from the cytosol to the periplasm while ABC transporters throw out metals present in the periplasm; thus, the ABC class of efflux pump removes metal ions transported by other ATPases, transporters, or metals that have entered to the periplasm from the outer environment. This suggests that P-type ATPases and CDF transporters can alter each other but they cannot replace ABC transporters and vice versa [130]. The Czc determinant needs the sensor CzcD for regulation. It resembles to four predicted proteins and one of the open reading frame fragments found in bacteria and yeasts. However, *Bacillus caldolyticus,* and *Bacillus stearothermophilus* also contain an incomplete ORF and ORF2, respectively, but their function is not known [131]. Hexavalent chromium makes an entry into the bacterial cell by mechanisms involved in sulfate uptake as $CrO_4{}^{2-}$ is a structural analog of $SO_4{}^{2-}$. Chromate reductase present in the cytosol reduces Cr(VI) to Cr(III), but when high levels of Cr(VI) accumulate inside the cell, the chr operon is activated. chrA encodes a chromate efflux pump that expels chromate out of the cell and protects the cell from Cr toxicity [132]. ChrA has been reported from thermophilic bacteria, viz. [133]. *G. kaustophilus*, *Geobacillus* spp. C56-T3, *G. thermoglucosidasius*, *Thermosipho melanesiensis* [14]. Resistance to arsenicals (As^{3+} and As^{5+}) and antimonials (Sb^{3+}) is provided by proteins encoded by the ars operon reported widely in gram-positive and gram-negative bacteria, implying its ancient presence in the microbial world. As^{5+} is first converted to As^{3+} by arsenate reductase, ArsC, and thereafter ArsB, which is a chemiosmotic As^{3+} and Sb^{3+} efflux protein that ejects As3+/Sb3+ outside the bacterial cell. Sometimes, a dimer ArsA is also encoded that binds to ArsB, provides energy through ATP and enhances the function of efflux pump in expelling the high level of As^{3+} from the cell [134]. The arsB gene has been reported from *S. thermosul fi dooxidans*, isolated from a deep high-temperature mine heap [135]. The presence of ArsB in *Sulfobacillus* spp. strengthens its response to As and permits detoxification of H3AsO4-As (V) [135]. Czr and Czc proteins of the mesophilic bacterial species *E. coli* share arrangement similarity and structural conservation with proteins of the thermophilic bacteria *Thermus thermophilus*. The expression of the gene of *T. thermophilus* in *E. coli* resulted in the improvement in cell physiology, delay in cell bursting, accumulation of elevated amounts of recombinant proteins, high level of resistance to Zn^{2+} and Cd^{2+}, and lesser resistance to Co^{2+} because of the efflux of these metals outside the cell. This was a phage display-based analysis that led to the identification of metal efflux pump in the genome of *T. thermophilus* when *T. thermophilus* genes were expressed in *E. coli* [136].

The thermophilic archaea *A. fulgidus* possess P-type Ag^+/Cu^+-ATPase. It is a class of P1B-type heavy metal transport ATPases encoded by two genes: AF0473 and AF0152. AF0473 encodes a thermostable ATPase-CopA, which was active at 75 °C and exhibited no activity at 37 °C. CopA interacts with Cu+ with high affinity, but activation is determined by Ag+. However, cysteine in millimolar concentrations has a profound effect on the activation of CopA by Ag+ or Cu+. Metals initiate the synthesis of phosphoenzyme in the presence of ATP and CopA, which has functional characters similar to P-type ATPases and starts expelling metal outside the cell. Thus, the characterization of this P-type Ag^+/Cu^+-ATPase indicates the efficient functioning of thermophilic ion transporters [137]. In some microorganisms, metal-phosphate complexes participate in acquiring metal tolerance and aid in detoxification of metals by hydrolysis of inorganic polyphosphate (polyP) [138]. Remonsellez et al. [139] studied the presence of polyP in the archaea *S. acidocaldarius*, *S. metallicus,* and *S. solfataricus*. The accumulation of a large amount of polyP in *S. metallicus* allowed it to tolerate a high concentration of Cu^{2+} up to 200 mM while other species of Sulfolobus were unable to tolerate >1–5 mM Cu^{2+}. In *S. metallicus*, exopolyphosphatase (PPX) activity was readily increased, when a concentration of Cu^{2+} was 100 mM, which, in turn, brought polyP to low levels and activated phosphate efflux pump. This strongly recommends that the polyP metal resistance mechanism is highly active in *Sulfolobus* spp. as accumulation and hydrolysis of polyP plays an indispensable role in the survival of *Sulfolobus* spp. in a harsh environment and also makes it a good candidate for biomining processes [139].

7.4 Future Perspectives

Bioremediation of metal-contaminated soils, sediments, and waters is a multifaceted business that exploits diverse biological activities ranging from single molecules to entire microorganisms [140]. To make it more attractive, this technology must be adapted with specific contamination problems in specific environments. Advances of *in-situ* treatment methods, treatment of mixed wastes or radioactive wastes, particularly in high temperature environments, are some of the main targets. Knowledge of microbial processes that are operative at high temperature can be employed in metal decontamination under such conditions. Current observations on multiple metal and radionuclide [Cr(VI), Co(III), U(VI), and Tc(VII)] reduction abilities of hyperthermophilic bacterium *P. islandicum* resulting in the formation of insoluble uranium deposits at temperature nearly 100 °C in hydrothermal vents provide evidence about the extraordinary capability of thermophiles, which could be relevant for metal bioremediation. Further study in this direction may result in novel strategies for *in-situ* treatment of metal-contaminated environments [121]. The distribution and diversity of thermophilic microorganisms inhabiting contaminated sites and their genes that code for the attributes required to cope with harsh environments having elevated concentrations of metals should be taken into account for future research as these insights are invaluable in metal bioremediation. Metagenomic investigations can provide key information of phylogenic and functional diversity of thermophiles in specific environments. Also, such studies will provide direct insights into the metabolic potentials of these indigenous microorganisms and on their genetic determinants relevant to their functions in the natural habitat. Metagenome sequencing from high-temperature geothermal environments provides an excellent tool for accessing and characterizing the predominant members of such microbial communities as well as their functional attributes relevant for survival and growth of these extremophiles [141].

Intensive research should be focused on the various modes of metal resistance and interaction in thermophilic microorganism, which may contribute to improvement in bioremediation strategies. Studies on some chemolithotrophic, thermophilic, and anaerobic bacteria, or on thermophilic bacterial heavy metal resistance/accumulation/transformation characteristics, or on the metal efflux/transformation proteins/enzymes of such microorganisms may contribute to the progress in bioremediation [75]. Development of recombinant thermophilic strains incorporating desirable genes for metal bioremediation from suitable mesophilic or other extremophilic microorganism may open up new possibilities. Engineering extremely radiation-resistant thermophilic bacterium *D. geothermalis* has generated a Hg(II)-resistant strain capable of reducing Hg(II), Fe(III)-nitrilotriacetic acid, U(VI), and Cr(VI) at elevated temperatures and in the presence of ionizing radiation [72].

Biosurfactants or other extracellular polymers from thermophilic microorganisms can also be used in solubilization and mobilization of metals from polluted soils and sediments. These molecules are competitive in soil treatments as they have excellent surface properties and are more environment friendly than synthetic compounds [140]. Polymer engineering through the introduction of required genes of the biosynthetic pathways may further enhance the promising possibilities of these polymers in metal bioremediation [142]. The metal-microbe interaction processes taking place in the cell microenvironment need to be better elucidated so that such well-characterized systems can be used efficiently. Similar to mesophilic organisms, genetic engineering of thermophilic bacterial biosorbents and exploration of novel-binding molecules, such as designed phytochelatin analogues and selected polypeptides with new affinities, may further enhance metal bioaccumulation/biosorption [121]. Genomic and proteomics approaches will provide us access to new and enticing activities of such extremophiles. The availability of genome sequences of some of the thermophilic microorganisms inhabiting metal-polluted environments will be very advantageous for further research [143].

Acknowledgments

The fellowship provided by the Ministry of Environment, Forest & Climate Change (MoEF&CC), the Government of India along with G.B. Pant National Institute of Himalayan Environment & Sustainable Development under the National Mission on Himalayan Studies is gratefully acknowledged.

References

1 Rawat, S., Agarwal, P.K., Chaudhary, D.K. et al. (2005). Microbial diversity and community dynamics of mushroom compost ecosystem. In: *Microbial Diversity: Current Perspectives and Potential Application* (eds. T. Satyanarayana and B.N. Johri), 181–206. New Delhi: I. K. International Pvt. Ltd.

2 Odlare, M. ed. (2005). Organic residues: a resource for arable soils. PhD thesis, Swedish University of Agricultural Sciences Uppsala 51.

3 Steger, K., Sjogren, M.A., Jarvis, A. et al. (2007). Development of compost maturity and Actinobacteria populations during full-scale composting of organic household waste. *Journal of Applied Microbiology* 103 (2): 487–495.

4 Peters, S., Koschinsky, S., Schwieger, F. et al. (2000). Succession of microbial communities during hot composting as detected by PCR–single-strand-conformation polymorphism-based genetic profiles of small-subunit rRNA genes. *Applied Environmental Microbiology* 66: 930–936.

5 Purohit, H.J. and Kapley, A. (2005). Dynamics of microbial community: an environmental perspective. In: *Microbial Diversity: Current Perspectives and Potential Applications* (eds. T. Satyanarayana and B.N. Johri), 559–572. New Delhi: I. K. International Pvt. Ltd.

6 Klamer, M., Baath, E. et al. (1998). Microbial community dynamics during composting of straw material studied using phospholipid fatty acid analysis. *FEMS Microbial Ecology* 27: 9–20.

7 Boggs, L.C., Kennedy, A.C., and Reganold, J.P. (1998). Use of phospholipid fatty acids and carbon source utilization patterns to track microbial community succession in developing compost. *Applied Environmental Microbiology* 64: 4062–4064.

8 Kowalchuk, G.A., Naovmenko, Z.S., Derikx, P.J.L. et al. (1999). *Applied and Environmental Microbiology* 65: 396–403.

9 Watkin, E.L.J., Keeling, S.E., Perrot, F.A. et al. (2009). Metals tolerance in moderately thermophilic isolates from a spent copper sulfide heap, closely related to*Acidithiobacillus caldus, Acidimicrobium ferrooxidans* and *Sulfobacillus thermosulfidooxidans. Journal of Industrial Microbiology and Biotechnology* 36: 461–465.

10 Llanos, J., Capasso, C., Parisi, E. et al. (2000). Susceptibility to heavy metals and cadmium accumulation in aerobic and anaerobic thermophilic microorganisms isolated from deep-sea hydrothermal vents. *Current Microbiology* 41: 201–205.

11 Dopson, M., Baker-Austin, C., Koppineedi, P.R. et al. (2003). Growth in sulfidic mineral environments: metal resistance mechanisms in acidophilic micro-organisms. *Microbiology* 149: 1959–1970.

12 Al-Daghistani, H.I. (2012). Bio-remediation of Cu, Ni and Cr from rotogravure wastewater using immobilized, dead, and live biomass of indigenous Thermophilic bacillus species. *Internet Journal of Microbiology* 10: 1.

13 Al-Daghistani, H. (2012). Bio-remediation of Cu, Ni And Cr from rotogravure wastewater using immobilized, dead, and live biomass of indigenous thermophilic bacillus species. *The Internet Journal of Microbiology* 10 (1): 2012 02 02.

14 Sar, P., Kazy, S., Paul, B. et al. (2013). Metal bioremediation by thermophilic microorganisms. In: *Thermophilic Microbes in Environment and Industrial Biotechnology: Biotechnology of Thermophiles* (ed. T. Satyanarayan). Dordrecht: Springer Science.

15 Ahring, B.K. and Westermann, P. (1985). Sensitivity of thermophilic methanogenic bacteria to heavy metals. *Current Microbiology* 12: 273–276.

16 Chatziefthimiou, A.D., Crespo-Medina, M., Wang, Y. et al. (2007). The isolation and initial characterization of mercury resistant chemolithotrophic thermophilic bacteria from mercury rich geothermal springs. *Extremophiles* 11: 469–479.

17 Schelert, J., Dixit, V., Hoang, V. et al. (2004). Occurrence and characterization of mercury resistance in the hyperthermophilic archaeon Sulfolobus solfataricus by use of gene disruption. *Journal of Bacteriology* 186: 427–437.

18 Schelert, J., Drozda, M., Dixit, V. et al. (2006). Regulation of mercury resistance in the crenarchaeote Sulfolobus solfataricus. *Journal of Bacteriology* 88: 7141–7150.

19 Barkay, T., Miller, S.M., and Summers, A.O. (2003). Bacterial mercury resistance from atoms to ecosystems. *FEMS Microbiology Reviews* 27 (2–3): 355–384.

20 Aravind, L., Anantharaman, V., Balaji, S. et al. (2005). The many faces of the helix-turn-helix domain: Transcription regulation and beyond. *FEMS Microbiology Reviews* 29 (2): 231–262.

21 Simbahan, J., Kurth, E., Schelert, J. et al. (2005). Community analysis of a mercury hot spring supports occurrence of domain-specific forms of mercuric reductase. *Applied and Environmental Microbiology* 71: 8836–8845.

22 Schiering, N., Kabsch, W., Moore, M.J. et al. (1991). Structure of the detoxification catalyst mercuric ion reductase from *Bacillus* sp. strain RC607. *Nature* 352: 168–172.

23 Endo, G. and Silver, S.L. (1995). CadC, the transcriptional regulatory protein of the cadmium resistance system of *Staphylococcus aureus* plasmid pi258. *Journal of Bacteriology* 177 (15): 4437–4441.

24 Schirawski, J., Hagens, W., Fitzgerald, G.F. et al. (2002). Molecular characterization of cadmium resistance in *Streptococcus thermophilus* strain 4134: an example of lateral gene transfer. *Applied and Environmental Microbiology* 68: 5508–5516.

25 Saw, J.H., Mountain, B.W., Feng, L. et al. (2008). Encapsulated in silica. *Genome Biology* 9: 11–25.

26 Saier, M.H., Tam, R., Reizer, A. et al. (1994). Two novel families of bacterial membrane proteins concerned with nodulation, cell division and transport. *Molecular Microbiology* 11: 841–847.

27 Nies, D.H. (1999). Microbial heavy-metal resistance. *Applied Microbiology and Biotechnology* 51: 730–750.

28 Alvarez, A.H., Moreno-Sanchez, R., and Cervantes, C. (1999). Chromate efflux by means of the ChrA chromate resistance protein from *Pseudomonas aeruginosa*. *Journal of Bacteriology* 181: 7398–7400.

29 Takami, H., Takaki, Y., Chee, G.J. et al. (2004). Thermoadaptation trait revealed by the genome sequence of thermophilic *Geobacillus kaustophilus*. *Nucleic Acids Research* 32: 6292–6303.

30 Oremland, R.S. and Stoltz, J.F. (2003). The ecology of arsenic. *Science* 300: 939–944.

31 Silver, S. and Phung, L.T. (1996). Bacterial heavy metal resistance. *Annual Review of Microbiology* 50: 753–789.

32 Mukhopadhyay, R. and Rosen, B.P. (2002). Arsenate reductases in prokaryotes and eukaryotes. *Environmental Health Perspectives* 110: 745–748.

33 Angelov, A., Liebl, S., Ballschmiter, M. et al. (2010). Genome sequence of the polysaccharide-degrading, thermophilic anaerobe Spirochaeta thermophila DSM 6192. *Journal of Bacteriology* 192: 6492–6493.

34 Francisco, S.M.J.D., Hope, C.L., Owolabi, J.B. et al. (1990). *Nucleic Acids Research* 18: 619.

35 Rosen, B.P. (1999). Families of arsenic transporters. *Trends in Microbiology* 7: 207–212.

36 Wu, J., Tisa, L.S., and Rosen, B.P. (1992). Membrane topology of the ArsB protein, the membrane subunit of an anion-translocating ATPase. *Journal of Biological Chemistry* 267: 12570–12576.

37 Saltikov, C.W. and Newman, D.K. (2003). Genetic identification of a respiratory arsenate reductase. *Proceedings of the National Academy of Science, USA.* 100: 10983–10988.

38 Richey, C., Chovanec, P., Hoeft, S.E. et al. (2009). Respiratory arsenate reductase as a bidirectional enzyme. *Biochemical and Biophysical Research Communications.* 382: 298–302.

39 Afkar, E., Lisak, J., Saltikov, C. et al. (2003). The respiratory arsenate reductase from *Bacillus selenitireducens* strain MLS10. *FEMS Microbiology Letters* 226: 107–112.

40 Takai, K., Kobayashi, H., Nealson, K.H. et al. (2003). *Deferribacter desulfuricans* sp. nov., a novel sulfur-, nitrate- and arsenate-reducing thermophile isolated from a deep-sea hydrothermal vent. *International Journal of Systemic and Evolutionary Microbiology* 53: 839–846.

41 Wheaton, G., Counts, J., Mukherjee, A. et al. (2015). The confluence of heavy metal biooxidation and heavy metal resistance: implications for bioleaching by extreme thermoacidophiles. *Minerals* 5: 397–451.

42 Muller, D., Lievremont, D., Hubert, J.-C. et al. (2003). Arsenite oxidase (aox) genes from a metal-resistant β-proteobacterium. *Journal of Bacteriology* 185: 135–141.

43 Okibe, N., Koga, M., Sasaki, K. et al. (2013). Simultaneous oxidation and immobilization of arsenite from refinery waste water by thermoacidophilic iron-oxidizing archaeon, Acidianus brierleyi. *Minerals Engineering* 48: 126–134.

44 Lebrun, E., Brugna, M., Bayman, F. et al. (2003). Arsenite oxidase, an ancient bioenergetic enzyme. *Molecular Biology and Evolution* 20: 686–693.

45 Van Lis, R., Nitschke, W., Duval, S. et al. (2013). Arsenics as bioenergetic substrates. *Biochimica et Biophysica Acta* 1827: 176–188.

46 Lett, M.-C., Muller, D., Lièvremont, D. et al. (2012). Unified nomenclature for genes involved in prokaryotic aerobic arsenite oxidation. *Journal of Bacteriology* 194: 207–208.

47 Tang, K., Barry, K., Chertkov, O. et al. (2011). Complete genome sequence of the filamentous anoxygenic phototrophic bacterium *Chloroflexus aurantiacus*. *BMC Genomics* 12: 334.

48 Yamamura, S. and Amachi, S. (2014). Microbiology of inorganic arsenic: from metabolism to bioremediation. *Journal of Bioscience and Bioengineering* 118 (1): 1–9.

49 Heinrich-Salmeron, A., Cordi, A., Brochier-Armanet, C. et al. (2011). Unsuspected diversity of arsenite oxidizing bacteria as revealed by widespread distribution of aoxB gene in prokaryotes. *Applied Microbiology and Biotechnology* 77: 4685–4692.

50 Konishi, Y., Yoshida, S., and Asai, S. (1995). Bioleaching of pyrite by acidophilic thermophile Acidianus brierleyi. *Biotechnology and Bioengineering* 48 (6): 592–600.

51 Bathe, S. and Norris, P.R. (2007). Ferrous iron- and sulfur-induced genes in Sulfolobus metallicus. *Applied Environmental Microbiology* 73: 2491–2497.

52 Mukherjee, A., Wheaton, G.H., Blum, P.H. et al. (2012). Uranium extremophily is an adaptive, rather than intrinsic, feature for extremely thermoacidophilic Metallosphaera species. *Proceedings of the National Academy of Sciences, USA* 109: 16702–16707.

53 Kozubal, M.A., Dlakic, M., Macur, E. et al. (2011). Terminal oxidase diversity and function in BMetallosphaera yellowstonensis: gene expression and protein modeling suggest mechanism of Fe (II) oxidation in the Sulfolobales. *Applied and Environmental Microbiology* 77: 1844–1853.

54 Castresana, J., Lübben, M., and Saraste, M. (1995). New archaebacterial genes coding for redox proteins: implications for the evolution of aerobic metabolism. *Journal of Molecular Biology* 250 (2): 202–210.

55 Blake, R.C., Shute, E.A., Greenwood, M.M. et al. (1993). Enzymes of aerobic respiration on iron. *FEMS Microbiology Reviews* 11: 9–18.

56 Kappler, U., Sly, L.I., and McEwan, A.G. (2005). Respiratory gene clusters of Metallosphaera sedula–differential expression and transcriptional organization. *Microbiology* 151: 35–43.

57 Slobodkin, A.I. (2005). Thermophilic microbial metal reduction. *Microbiology* 74 (5): 581–595.

58 Glasauer, S., Langley, S., and Beveridge, T.J. (2002). Intracellular iron minerals in a dissimilatory iron-reducing bacterium. *Science* 295: 117–119.

59 Glasauer, S., Langley, S., and Beveridge, J. (2004). Intracellularmanganese granules formed by subsurface bacterium. *Environmental Microbiology* 6: 1042–1048.

60 Schroeder, I., Johnson, E., and de Vries, S. (2003). Microbial ferric iron reductases. *FEMS Microbiology Reviews* 27: 427–447.

61 Lauwerys, R., Haufroid, V., Hoet, P. et al. (2007). *Toxicologie Industrielleet Intoxications Professionnelles*, 5e. Issy-les-Moulineaux: Elsevier.

62 Sobol, Z. and Schiestl, R.H. (2012). Intracellular and extracellular factors influencing Cr(VI) and Cr(III) genotoxicity. *Environmental and Molecular Mutagenesis* 53: 94–100.

63 Sau, G.B., Chatterjee, S., Sinha, S. et al. (2008). Isolation and characterization of a Cr (VI) reducing Bacillus firmus strain from industrial effluents. *Polish Journal of Microbiology* 57: 327–332.

64 Cheung, K.H., Lai, H.Y., and Gu, J.-D. (2006). Membrane-associated hexavalent chromium reductase of Bacillus megaterium TKW3 with induced expression. *Journal of Microbiology and Biotechnology* 16: 855–862.

65 Elsevier-Masson, P., Thatoi, H., Mishra, J. et al. (2014). Bacterial chromate reductase, a potential enzyme for bioremediation of hexavalent chromium: a review. *Journal of Environmental Management* 146: 383–399.

66 Patra, R.C., Malik, B., Beer, M. et al. (2010). Molecular characterization of chromium (VI) reducing potential in gram positive bacteria isolated from contaminated sites. *Soil Biology and Biochemistry* 42 (10): 1857–1863.

67 Ramirez-Diaz, M.I., Díaz-Pérez, C., Vargas, E. et al. (2008). Mechanisms of bacterial resistance to chromium compounds. *Biometals* 21: 321–332.

68 Cervantes, C., Garcia, J.C., Devars, S. et al. (2001). Interactions of chromium with microorganisms and plants. *FEMS Microbiology Reviews* 25: 335–347.

69 McLean, J. and Beveridge, T.J. (2001). Chromate reduction by a pseudomonad isolated from a site contaminated with chromate copper arsenate. *Applied Environmental Microbiology* 67: 1076–1084.

70 Gursahani, Y.H. and Gupta, S.G. (2015). Hexavalent chromium reduction by Anoxybacillus rupiensis isolated from hot water spring of Dhapoli, Maharashtra, India. *Journal of Petroleum and Environmental Biotechnology* 6 (4): 1–5.

71 Opperman, D.J. and van Heerden, E. (2008). A membrane-associated protein with Cr(VI)-reducing activity from Thermus scotoductus SA-01. *FEMS Microbiology Letters* 280 (2): 210–218.

72 Brim, H., Venkateswaran, A., Kostandarithes, H.M. et al. (2003). Engineering Deinococcus geothermalis for bioremediation of high-temperature radioactive waste environments. *Applied Environmental Microbiology* 69: 4575–4582.

73 Perry, J.J., Perman, J.A., and Zaworotko, M.J. (2009). Design and synthesis of metal-organic frameworks using metal-organic polyhedral as supermolecular building blocks. *Chemical Society Reviews* 38 (5): 1400–1417.

74 Barkay, T. and Wagner-Dobler, I. (2005). Microbial transformations of mercury: potentials, challenges, and achievements in controlling mercury toxicity in the environment. *Advances in Applied Microbiology* 57: 1–52.

75 Barkay, T., Kritee, K., Boyd, E. et al. (2010). A thermophilic bacterial origin and subsequent constraints by redox, light and salinity on the evolution of the microbial mercuric reductase. *Environmental Microbiology* 12 (11): 2904–2917.

76 Plum, L.M., Rink, L., and Haase, H. (2010). The essential toxin: impact of zinc on human health. *International Journal of Environmental Research and Public Health* 7 (4): 1342–1365.

77 Dhanjal, S. and Cameotra, S.S. (2010). Aerobic biogenesis of selenium nanospheres by Bacillus cereus isolated from coalmine soil. *Microbial Cell Factories* 9: 52.

78 Correa-Llantén, D.N., Munoz-Ibacache, S.A., and Blamey, J. (2014). Enzyme involvement in the biosynthesis of selenium nanoparticles by Geobacillus wiegelii strain Gwe1 isolated from a drying oven. *International Journal of Biological, Biomolecular, Agricultural, Food and Biotechnological Engineering* 8 (6): 637–641.

79 Oremland, R., Herbel, M., Blum, J. et al. (2004). Structural and spectral features of selenium nanospheres produced by Se-respiring bacteria. *Applied and Environmental Microbiology* 70 (1): 52–60.

80 Tejo, P., Sharma, N., Prakash, R. et al. (2009). Aerobic microbial manufacture of nanoscale selenium: exploiting nature's bio-nanomineralization potential. *Biotechnology Letters* 31 (12): 1857–1862.

81 Kaur, G., Iqbal, M., and Bakshi, M. (2009). Biomineralization of fine selenium crystalline rods and amorphous spheres. *The Journal of Physical Chemistry C* 113 (31): 13670–13676.

82 Ip, C. ed. (2006). Selenium and ER stress response: implication and exploitation for cancer therapy. Proceedings of the International Conference on SBM (July 25–30, 2006) University of Wisconsin-Madison: 63.

83 Wang, H., Zhang, J., and Yu, H. (2007). Elemental selenium at nano size possesses lower toxicity without compromising the fundamental effect on selenoenzymes: comparison with selenomethionine in mice. *Free Radical Biology and Medicine* 42 (10): 1524–1153.

84 Peng, D., Zhang, J., Liu, Q. et al. (2007). Size effect of elemental selenium nanoparticles (nano-Se) at supranutritional levels on selenium accumulation and glutathione S-transferase activity. *Journal of Inorganic Biochemistry* 101 (10): 1457–1463.

85 Dobias, E., Suvorova, R., and Bernier, L. (2011). Role of proteins in controlling selenium nanoparticle size. *Nanotechnology* 22 (19): 1–9.

86 Ghalib, A.K., Yasin, M., and Faisal, M. (2014). Characterization and metal detoxification potential of moderately thermophilic Bacillus cereus from geothermal springs of Himalayas. *Brazilian Archives of Biology and Technology* 57 (4): 554–560.

87 Wright, M.H., Patel, B.K.C., and Greens, A.C. (2012). Thermophilic bacteria from Paralana hot springs in the Northern Flinders ranges of South Australia Conference paper. https://doi.org/10.13140/RG.2.1.2765.5525

88 Francis, A.J. and Dodge, C.J. (2008). Bioreduction of uranium (VI) complexed with citric acid by Clostridia affects its structure and solubility. *Environmental Science and Technology* 42: 8277–8282.

89 Lovley, D.R. and Phillips, E.J. (1992). Reduction of uranium by Desulfovibrio desulfuricans. *Applied Environmental Microbiology* 58 (3): 850–885.

90 Khijniak, T.V., Slobodkin, A.I., Coker, V. et al. (2005). Reduction of uranium (VI) phosphate during growth of the thermophilic bacterium Thermoterrabacterium ferrireducens. *Applied Microbiology and Biotechnology* 71 (10): 6423–6426.

91 Cason, E.D., Piater, L.A., and van Heerden, E. (2012). Reduction of U(VI) by the deep subsurface bacterium, Thermus scotoductus SA-01, and the involvement of the ABC transporter protein. *Chemosphere* 86 (6): 572–527.

92 Brock, T.D. and Gustafson, J. (1976). Ferric iron reduction by sulfur- and iron-oxidizing bacteria. *Applied and Environmental Microbiology* 32 (4): 567–571.

93 Sokolova, T.G., Gonzalez, J.M., Kostrikina, N.A. et al. (2004). Thermosinus carboxydivorans gen. nov., sp. nov., a new anaerobic thermophilic carbon monoxide oxidizing hydrogenogenic bacterium from a hot pool of Yellowstone National Park. *International Journal of Systemic and Evolutionary Microbiology* 54: 2353–2359.

94 Bridge, T. and Johnson, D.B. (1998). Reduction of soluble iron and reductive dissolution of ferric iron-containing minerals by moderately thermophilic iron-oxidizing bacteria. *Applied Environmental Microbiology* 64: 2181–2186.

95 Miroshnichenko, M.L., Slobodkin, A.I., Kostrikina, N.A. et al. (2003). Deferribacter abyssi sp. nov., an anaerobic thermophile from deep-sea hydrothermal vents of the Mid-Atlantic Ridge. *International Journal of Systemic and Evolutionary Microbiology* 53: 1637–1641.

96 Kashefi, K., Holmes, D.E., Baross, J.A. et al. (2003). Thermophily in the Geobacteraceae: Geothermobacter ehrlichii gen. nov., sp. nov., a novel thermophilic member of the Geobacteraceae from the "Bag City" hydrothermal vent. *Applied and Environmental Microbiology* 69 (5): 2985–2993.

97 Balkwill, D.L., Kieft, T.L., Tsukuda, T. et al. (2004). Identification of iron-reducing Thermus strain as Thermus scotoductus. *Extremophiles* 8: 37–44.

98 Slobodkin, A.I., Chistyakova, N.I., and Rusakov, V.S. (2004). High-temperature microbial sulfate reduction can be accompanied by magnetite formation. *Mikrobiologiya* 73: 553–557.

99 Zachara, J.M., Kukkadapu, R.K., Fredrickson, J.K. et al. (2002). Biomineralization of poorly crystalline Fe(III) oxides by dissimilatory metal reducing bacteria. *Geomicrobiology Journal* 19: 179–207.

100 Childers, S.E. and Lovley, D.R. (2001). Differences in Fe(III) reduction in the hyperthermophilic archaeon Pyrobaculum islandicum versus mesophilic Fe(III)-reducing bacteria. *FEMS Microbiology Letters* 195: 253–258.

101 Schuler, D. and Frankel, R.B. (1999). Bacterial magnetosomes: microbiology, biomineralization and biotechnological applications. *Applied Microbiology and Biotechnology* 52: 464–473.

102 Ayangbenro, A.S. and Babalola, O.O. (2017). A new strategy for heavy metal polluted environments: a review of microbial biosorbents. *International Journal of Environmental Research and Public Health* 14: 94.

103 Hassan, S.H.A., Awad, Y.S., Kabir, M.H. et al. (2010). Bacterial biosorption of heavy metals. In: *Biotechnology: Cracking New Pastures* (eds. C.P. Malik and A. Verma), 79–110. Delhi: MD Publications Pvt Ltd.

104 Karna, R.R., Uma, L., Subramanian, G. et al. (1999). Biosorption of toxic metal ions by alkali-extracted biomass of a marine cyanobacterium, Phormidium valderianum BDU 30501. *World Journal of Microbiology and Biotechnology* 15: 729–732.

105 Hassan, S.H.A., Kim, S.J., Jung, Y.A. et al. (2009). Biosorptive capacity of Cd(II) and Cu(II) by lyophilized cells of Pseudomonas stutzeri. *The Journal of General and Applied Microbiology* 55: 27–34.

106 Hetzer, A., Daughney, C.J., and Morgan, H.W. (2006). Cadmium ion biosorption by the thermophilic bacteria Geobacillus stearothermophilus and G. thermocatenulatus. *Applied Microbiology and Biotechnology* 72: 4020–4027.

107 Özdemir, S., Kılınc, E., Poli, A. et al. (2013). Biosorption of heavy metals (Cd2+, Cu2+, Co2+, and Mn2+) by thermophilic bacteria *Gebacillus thermantarcticus* and *Anoxybacillus amylolyticus*: equilibrium and kinetics. *Bioremediation Journal* 17 (2): 86–96.

108 Alkan, H., Gul-Guven, R., Guven, K. et al. (2015). Biosorption of Cd2+, Cu2+ and Ni2+ ions by a thermophilic haloalkalitolerant bacterial strain (KG9) immobilized on Amberlite XAD-4. *Polish Journal of Environmental Studies* 24 (5): 1903–1910.

109 Pinto, G., Albertano, P., Ciniglia, C. et al. (2003). Comparative approaches to the taxonomy of the genus Galdieria merola (Cyanidiales, Rhodophyta). *Cryptogamie. Algologie* 24: 13–32.

110 Babàk, L., Šupinova, P., Zichová, M. et al. (2012). Biosorption of Cu, Zn and Pb by thermophilic bacteria—effect of biomass concentration on biosorption capacity. *Acta Universitatis Agriculturae et Silviculturae Mendelianae Brunensis* 5: 9–18.

111 Zhang, D., Wang, J., and Pan, X. (2006). Cadmium sorption by EPSs produced by anaerobic sludge under sulfate-reducing conditions. *Journal of Hazardous Materials B* 38: 589–593.

112 Sutherland, I.W. (1983). Extracellular polysaccharides. In: *Biotechnology: Biomass, Microorganisms for Special Applications, Microbial Products I, Energy From Renewable Resources* (eds. H.J. Rehm and G. Reed), 531–574. Wienheim: Chemie.

113 Choi, S.B. and Yun, Y.S. (2006). Biosorption of cadmium by various types of dried sludge: an equilibrium study and investigation of mechanisms. *Journal of Hazardous Materials* 138 (2): 378–383.

114 Zhao, S., Cao, F., Zhang, H. et al. (2014). Structural characterization and biosorption of exopolysaccharides from Anoxybacillus sp. R4-33 isolated from radioactive radon hot spring. *Applied Biochemistry and Biotechnology* 172 (5): 2732–2746.

115 Lapaglia, C. and Hartzell, P.L. (1997). Stress-induced production of biofilm in the hyperthermophile Archaeoglobus fulgidus. *Applied and Environmental Microbiology* 63 (8): 3158–3163.

116 Kambourova, M., Mandeva, R., Dimova, D. et al. (2009). Production and characterization of amicrobial glucan, synthesized by Geobacillus tepidamans V264 isolated from Bulgarian hot spring. *Carbohydrate Polymers* 77 (2): 338–343.

117 Mancuso, N.C., Lardière, S.G., Bowman, J.P. et al. (2005). Chemical characterization of exopolysaccharides from Antarctic marine bacteria. *Microbial Ecology* 49: 578–589.

118 Almeida, M.A.N. and de Franca, F.P. (1999). Thermophilic and mesophilic bacteria in biofilms associated with corrosion in a heat exchanger. *World Journal of Microbiology and Biotechnology* 154: 39–442.

119 Lata, S., Sharma, C., and Singh, A.K. (2012). Microbial influenced corrosion by thermophilic bacteria. *Central European Journal of Engineering* 2 (1): 113–122.

120 Dogan, N.M., Doganli, G.A., Doğan, G. et al. (2015). Characterization of extracellular polysaccharides (eps) produced by thermal Bacillus and determination of environmental conditions affecting exopolysaccharide production. *International Journal of Environmental Research* 9 (3): 1107–1116.

121 Valls, M. and de Lorenzo, V. (2002). Exploiting the genetic and biochemical capacities of bacteria for the remediation of heavy metal pollution. *FEMS Microbiology Reviews* 26: 328–338.

122 Blindauer, C.A. and Harrison, M.D. (2002). Multiple bacteria encode metallothioneins and Smt-like zinc fingers. *Molecular Microbiology* 45 (2): 1421–1432.

123 Olafson, R.W., McCubbin, W.D., and Kay, C.M. (1988). Primary- and secondary-structural analysis of a unique prokaryotic metallothionein from a Synechococcus sp. cyanobacterium. *Biochemical Journal* 251: 691–696.

124 Shelake, R.M., Hayashi, H., and Morita, E.H. (2016). Structural analysis and homology modeling of members of smt-like operon from thermophilic cyanobacterium Thermosynechococcus elongatus BP-1. *Journal of Proteins and Proteomics* 7 (3): 221–230.

125 Blencowe, D.K. and Morby, A.P. (2003). Zn(II) metabolism in prokaryotes. *Microbiology Reviews* 27: 291–311.

126 Noll, M., Petrukhin, K., and Lutsenko, S. (1998). Identification of a novel transcription regulator from Proteus mirabilis, PMTR, revealed a possible role of YJAI protein in balancing zinc in Escherichia coli. *Journal of Biological Chemistry* 273: 21393–21401.

127 Leal, S.S. and Gomes, C.M. (2007). Studies of the molten globule state of ferredoxin: structural characterization and implications on protein folding and iron-sulfur center assembly. *Proteins* 68 (3): 606–616.

128 Srivastava, P. and Kowshik, M. (2013). Mechanisms of metal resistance and homeostasis in Haloarchea. *Archaea* 732864: 1–16.

129 Paulsen, I.T., Park, J.H., Choi, P.S. et al. (1997). A family of gram-negative bacterial outer-membrane factors that function in the export of proteins, carbohydrates, drugs and heavy metals from gram-negative bacteria. *FEMS Microbiology Letters* 156: 1–8.

130 Scherer, J. and Nies, D.H. (2009). CzcP is a novel efflux system contributing to transition metal resistance in Cupriavidus metallidurans CH34. *Molecular Microbiology* 73: 601–621.

131 Kiel, J.A.K.W., Boels, J.M., Beldman, G. et al. (1991). The glgB gene from the thermophile Bacillus caldolyticus encodes a thermolabile branching enzyme. *DNA Sequence* 3: 221–232.

132 Joutey, N.T., Sayel, H., Bahafid, W. et al. (2015). Mechanism of hexavalent chromium resistance and removal by microorganisms. In: *Reviews of Environmental Contamination and Toxicology*, vol. 233, 45–69.

133 Berlin Silver, S. and le Phung, T. (2005). A bacterial view of the periodic table: genes and proteins for toxic inorganic ions. *Journal of Industrial Microbiology and Biotechnology* 32: 587–605.

134 Rosen, B.P., Bhattacharjee, H., Zhou, T. et al. (1999). Mechanism of the ArsA-ATPase. *Biochimica et Biophysica Acta* 1461: 207–215.

135 Casas-Flores, S., Gomez-Rodriguez, E.Y., and García-Meza, J.V. (2015). Community of thermoacidophilic and arsenic resistant microorganisms isolated from a deep profile of mine heaps. *AMB Express* 5 (54): 1–12.

136 Spada, S., Pembroke, J.T., and Wall, G. (2002). Isolation of a novel Thermus thermophilus metal efflux protein that improves Escherichia coli growth under stress conditions. *Extremophiles* 6: 301–308.

137 Mandal, A.K., Cheung, W.D., and Argüellos, J.M. (2002). Characterization of a thermophilic P-type Ag+/Cu+-ATPase from the extremophile Archeoglobus fulgidus. *Journal of Biological Chemistry* 277 (9): 7201–7208.

138 Kornberg, R.N.N. and Ault-Riche, D. (1999). Inorganic polyphosphate: a molecule of many functions. *Annual Review of Biochemistry* 68: 89–125.

139 Remonsellez, F., Orell, A., and Jerez, C.A. (2006). Copper tolerance of the thermoacidophilic archaeon Sulfolobus metallicus: possible role of polyphosphate metabolism. *Microbiology* 152: 59–66.

140 Tabak, H.H., Lens, P., van Hullesbusch, E.D. et al. (2005). Developments in bioremediation of soils and sediments polluted with metals and radionuclides. *Reviews in Environmental Science and Bio/Technology* 4: 115–156.

141 Inskeep, W.P., Rusch, D.B., Jay, Z.J. et al. (2010). Metagenomes from high-temperature chemotrophic systems reveal geochemical controls on microbial community structure and function. *PLoS One* 5: e9773. https://doi.org/10.1371/journal.pone.0009773.

142 Gutnick, D.L. and Bach, H. (2000). Engineering bacterial biopolymers for the biosorption of heavy metals. *Applied Microbiology and Biotechnology* 54: 451–460.

143 Auernik, K.S., Maezato, Y., Blum, P.H. et al. (2008). The genome sequence of the metal-mobilizing, extremely thermoacidophilic archaeon *Metallosphaera sedula* provides insights into bioleaching-associated metabolism. *Applied and Environmental Microbiology* 74: 682–692.

8

Ecological Perspectives of Halophilic Fungi and their Role in Bioremediation

Shekhar Jain[1], Devendra Kumar Choudhary[2], and Ajit Varma[2]

[1] *Faculty of Life Sciences, Mandsaur University, Mandsaur, Madhya Pradesh, India*
[2] *Amity Institute of Microbial Technology (AIMT), Amity University Campus, Noida, Uttar Pradesh, India*

8.1 Introduction

Any hostile change in the surrounding results in a challenge to normal life. Extreme conditions are formed in the environment due to a change in physicochemical parameters, which get close to the lower or upper limits that sustain life [1]. Being a high salty condition, hypersaline environments provide an extreme situation and may limit the growth of organisms along with the help of some additional growth delimiting factors that include temperature, pH, pressure, oxygen, nutrient availability, and solar radiation [2, 3]. Hypersaline environments include hypersaline waters and soils. It is usually considered hypersaline if the salt concentrations are higher than that of seawater [3]. However, the aforesaid condition may differ based on the origin and the salt composition. Furthermore, hypersaline water habitats have been classified into two types depending on the salt level and composition. First, the thalassohaline environment, where the relative amounts of the different inorganic salts are just about equal to those present in seawater, typically chloride types and the pH is near neutral to slightly alkaline. The Great Salt Lake or the solar salterns used for the industrial production of marine salt by evaporation of seawater are considered to be thalassohaline environments. Brines contain higher Mg^{2+} and Cl^{-} ions and are more acidic than seawater due to halite precipitation [2, 4]. pH is also one of the important growth-limiting factors of hypersaline water apart from the total salinity and ionic composition. Brines derived from seawater contain relatively high concentrations of Ca^{2+} and can remain approximately neutral even after extensive evaporation

Soil Bioremediation: An Approach Towards Sustainable Technology, First Edition.
Edited by Javid A. Parray, Abeer Hashem Abd Elkhalek Mahmoud, and Riyaz Sayyed.

because the molarity of Ca^{2+} always exceeds that of $CO_3{}^{2-}$ and it removes the alkaline $CO_3{}^{2-}$ through precipitation of insoluble calcite [4].

The second class is the athalassohaline environment, in which the proportions of the different salts are noticeably different to those of seawater and are considered as more heterogeneous. This type environment is noticeably influenced by the geology of an area and are generally found in the Dead Sea, the Wadi Natrun, Lake Magadi, and several other soda lakes [4]. A lack of Ca^{2+} and Mg^{2+} results in pH values ≈11 vis a vis development of an alkaline lake [5]. In the Dead Sea, due to Mg^{2+} precipitation, H^+ ions get released and generate slightly acidic water conditions [6]. The Dead Sea is considered to be one of the most extreme and hostile environments on the planet due to low water activity (<0.669) and the very limited amounts of organic material [7]. The environment of the Great Salt Lake has 86% NaCl of the total salts, close to the composition of seawater [2]. Hypersaline soil is more complicated and not even a clear demarcation is there to categorize it as hypersaline. Because most soils contain small amounts of soluble salts, a soil would be considered as hypersaline when its salt concentration is above a certain threshold [3]. According to Kaurichev [8], soils containing more than 0.2% (w/v) soluble salt should be considered saline soils. Salinization is one of the most serious land degradation problems to confront of world. It results in deprived plant growth and low soil microbial activity due to osmotic stress and toxic ions [9]. Soil salinization consists of an accumulation of water-soluble salts such as potassium (K^+), magnesium (Mg^{2+}), calcium (Ca^{2+}), chloride (Cl^-), sulfate ($SO_4{}^{2-}$), carbonate ($CO_3{}^{2-}$), bicarbonate (HCO^{3-}), and sodium (Na^+) ions in the soil [10]. Salinity can be changed in sodicity due to a higher sodium accumulation over the time, which results in destruction of the soil structure due to a lack of oxygen and soil becomes incapable of assuring plant growth and animal life.

Soil microorganisms are very important for plant growth in soil as they play a pivotal role through mineralization of organic matter into plant available nutrients so it is important to maintain high microbial activity. The presence of high concentrations of soluble salts in the hypersaline soil affects microbial growth through increases in the osmotic potential (more negative) of the soil water, that results in the death of microbes via drawing water out of the cells and roots through plasmolysis. Low osmotic potential also makes it more difficult for roots and microbes to remove water from the soil [11]. The requirement of large energy input to adapt to such a saline environment leads to retarded plant growth and decreased microbial activity [12]. Over a long period, several research studies have shown that salinity reduces microbial activity, microbial biomass, and changes microbial community structure through drying and lysis of cells [13–17]. In the case of fungi, spore germination potential get decreased due to low osmotic potential and the growth of hyphae is reduced. Additionally, morphology and gene expression also get changed resulting in the formation of spores with thick walls [18–20]. Water activity (a_w), i.e. availability of water, is a major growth-limiting factor in hypersaline environments. The addition of concentrations of solutes, a_w of pure water decreases gradually, this is 1.00 for the purest water. Saturated NaCl solution possesses water activity (a_w) of approximately 0.75 or even less such as in dried foods. Due to the semi-permeable nature of biological membranes microorganism that live in high salt concentrations need to keep the cytoplasm hyperosmotic to the extracellular environment, to build up the necessary turgor pressure [2].

8.2 Hypersaline Inhabitant Fungi

Although the hypersaline environment is challenging for the life, microorganism can survive over the entire array of salt concentrations that are found in natural habitats, from freshwater environments to hypersaline lakes, saltern crystallizer ponds, and other environments that are saturated with sodium chloride [2]. These microorganisms possess the capability to adapt to the extreme concentrations of NaCl, and also often to high concentrations of other ions, to high ultraviolet (UV) irradiation and, in some cases, to extremes of pH [21]. Even though, microorganisms of many phylogenetic groups can grow in salty condition, based on the salt concentration required for their optimal growth, and can be divided into the following groups: (i) nonhalophiles (require less than 1% NaCl); (ii) halotolerant (nonhalophilic, can tolerate high salt concentrations up to 25% NaCl); (iii) slight halophiles (grow best in media with 1–3% NaCl); (iv) moderate halophiles (grow best in media with 3–15% NaCl); and (v) extreme halophiles (grow best in media with 15–25% NaCl). Extreme halophiles are able to grow even at saturated salt concentrations [22]. All three domains of life i.e. Archaea, Bacteria, and Eukarya include members that can tolerate/require higher salt condition for growth [23].

Until the isolation as an active inhabitant from the solar salterns, fungi were not considered to be the part of hypersaline environments [21]. Many species including *Aspergillus*, *Cladosporium*, *Pencillium*, and some yeast including those that were previously reported only to be food contaminants, have been isolated from hypersaline habitats [6, 24–26]. Kirk et al. [27] stated that out of a total existing 106 orders of fungi, 10 orders of fungi were demonstrated for tolerance to low a_w. Halophilic/halotolerant groups were found in the three orders namely Wallemiales, Trichnosporales, and Sporidiales of Basidiomycota. Whereas halotolerance is spread in several genera of these orders, a few obligate halophilic genera are described. Yeast and yeast-like *Rhodotorula*, *Debaryomyces*, *Aureobasidium*, *Trichosporum* as well as filamentous fungi *Cladosporium*, *Scopulariopsis*, *Alternaria*, *Aspergillus* spp., *Penicillium* spp. are described as halotolerant. Halophilic species are found in the genus *Wallenia*, *Hortea*, *Phaetotheca*, *Trimmatostroma* [28, 29]. Abundant and consistent occurrence of several fungi was found in studies of fungal populations in hypersaline water and other hypersaline environments on several continents [21]. Members of the different order of fungal class Ascomycota (Dothideales, Capnodiales, Eurotiales) and Basidiomycota (Wallemiales and Tremellales) were found halotolerant but black yeasts are regarded as the best-represented fungal group in the hypersaline waters of the salterns [30]. It possesses melanized cell walls and can withstand with high salt stress conditions [31].

Among the black yeasts, *Hortaea werneckii* is the most successful and found in whole range of saline environments, optimally at 3.0–4.5 M NaCl [32]. It shows an isodiametric type of thallus expansion in the highest salinity condition, which helps its survival in endogenous conditions by making highly resistant, meristematic cell clumps while at lower salinity, based on the substrate, it will grow either as a yeast cell or as hyphae [33]. This polymorphic character is typical for extremophilic species within the Capnodiales [34]. *H. werneckii* is considered as a fitting model organism for studying salt adaptation mechanisms of fungi and other eukaryotes due to its copious and consistent occurrence in several hypersaline environments. To ameliorate the harmful effect of the hyper osmolar environment it uses a compatible-solute strategy and keeps the cytoplasm concentrations of inorganic ions to a minimum [35]. Basidiomycetes fungus *Wallemia ichthyophaga* are another important fungus, which is halophilic in nature. It prefers high concentrations of NaCl compared to nonionic solutes and requires at least 1.5 M NaCl for growth *in vitro* [26]. According to Zajc et al. [36], it possesses the capability to grow even in saturated NaCl solution but the optimum *in vitro* growth can be obtained in between 2.6 and 3.5 M NaCl.

Like *H. werneckii*, it also forms multicellular meristematic clumps at all salinities, which are covered in a salt-dependent manner with an extracellular polymeric substance matrix [29]. *Debaryomyces hansenii* (anamorphic state – *Candida famatais*) is a halotolerant yeast having a role in meat fermentation, it is present in all kinds of cheeses, and produces lytic enzymes and alditols [37]. This species is a universal inhabitant of the world's oceans [38]. Apart from oceans, it has also been isolated from hypersaline waters in natural hypersaline habitats, and particularly from the salterns of the Namibian Skeleton coast and the Great Salt Lake, both of which are exposed to low seasonal temperatures [7].

8.3 Strategies Against the Hypersaline Environment

8.3.1 Ion Homeostasis

In order to survive in the hypersaline environment cells maintain a lower water potential than their surroundings. Halotolerant fungi *D. hansenii* have been studied considerably for ion homeostasis and it was reported that it is the fungi least affected by high NaCl concentrations and that K^+/Na^+ fluxes have roles in this process [38, 39]. Further, the nontoxic nature of Na^+ was observed for *D. hansenii* and more accumulation of Na^+ was also reported compared to *Saccharomyces cerevisiae* for better growth [40, 41]. Additionally, the presence of Na^+ can defend the cells against added stress factors [42, 43]. Looking at the intracellular Na^+ and K^+ concentrations *D. hansenii* and halophilic *W. ichthyophaga* are considered as Na^+-intruder fungus. Compare to extremely halotolerant *H. wernecki*, which apparently excludes this cation, these two fungi can accumulate higher amounts of Na^+, which shows the independent evolution of the ion homeostasis mechanisms of these fungi [44]. Two genes DhENA1 and DhENA2 encode Na^+-ATPases, the ENA ATPases in *D. hansenii* are responsible for salt extrusion. The ENA ATPases are one of the most important salt-tolerance determinants in *Saccharomyces.* DhENA1 is more expressed in the presence of high NaCl concentrations, while the expression of DhENA2, as an alternative, requires a high pH. Expression of these genes recovered the ability to tolerate salt stress and extrude Na^+ in *S. cerevisiae* mutant (ena) even though this recovered tolerance was still far from the tolerance level of *D. hansenii.* Therefore, Na^+ independent extrusion is not sufficient to explain the high salt tolerance of *D. hansenii* [45]. The existence of low and high affinity K^+-influx transport and Na^+-efflux processes that are somehow similar, but also considerably different to those of *S. cerevisiae*, has been shown in *D. hansenii.* Two interesting characteristics are seen in the cation/cation exchange system and the sodium–glycerol symporter of *D. hansenii* [41].

Furthermore, intracellular H^+ homeostasis also has an important role in salt tolerance, and a study on the two strains of *D. hansenii* have shown that the nearer a strain can maintain its pK(a, i) to its pHi homeostasis level, the better it manages NaCl stress [46]. In addition, several K^+/Na^+ transporters located in the plasma membrane or in internal membranes have been identified in *D. hansenii* [47–50]. Prista et al. [51] reported that DhPMA1, an orthologue of the gene in *S. cerevisiae* that encodes the H^+-ATPase, has been found in *D. hansenii* and can generate the membrane potential that is needed to drive the K^+/Na^+ exchange of these cells. In another salt tolerant fungi *H. werneckii*, two novel salt-responsive ENA-like P-ATPases, HwENA1 and HwENA2, have been identified, which are responsive to increased salt concentrations and increased pH conditions in *H. werneckii.* [52]. Similar to *D. hansenii*, one gene is expressed more at high salt concentrations (HwENA2), with greater expression of the other (HwENA1) at high pH [29].

8.3.2 Accumulation of Compatible Solutes

Compatible solutes or osmolytes are components that can be quickly accumulated and degraded inside fungal cells. They do not disturb the functioning of proteins and protect the cell under adverse conditions. Through accumulation of compatible solutes, fungi maintain their intracellular concentrations of Na^+ below the toxic levels for the cells. Polyols, free amino acids, and their derivatives are the most common osmolytes accumulated by the fungi. While the salt-sensitive *S. cerevisiae* almost exclusively uses glycerol as an osmolyte, other yeast and fungi are known to produce and/or accumulate polyols from the environment, such as erythritol, ribitol, arabinitol, xylitol, sorbitol, mannitol, and galacticol [53]. *D. hansenii* accumulates glycerol as it is a main compatible solute along with some additional osmolytes such as arabitol, trehalose, glutamic acid, and alanine [54]. A correlation in the accumulation of glycerol and trehalose was observed based on the salt concentration in this yeast. It accumulates more glycerol than trehalose in the presence of high salt concentrations (2.0–3.0 M salt), although under moderate NaCl stress, these cells accumulated more trehalose than glycerol [55]. Glycerol synthesis in *D. hansenii* is stimulated in the presence of high Na^+, partly due to increased expression of the GPD1 and GPP2 genes coding for a glycerol-3-phosphate dehydrogenase and a glycerol-3-phosphatase [56, 57].

In addition to glycerol, glutamate might also work as an additional salt-tolerance determinant in *D. hansenii,* as it exerts a protective effect against salt stress, and NADP-glutamate dehydrogenase activity is increased when cells are grown in high salt [58]. Like *D. hansenii,* glycerol is also the most important compatible solute in *H. werneckii,* [59]. Upon growth on the hypersaline media, as well as glycerol, *H. werneckii* also accumulates a mixture of organic compounds, including the polyols erythritol, arabitol, and mannitol [60]. Their amounts vary both with the salinity of the growth medium and with the growth phase of the fungal culture. However, the total amount of polyols that correlate to increasing salinity is mostly due to glycerol during all growth phases. It accumulates largely during the exponential growth phase and diminishes abruptly during the stationary phase while the amount of erythritol increases gradually during the exponential growth phase and reaches its highest level during the stationary phase. Remaining compatible solutes stay low, which is why the total amount of polyols decreases during the stationary phase. Apart from polyols, *H. werneckii* also accumulates Mycosporines steeply up to 1.0 M NaCl, and decrease at higher NaCl concentrations [61, 62]. Halophilic fungi, *W. ichthyophaga* also accumulate glycerol as the main compatible solute, and additionally arabitol and mannitol, the latter being a typical compatible solute for basidiomycetes fungi. Recently Turk and Gostinčar [63] studied glycerol metabolism in polyextremotolerant black yeasts *Aureobasidium pullulans* and *Aureobasidium subglaciale* and found increased accumulation with increasing salinity. Summing up, all three-model organisms synthesize and accumulate glycerol as the main compatible solute for protection against salt stress [29].

8.3.3 Maintaining Plasma Membrane Fluidity

Plasma membrane fluidity is one of the important properties that can change according to a situation in order to adapt to the surroundings. Toh et al. [64] reported that in *S. cerevisiae*, deletion of the main glycerol channel, Fps1p, is associated with a decrease in membrane ergosterol content and with changes in the flux of glycerol across the plasma membrane. The lipid composition of the plasma membrane has an important role in glycerol retention; a decrease in the sterol-to-phospholipid ratio and higher plasma membrane fluidity over a wide range of NaCl concentrations was observed in *H. werneckii* compared to the salt-sensitive *S. cerevisiae* [65]. Two copies of each gene encoding for three fatty-acid-modifying enzymes, D9-saturase, D12-desaturase, and an elongase, have been identified in the genome of *H. werneckii* [31]. Differential expressions of these genes have been reported based on different salinity conditions. The lowest expression at optimal salinity infers that changes in the expression of fatty-acid-modifying enzymes might be a part of the mechanisms that facilitate precise regulation of membrane fluidity, which is in turn needed for the correct functioning of cells in a constantly changing osmotic environment. The major fatty acid in the plasma membrane of *D. hansenii* is oleic acid (cis-9-octadecenoic acid). Like *H. werneckii*, the presence of greater amounts of salt show a concomitant increased in fatty acid unsaturation in *D. hansenii* [66]. In contrast, increased fatty-acid shortening and saturation has been reported in *S. cerevisiae*, showing once again clear differences between salt-sensitive and halotolerant organisms [65].

8.3.4 High Osmolarity Glycerol (HOG) Signaling Pathway

The ability to sense the changes in surrounding Na^+ concentrations is of vital importance for cell survival. The high osmolarity glycerol (HOG) signaling pathway, which is a mitogen-activated protein kinase (MAPK) pathway plays a central role in *S. cerevisiae* survival at low environmental water potential. On exposure to high osmolarity in the surrounding areas, a loss of turgor, due to intracellular water loss, leads to shrinkage of the cells and this serves as a signal for activation of the HOG pathway [53]. After exposure to such medium for only 20 s, when the cell volume reaches its minimum, phosphorylated Hog1 is already maximally enriched in the nucleus and glycerol synthesis increases [67]. Earlier studies confirmed the existence of HOG pathway in the *H. werneckii* was similar to that of *S. cerevisiae* by the identification of the homolog of MAP kinase, HwHog1 but some additional features were found relating to sensing and responding to increases in NaCl in the form of a novel motif that suggests its unique mechanisms of adaptation [68–71]. Unlike *S. cerevisiae* Hog1, HwHog1 are only activated in extremely high salt concentrations [71]. Two copies of histidine kinase group 7 genes (HK7) i.e., HwHHK7A and HwHHK7B were also reported in the *H. werneckii* genome, play an important role in osmosensing [69]. Location-wise this HK7 protein is present in the cytosol of *H. werneckii* whereas *S. cerevisiae* histidine kinases Sln1 are membrane bound. This is a prominent difference between these two microorganisms. It is reported that 36 genes out of a total of 95 salt-dependent differentially expressed genes in *H. werneckii* were dependent on Hog1 for long-term adaptation to extreme environments, which shows HOG pathway control over expression of osmo-responsive genes [53, 72–74]. A correlation was found between the salinity regulation and mitochondrial activity, as more than 20% of the differentially expressed genes in *H. werneckii*, following adaptation to 4.5 M NaCl, were assigned to mitochondrial metabolism and biogenesis [75]. Sequence homology analysis has shown that the presence of Hog1 gene conserves sequences in the halotolerant *A. pullulans*, the halophilic *Phaeotheca triangularis*, and *W. ichthyophaga*. Exceptionally, WiHog1 of the *W. ichthyophaga* reported with some sequence motif that is more similar to the mammalian p38 isoforms than to the fungal Hog1 orthologues. Expression of halotolerant fungus *A. glaucus*Hog1 gene in the *S. cerevisiae* hog1D mutant restored its growth in high osmotic stress and the intracellular glycerol content in the transformants increases to much higher levels than those in the mutant during salt-stress conditions [76].

8.4 Ecological Perspectives of Hypersaline Fungi

Being adaptable/native of the hypersaline environment, halophiles and halotolerant fungi have a significant role in various fields. Various agriculture chemicals such as herbicides and pesticides were used to control weed and pests in crop fields. The increasing in use of agricultural chemicals results in the accumulation of these compounds and their derivatives in soil and water. Diuron is a phenyl urea herbicide (N-(3, 4-dichlorophenyl)-N, N-dimethylurea) and it is one of the most often employed agrochemicals for controlling weeds in sugarcane, citrus, and coffee crops. It has high mobility in soils, low susceptibility to natural attenuation, and strong toxicity. Additionally, more toxic metabolites with genotoxic and teratogenic actions such as 3, 4-dichloro anilines, N-3, 4-dichloro phenyl urea and N-(3, 4-dichloro phenyl)-N-methylurea can be generated from diuron by biotic and abiotic reactions [77–79]. The extravagance of phenyl urea herbicides in the environment was reported as a primarily biological process, suggesting that the restoration of polluted sites through bioremediation is a promising tool [80, 81].

Bioremediation of toxic material such as herbicides, pesticides, or any other chemical by using microorganism is an ecofriendly approach. However, in the salty conditions such as soil or industrial effluent microbe's growth is retarded and their enzymes are also inhibited due to low water potential. Crop soil, where herbicides are frequently used, can have its salinity increased by irrigation or by the use of chemical fertilizers. Considering these facts, Arakaki et al. [82] have grown *Dacryopinax elegans*, a halophilic fungus, which required salt for growth on the culture media sodium chloride, mannitol, glycerol, and diuron, and its ability to degrade herbicide diuron was evaluated. *D. elegans* was able to degrade the diuron in all the culture media used. The highest degradation of the herbicide was observed in the medium with glycerol (38%). Another important aspect is the presence of heavy metals in the environment that poses a serious threat to human health. The use of microbes for the remediation of this problem has been widely researched in order to find a sustainable solution. Recently, Bano et al. [83] have used obligate halophilic fungi comprising *Aspergillus flavus*, *Acanthorhodeus gracilis*, *Aspergillus penicillioides* (sp. 1), *A. penicillioides* (sp. 2), *Aspergillus restrictus*, and *Sterigmatomyces halophilus* for the biosorption of cadmium, copper, ferrous, manganese, lead, and zinc. The metals were supplemented as salts in potato dextrose broth for the growth of obligate halophilic fungi and incubated for 14 days and the biosorption was screened by atomic absorption spectroscopy. All tested fungi showed moderate to high adsorption of heavy metals, among which *A. flavus* and *S. halophilus* showed the best average adsorption of all heavy metals studied, with averages of 86 and 83%, respectively. On average, Fe and Zn are best removed from the liquid media of obligate halophilic fungi, with averages of 85 and 84%, respectively. This pioneering study of biosorption by obligate halophilic fungi using inexpensive media in stagnant conditions provides a cost-effective environmental solution for the removal of heavy metals. Bioremediation of hypersaline environments is also becoming an increasingly important topic, together with the need to treat oil-polluted salt marshes and salty wastewaters that result from many industrial processes [84–86].

Ali et al. [87] have isolated six obligate halophilic fungi namely *A. flavus, A. gracilis, A. penicillioides* (2 strains), and *A. restrictus* from a man-made solar saltern, extreme hypersaline habitat, located in Phetchaburi province, Thailand and used for their antibacterial, antioxidant, and enzymatic production potential. All of the isolates in this study were found to have at least one of the applications screened. *A. flavus, A. gracilis*, and *A. penicillioides* (1) were found to have high antibacterial potential. *A. flavus* and *A. penicillioides* (2) showed the most antioxidant activity while *A. gracilis* and *A. restrictus* were found to be most promising candidates for enzymes (of the five enzymes tested) having activity under saline conditions. The biotechnological applications of hypersaline microbes and their components are one of the main reasons for studying them [1]. Hypersaline microbes produce various industrially important enzymes and fungi in particular are rich sources of industrially important extracellular enzymes

[88, 89]. In the fermentation processes such as production of soy sauce and fish sauce halophiles have been used for centuries, although they can at the same time cause significant damage by spoiling salt-cured and sugar-cured food. Furthermore, in a few modern industrial processes, such as for the production of b-carotene and ectoine they have been exploited successfully [90, 91]. Additionally, hypersaline fungi are also important for the production of biorhodopsin for optical computing, biosurfactants, and exopolysaccharides for microbially enhanced oil recovery, pigments for food coloring, and compatible solutes as stress protectants [85, 92].

Halotolerant fungi also represent a valuable genetic resource. The problem of water scarcity and increasing salinity in soil makes halophilic fungi excellent candidates to acquire the genes for increasing the halophilic and xerophilic tolerant characteristics in food-producing crops [35]. Abiotic stress can significantly decrease the yields of industrial important microorganisms. Industrial processes such as production of biofuel and fermentation of dough all need improved multi-stress tolerance of *S. cerevisiae* and other microorganisms as they face stress conditions that reduces biomass as well as enzyme production [93–95]. For the food compensation of a growing world population, irrigation of agricultural land is crucial but it leads to increased salinity after evaporation. Further, this increased salt concentration inhibits plant growth [96]. Apart from these factors, water scarcity is also a main reason of increasing salinity, and it is one of the major causes of poor plant performance and limited crop yields worldwide. Indeed, it is the single most common cause of severe food shortage in developing countries [97, 98]. The ultimate resolution for this problem is the production of drought and salt-tolerant crops, which reduce the irrigation need and can be grown in saline soil. However, although substantial efforts have been made over the past century to develop salt-tolerant lines/cultivars of various crops using conventional plant-breeding methods as well as genetic engineering, they still have not yielded the desired results [99]. Consequently, novel genes sources are required for improving salt and drought tolerance crops [99–101]. Halotolerant and halophilic fungi possess these types of genes but so far they have been largely neglected for genetic resources, despite some of their important advantages: (i) they have evolved specialized mechanisms for salt tolerance that are lacking in salt-sensitive gene donors (which have often been used for improving plants in the past); (ii) they are more closely related to plants than the structurally different and phylogenetically distant prokaryotes; and finally, (iii) as microorganisms, for studies of salt tolerance, they are more convenient model organisms than the higher plants. Thus, their genetic resources may be applied to develop salt-tolerant crop cultivars.

8.5 Conclusions

Salinity is one of the most brutal environmental factors limiting the productivity of crop plants because most crop plants are sensitive to salinity caused by high concentrations of salts in the soil, and the area of saline and hypersaline soils is increasingly abundant as a consequence of irrigation and desertification processes. The study of the microbiological activities in these habitats yields results of great interest. Halophilic and halotolerant microorganisms are the most important microbial population of the hypersaline environment. Compared to hypersaline aquatic habitats, very little information exists thus far regarding the diversity of halophilic and halotolerant microorganisms isolated from saline soils, and much work needs to be done on this subject. Improved knowledge about hypersaline ecosystems will lead to important advances in the understanding of microbial adaptation mechanisms, and facilitate the design of ecological and biotechnological applications of microbes and microbial compounds to function under hypersaline conditions. Halophilic and halotolerant fungi, studied so far, found an important role in amelioration of ecological problems such as the accumulation of herbicide, pesticide, and heavy metals along with biotechnological applications in extreme saline condition. However, more extensive studies on the ecology, biochemistry, and physiology of these organisms present in the hypersaline aquatic and soil ecosystem is required to explore its capability to thrive in these environments.

References

1 Amils, R., Blix, A., Danson, M. et al. (2007). *Investigating Life in Extreme Environments: A European Perspective* (ed. N. Walter). Strasbourg: ESF.

2 Oren, A. (2002). *Halophilic microorganisma and their environments, COLE Series 5*, 575. Dordrecht, Boston, London: Kluwer Academic Publishers.

3 Rodríguez-Valera, F. (1988). Characteristics and microbial ecology of hypersaline environments. In: *Halophilic Bacteria*, vol. 1 (ed. F. Rodríguez-Valera), 3–30. Boca Raton, FL: CRC Press.

4 Grant, W.D. (2004). Life at low water activity. *Philosophical Transactions of the Royal Society London B* 359: 1249–1269.

5 Jones, B.E., Grant, W.D., Duckworth, A.W. et al. (1998). Microbial diversity of soda lakes. *Extremophiles* 2: 191–200.

6 Perl, T., Kis-Papo, T., and Nevo, E. (2017). Fungal biodiversity in the hypersaline Dead Sea: extinction and evolution. *Biological Journal of the Linnean Society* 121 (1): 122–132.

7 Gunde-Cimerman, N., Frisvad, J.C., Zalar, P. et al. (2005). Halotolerant and halophilic fungi. In: *Biodiversity of Fungi – Their Role in Human Life* (eds. S.K. Deshmukh and M.K. Rai), 69–128. New Delhi: Oxford & IBH Publishing.

8 Kaurichev, I.S. (1980). *Practicas de Edafologia*. Moscow: Mir.

9 Yan, N., Marschner, P., Cao, W. et al. (2015). Influence of salinity and water content on soil microorganisms. *International Soil and Water Conservation Research* 3 (4): 316–323.

10 Shi, D. and Wang, D. (2005). Effects of various salt-alkaline mixed stresses on *Aneurolepidium chinense* (Trin.). Kitag. *Plant and Soil* 271: 15–26.

11 Chowdhury, N., Marschner, P., and Burns, R.G. (2011). Soil microbial activity and community composition: impact of changes in matric and osmotic potential. *Soil Biology and Biochemistry* 43 (6): 1229–1236.

12 Wichern, J., Wichern, F., and Joergensen, R.G. (2006). Impact of salinity on soil microbial communities and the decomposition of maize in acidic soils. *Geoderma* 137 (1–2): 100–108.

13 Andronov, E.E., Petrova, S.N., Pinaev, A.G. et al. (2012). Analysis of the structure of microbial community in soils with different degrees of salinization using T-RFLP and real-time PCR techniques. *Eurasian Soil Science* 45 (2): 1a47–1a56.

14 Rietz, D.N. and Haynes, R.J. (2003). Effects of irrigation-induced salinity and sodicity on soil microbial activity. *Soil Biology and Biochemistry* 35 (6): 845–854.

15 Rousk, J., Elyaagubi, F.K., Jones, D.L. et al. (2011). Bacterial salt tolerance is unrelated to soil salinity across an arid agroecosystem salinity gradient. *Soil Biology and Biochemistry* 43 (9): 1881–1887.

16 Setia, R., Marschner, P., Baldock, J. et al. (2011). Relationships between carbon dioxide emission and soil properties in salt-affected landscapes. *Soil Biology and Biochemistry* 43 (3): 667–674.

17 Yuan, B.C., Li, Z.Z., Liu, H. et al. (2007). Microbial biomass and activity in salt affected soils under arid conditions. *Applied Soil Ecology* 35 (2): 319–328.

18 Juniper, S. and Abbott, L.K. (2006). Soil salinity delays germination and limits growth of hyphae from propagules of arbuscular mycorrhizal fungi. *Mycorrhiza* 16 (5): 371–379.

19 Liang, Y., Chen, H., Tang, M.J. et al. (2007). Proteome analysis of an ectomycorrhizal fungus *Boletus edulis* under salt shock. *Mycological Research* 111 (8): 939–946.

20 Mandeel, Q.A. (2006). Biodiversity of the genus *Fusarium* in saline soil habitats. *Journal of Basic Microbiology* 46 (6): 480–494.

21 Gunde-Cimerman, N., Zalar, P., de Hoog, S. et al. (2000). Hypersaline waters in salterns – natural ecological niches for halophilic black yeasts. *FEMS Microbiology Ecology* 32: 235–240.

22 Kushner, D.J. and Kamekura, M. (1988). Physiology of halophilic eubacteria. In: *Halophilic Bacteria*, vol. 1 (ed. F. Rodríguez-Valera), 87–103. Boca Raton, FL: CRC Press.

23 de Hoog, G.S., Zalar, P., Gerrits van den Ende, B. et al. (2005). Relation of halotolerance to human-pathogenicity in the fungal tree of life: an overview of ecology and evolution under stress. In: *Adaptation to Life at High Salt Concentrations in Archaea, Bacteria and Eukarya* (eds. N. Gunde-Cimerman, A. Oren and A. Plemenita,š), 453–470. Dordrecht, The Netherlands: Springer COLE.

24 Butinar, L., Zalar, P., Frisvad, J.C., and Gunde-Cimerman, N. (2005). The genus *Eurotium*-members of indigenous fungal community in hypersaline waters of salterns. *FEMS Microbiology Ecology.* 51: 155–166.

25 Cantrell, S.A. and Baez-Félix, C. (2010). Fungal molecular diversity of a Puerto Rican subtropical hypersaline microbial mat. *Fungal Ecology* 3: 402–405.

26 Zalar, P., de Hoog, G.S., Schroers, H.-J. et al. (2005). Taxonomy and phylogeny of the xerophilic genus Wallemia (*Wallemiomycetes* and *Wallemiales*, cl. et ord. nov.). *Antonie Van Leeuwenhoek* 87: 311–328.

27 Kirk, P.M., Cannon, P.F., David, J.C. et al. (2001). *Ainsworth and Bisby's Dictionary of the Fungi*, 9e. Oxford, UK: CAB International.

28 Cantrell, S.A., Casillas-Martinez, L., and Molina, M. (2006). Characterization of fungi from hypersaline environments of solar salterns using morphological and molecular techniques. *Mycological Research* 110: 962–970.

29 Gunde-Cimerman, N., Ramos, J., and Plemenitaš, A. (2009). Halotolerant and halophilic fungi. *Mycological Research* 113 (11): 1231–1241.

30 Gunde-Cimerman, N., Zalar, P., Petrovič, U. et al. (2004). Fungi in the salterns. In: *Halophilic Microorganisms* (ed. A. Ventosa), 103–111. Berlin: Springer-Verlag.

31 Gostinčar, C., Turk, M., Plemenitas, A. et al. (2009). The expressions of D9-, D12-desaturases and an elongase by the extremely halotolerant black yeast *Hortaea werneckii* are salt dependent. *FEMS Yeast Research* 9: 247–256.

32 Butinar, L., Santos, S., Spencer-Martins, I. et al. (2005a). Yeast diversity in hypersaline habitats. *FEMS Microbiology Letters* 244: 229–234.

33 Plemenitaš, A., Vaupotič, T., Lenassi, M. et al. (2008). Adaptation of extremely halotolerant black yeast *Hortaea werneckii* to increased osmolarity: a molecular perspective at a glance. *Studies in Mycology* 61: 67–75.

34 Crous, P.W., Schoch, C.L., Hyde, K.D. et al. (2009). Phylogenetic lineages in the Capnodiales. *Studies in Mycology* 64: 17–47.

35 Gostinčar, C., Grube, M., and Gunde-Cimerman, N. (2011). Evolution of fungal pathogens in domestic environments? *Fungal Biology* 115: 1008–1018.

36 Zajc, J., Kogej, T., Galinski, E.A. et al. (2014). Osmoadaptation strategy of the most halophilic fungus, *Wallemia ichthyophaga*, growing optimally at salinities above 15% NaCl. *Applied and Environmental Microbiology* 80: 247–256.

37 Breuer, U. and Harms, H. (2006). *Debaryomyces hansenii* – an extremophilic yeast with biotechnological potential. *Yeast* 23: 415–437.

38 Norkrans, B. (1966). Studies on marine occurring yeasts: growth related to pH, NaCl concentration and temperature. *Archives of Microbiology* 54: 374–392.

39 Norkrans, B. and Kylin, A. (1969). Regulation of the potassium to sodium ratio and of the osmotic potential in relation to salt tolerance in yeasts. *Journal of Bacteriology* 100: 836–845.
40 Gonzalez-Hernandez, J.C., Cardenas-Monroy, C.A., and Peña, A. (2004). Sodium and potassium transport in the halophilic yeast *Debaryomyces hansenii*. *Yeast* 21: 403–412.
41 Thomé-Ortiz, P.E., Peña, A., and Ramirez, J. (1998). Monovalent cation fluxes and physiological changes of *Debaryomyces hansenii* grown at high concentrations of KCl and NaCl. *Yeast* 14: 1355–1371.
42 Almagro, A., Prista, C., Castro, S. et al. (2000). Effects of salts on *Debaryomyces hansenii* and *Saccharomyces cerevisiae* under stress conditions. *International Journal of Food Microbiology* 56: 191–197.
43 Papouskova, K. and Sychrova, H. (2007). The co-action of osmotic and high temperature stresses results in a growth improvement of *Debaryomyces hansenii* cells. *International Journal of Food Microbiology* 118: 1–7.
44 Kogej, T., Ramos, J., Plemenitas, A. et al. (2005). The halophilic fungus *Hortaea werneckii* and the halotolerant fungus *Aureobasidium pullulans* maintain low intracellular cation concentrations in hypersaline environments. *Applied and Environmental Microbiology* 71: 6600–6605.
45 Almagro, A., Prista, C., Benito, B. et al. (2001). Cloning and expression of two genes coding for sodium pumps in the salt-tolerant yeast *Debaryomyces hansenii*. *Journey of Bacteriology* 183: 3251–3255.
46 Mortensen, H.D., Gori, K., Siegumfeldt, H. et al. (2006). Intracellular pH homeostasis plays a role in the NaCl tolerance of *Debaryomyces hansenii* strains. *Applied Microbiology and Biotechnology* 71: 713–719.
47 Carcía-Salcedo, R., Montiel, V., Calero, F. et al. (2007). Character-ization of DhKHA1, a gene coding for a putative Na(þ). trans-porter from *Debaryomyces hansenii*. *FEMS Yeast Research* 7: 905–911.
48 Montiel, V. and Ramos, J. (2007). Intracellular Na^+ and K^+ distribution in *Debaryomyces hansenii*. Cloning and expression in *Saccharomyces cerevisiae* of DhNHX1. *FEMS Yeast Research* 7: 102–109.
49 Prista, C., González-Hernández, J.C., Ramos, J. et al. (2007). Cloning and characterization of two K^+ transporters of *Debaryomyces hansenii*. *Microbiology* 153: 3034–3043.
50 Velkova, K. and Sychrova, H. (2006). The *Debaryomyces hansenii* NHA1 gene encodes a plasma membrane alkali-metal-cation anti-porter with broad substrate specificity. *Gene* 369: 27–34.
51 Prista, C., Loureiro-Dias, M.C., Montiel, V. et al. (2005). Mechanisms underlying the halotolerant way of *Debaryomyces hansenii*. *FEMS Yeast Research* 5: 693–701.

52 Gorjan, A. and Plemenitaš, A. (2006). Identification and characterization of ENA ATPases HwENA1 and HwENA2 from the halophilic black yeast *Hortaea werneckii*. *FEMS Microbiology Letters* 265: 41–50.

53 Hohmann, S. (2002). Osmotic stress signaling and osmoadaptation in yeasts. *Microbiology and Molecular Biology Reviews* 66: 300–372.

54 Jovall, P.A., Tunblad-Johanson, I., and Adler, L. (1990). ^{13}C-NMR analysis of production and accumulation of osmoregulatory metabolites in the salt-tolerant yeast *Debaryomyces hansenii*. *Archives of Microbiology* 154: 209–214.

55 González-Hernández, J.C., Jiménez-Estrada, M., and Peña, A. (2005). Comparative analysis of trehalose production by *Debaryomyces hansenii* and *Saccharomyces cerevisiae* under saline stress. *Extremophiles* 9: 7–16.

56 Gori, K., Mortensen, H.D., Arneborg, N. et al. (2005). Expression of the GPD1 and GPP2 orthologous and glycerol retention during growth of *Debaryomyces hansenii* at high NaCl concen-trations. *Yeast* 22: 1213–1222.

57 Thomé, P.E. (2005). Heterologous expression of glycerol 3-phosphate dehydrogenase gene (DhGPD1). from the osmotolerant yeast *Debaryomyces hansenii* in *Saccharomyces cerevisiae*. *Current Microbiology* 71: 713–719.

58 Alba-Lois, L., Segal, C., Rodarte, B. et al. (2004). NADP-glutamate dehydrogenase activity is increased under hyperosmotic conditions in the halotolerant yeast *Debaryomyces hansenii*. *Current Microbiology* 48: 68–72.

59 Petrovic, U., Gunde-Cimerman, N., and Plemenitas, A. (2002). Cellular responses to environmental salinity in the halophilic black yeast *Hortaea werneckii*. *Molecular Microbiology* 45: 665–672.

60 Plemenitaš, A., Lenassi, M., Konte, T. et al. (2014). Adaptation to high salt concentrations in halotolerant/halophilic fungi: a molecular perspective. *Frontiers in Microbiology* 5: 199.

61 Kogej, T., Gostinčar, C., Volkmann, M. et al. (2006). Mycosporines in extremophilic fungi – novel complementary osmolytes? *Environmental Chemistry* 3: 105–110.

62 Oren, A. and Gunde-Cimerman, N. (2007). Mycosporines and mycosporine-like amino acids: UV protectants or multipurpose secondary metabolites? *FEMS Microbiology Letters* 269: 1–10.

63 Turk, M. and Gostinčar, C. (2018). Glycerol metabolism genes in *Aureobasidium pullulans* and *Aureobasidium subglaciale*. *Fungal Biology* 122 (1): 63–73.

64 Toh, T.H., Kayingo, G., van der Merwer, M.J. et al. (2001). Implications of FPS1 deletion and membrane ergosterol content for glycerol efflux from *Saccharomyces cerevisiae*. *FEMS Yeast Research* 1: 205–211.

65 Turk, M., Mejanelle, L., Sentjurc, M. et al. (2004). Salt-induced changes in lipid composition and membrane fluidity of halophilic yeast-like melanized fungi. *Extremophiles* 8: 53–61.

66 Tunblad-Johansson, I., Andre, L., and Adler, L. (1987). The sterol and phospholipid composition of the salt-tolerant yeast *Debaryomyces hansenii* grown at various concentrations of NaCl. *Biochima et Biophysica Acta* 921: 116–123.

67 Mager, W.H. and Siderius, M. (2002). Novel insights into the osmotic stress response of yeast. *FEMS Yeast Research* 2: 251–257.

68 Kejžar, A., Grötli, M., Tamás, M.J. et al. (2015). HwHog1 kinase activity is crucial for survival of *Hortaea werneckii* in extremely hyperosmolar environments. *Fungal Genetics and Biology* 74: 45–58.

69 Lenassi, M. and Plemenitas, A. (2007). Novel group VII histidine kinase HwHhk7B from the halophilic fungi *Hortaea werneckii* has a putative role in osmosensing. *Current Genetics* 51: 393–405.

70 Lenassi, M., Vaupotič, T., Gunde-Cimerman, N. et al. (2007). The MAP kinase HwHog1 from the halophilic black yeast *Hortaea werneckii*: coping with stresses in solar salterns. *Saline Systems* 3: 3–14.

71 Turk, M. and Plemenitas, A. (2002). The HOG pathway in the halophilic black yeast *Hortaea werneckii*: isolation of the HOG1 homolog gene and activation of HwHog1p. *FEMS Microbiology Letters* 216: 193–199.

72 de Nada, E., Alepuz, P.M., and Posas, F. (2002). Dealing with osmostress through MAP kinase activation. *EMBO Reports* 3: 735–740.

73 Posas, F., Chambers, J.R., Heyman, J.A. et al. (2000). The transcriptional response of yeast to saline stress. *Journal of Biological Chemistry* 2752: 17249–17255.

74 Vaupotič, T. and Plemenitaš, A. (2007). Differential gene expression and Hog1 interaction with osmoresponsive genes in the extremely halotolerant black yeast *Hortaea werneckii*. *BMC Genomics* 8: 1–34.

75 Petrovic, U. (2006). Role of oxidative stress in the extremely salt tolerant yeast *Hortaea werneckii*. *FEMS Yeast Research* 6: 816–822.

76 Jin, Y., Weinin,g, S., and Nevo, E. (2005). A MAPK gene from Dead Sea fungus confers stress tolerance to lithium salt and freezing–thawing: prospects for saline agriculture. *Proceedings of the National Academy of Science, USA* 102: 18992–18997.

77 Dellamatrice, P.M. and Monteiro, R.T.R. (2004). Isolation of diuron degrading bacteria from treated soil. *Brazilian Archives of Biology and Technology* 47: 999–1003.

78 Giacomazzi, S. and Cochet, N. (2004). Environmental impact of diuron transformation: a review. *Chemosphere* 56: 1021–1032.

79 Tixier, C., Bogaerts, P., Sancelme, M. et al. (2000). Fungal biodegradation of a phenylurea herbicide, diurom: structure and toxicity of metabolites. *Pest Management Scence* 56: 455–462.

80 Cameron, M.D., Timofeevski, S., and Aust, S.D. (2000). Enzymology of Phanerochaete chrysosporium with respect to the degradation of recalcitrant compounds and xenobiotics. *Applied Microbiology and Biotechnology* 54: 751–758.

81 Sørensen, S.R., Bending, G.D., Jacobsen, C.S. et al. (2003). Microbial degradation of isoproturon and related phenylurea herbicides in and below agricultural fields. *FEMS Microbiology Reviews.* 45: 1–11.

82 Arakaki, R.L., Monteiro, D.A., Boscolo, M. et al. (2013). Halotolerance, ligninase production and herbicide degradation ability of basidiomycetes strains. *Brazilian Journal of Microbiology* 44 (4): 1207–1214.

83 Bano, A., Hussain, J., Akbar, A. et al. (2018). Biosorption of heavy metals by obligate halophilic fungi. *Chemosphere* 199: 218–222.

84 Chung, J., Shin, S., and Oh, J. (2009). Characterization of a microbial community capable of reducing perchlorate and nitrate in high salinity. *Biotechnology Letters* 31: 959–966.

85 DasSarma, S. and DasSarma, P. (2002). Halophiles. In: *Encyclopedia of Life Sciences* (ed. S. Robertson), xii. Chichester: Wiley.

86 Margesin, R. and Schinner, F. (2001a). Biodegradation and bioremediation of hydrocarbons in extreme environments. *Applied Microbiology and Biotechnology* 56: 650–663.

87 Ali, I., Siwarungson, N., Punnapayak, H. et al. (2014). Screening of potential biotechnological applications from obligate halophilic fungi, isolated from a man-made solar saltern located in Phetchaburi province, Thailand. *Pakistan Journal of Botany* 46: 983–988.

88 CAREX (2011). *Roadmap for Research on Life in Extreme Environments.* Strasbourg Cedex, France: CAREX Project Office.

89 Dalboge, H. (1997). Expression cloning of fungal enzyme genes: a novel approach for efficient isolation of enzyme genes of industrial relevance. *FEMS Microbiology* 21: 29–42.

90 Lentzen, G. and Schwarz, T. (2006). Extremolytes: natural compounds from extremophiles for versatile applications. *Applied Microbiology and Biotechnology* 72: 623–634.

91 Oren, A. (2010). Industrial and environmental applications of halophilic microorganisms. *Environmental Technology* 31: 825–834.

92 Margesin, R. and Schinner, F. (2001b). Potential of halotolerant and halophilic microorganisms for biotechnology. *Extremophiles* 5: 73–83.

93 Shima, J. and Takagi, H. (2009). Stress-tolerance of baker's-yeast (*Saccharomyces cerevisiae*). cells: stress-protective molecules and genes involved in stress tolerance. *Biotechnology and Applied Biochemistry* 53: 155–164.

94 Watanabe, T., Srichuwong, S., Arakane, M. et al. (2010). Selection of stress-tolerant yeasts for simultaneous saccharification and fermentation (SSF). of very high gravity (VHG). potato mash to ethanol. *Bioresour. Technol.* 101: 9710–9714.

95 Zheng, D.Q., Wu, X.C., Tao, X.L. et al. (2011). Screening and construction of *Saccharomyces cerevisiae* strains with improved multi-tolerance and bioethanol fermentation performance. *Bioresource Technology.* 102: 3020–3027.

96 Saysel, A.K. and Barlas, Y. (2001). A dynamic model of salinization on irrigated lands. *Ecological Modelling* 139: 177–199.

97 Bruinsma, J. (2009). The resource outlook to 2050: By how much do land, water and crop yields need to increase by 2050? FAO Expert Meeting on How to Feed the World in 2050, Rome.

98 Cominelli, E. and Tonelli, C. (2010). Transgenic crops coping with water scarcity. *Nature Biotechnology* 27: 473–477.

99 Ashraf, M. and Akram, N.A. (2009). Improving salinity tolerance of plants through conventional breeding and genetic engineering: an analytical comparison. *Biotechnology Advances* 27: 744–752.

100 Munns, R. (2002). Comparative physiology of salt and water stress. *Plant and Cell Environment* 25: 239–250.

101 Somvanshi, V.S. (2009). Patenting drought tolerance in organisms. *Recent Patents on DNA and Gene Sequences* 3 (1): 16–25.

9

Rhizobacteria-Mediated Bioremediation

Insights and Future Perspectives

Vijay Kant Dixit[1,2], Sankalp Misra[1], Shashank Kumar Mishra[1], Namita Joshi[2], and Puneet Singh Chauhan[1]

[1] *CSIR – National Botanical Research Institute, Rana Pratap Marg, Lucknow, Uttar Pradesh, India*
[2] *Department of Environmental Science, Kanya Gurukul Campus, Gurukul Kangri University, Haridwar, Uttarakhand, India*

9.1 Introduction

Clean air, soil, and water are the critical natural wealth for human society. But, the contamination of these natural resources is a challenging problem for the entire world. Environmental pollution keeps on expanding periodically due to uncontrolled urbanization, industrialization, technological advancements, and unsafe agricultural and pharmaceutical practices, which eagerly contaminate the environment. Various anthropogenic industrial activities like electroplating, leather tanning, mining, smelting, printing and textiles, pesticides, and chemical factories generate enormous amounts of solid and liquid waste that contains highly toxic organic and inorganic pollutants like toxic metals, phthalates, chlorinated compounds, xenobiotics compounds, phenols, tannins, dyes, and pathogenic organisms [1]. Discharge of such type of solid and liquid waste into the environment severely influences the physical and biological qualities of soil, air, and water with deleterious effects on their native flora and fauna. The natural resources with an effective concentration of these contaminants are key factors for serious health hazards to the ecosystem and are becoming an unavoidable issue in the provision of safe and healthy environments [2]. For the past two decades, various physical, chemical, and biological treatment technologies have been recognized for effective alleviation of contaminants from affected areas. But, in comparison to physico-chemical approaches, bioremediation is a convenient

Soil Bioremediation: An Approach Towards Sustainable Technology, First Edition.
Edited by Javid A. Parray, Abeer Hashem Abd Elkhalek Mahmoud, and Riyaz Sayyed.

approach for the proper remediation of contaminated sites [3]. Bioremediation is a vast term that deals with the degradation and detoxification of organic and inorganic pollutants through the use of microorganisms and plants [4]. In ancient times plants served as one of the important parts of our ecosystems. The technology that uses plants for the remediation of contaminated sites is known as phytoremediation [5]. In addition to bioremediation, phytoremediation is also a broad term in technology that involves mainly two components for the functioning of the rhizosphere colonizing microorganisms and the other by plants themselves.

The phytoremediation technology that involves the use of root colonizing microbes in association with plants is known as "rhizoremediation" [6]. Rhizoremediation is a convenient and economic technique for the remediation of toxic pollutants [7]. Most plants are symbiotic with soil bacteria and fungi, (forming ecto- and endo-mycorrhizal associations) [8]. These natural allies could be used to enhance plant survival on difficult sites.

However, the relationship between root-colonizing microbes and plants is still unexplored. Considering different remedial technologies, rhizoremediation is the most suitable technology for the complete mineralization of organic pollutants without generating any secondary pollutants or intermediate pollutants. Hence, this chapter explores the detailed knowledge and relationship of plant and rhizospheric microorganisms for the abatement of polluted lands and contaminated sites.

9.2 Bioremediation: Rescue Plan by Natural Agents

The term "bioremediation" was coined by scientists in the 1980s and defined as the process to remediate the environmental contaminants into non-/less-toxic forms by using living organisms (mainly microorganisms). Bioremediation is a vast term consists two parts: "bios" means life and refers to living organisms and "to remediate," which means to solve the problem. This is a cost-effective and environmentally sound technique, best suited for proper management of noxious waste contaminants in soil and wastewater [4]. In nature, bioremediation practices utilize the metabolic potential of microorganisms to reduce or remediate the hazardous substances by breaking them down or transforming them into nontoxic objects under *ex-situ* or *in-situ* conditions [9]. Microorganisms have various strategies to ameliorate the stress and toxicities associated with hazardous environments [10, 11]. They imply various strategies to detoxify the noxious environmental pollutant like heavy metal toxicity includes exclusion by using a permeability barrier, intracellular, and extracellular sequestration, active transport efflux pumps, enzymatic detoxification, and reductions in cellular sensitivity to metal ions [10, 11]. Bioremediation of heavy metals include transformation of oxidation states, (Cr(VI) to Cr(III), $SeO^{2-}{}_4$ to Se), biosorption (adsorb metal ions into their cell surface), bioaccumulation (sequester or accumulate heavy metals inside the cell), bioassimilation (by producing low molecular weight proteins that assimilates or chelates metal ions), extracellular chemical precipitation (e.g., by S^{2-} from sulfur-reducing bacteria), and volatilization (e.g., dimethylselenide, trimethylarsine, and Hg vapor). Apart from this, several strains of bacteria produce biosurfactants such as surfactin, rhamnolipids, sophorolipids, aescin, and saponin to enhance the solubilization of metals in soils [7]. Moreover, biotransformation, biosorption, and bioaccumulation are the main three kinds of microbial remediation processes that reduced heavy metal toxicity and transport, playing a critical role in microbial remediation of heavy metals. Incorporation of rhizobacteria for heavy metal remediation is cheap, noninfectious, and effective technology that can be executed on the polluted site together with physical and chemical treatments [12]. However, bioremediation technology has several limitations such as – it is longer process and has limited practical approach e.g., it is challenging to anticipate the bioremediation effect. Moreover related remediation mechanisms are also more complicated and not easy to explore. For heavy metal contaminated soils, there are only two different strategies (biomobilization and bioimmobilization) that may be efficient to restore the heavy metal contaminated soils sites [12]. Some rhizobacterial genera such as *Pseudomonas*, *Bacillus*, *Alcaligenes*, *Micrococcus*, *Sphingomonas*, and *Mycobacterium* are well reported for their potential degradation and removal of pesticides, hydrocarbon, polyaromatic compounds, and detoxification of toxic metals Cr, Pd, Cu, Ni, and As [13] (see Table 9.1).

Table 9.1 List of different microbes and host crops/plants evaluated for remediation of different metals.

S. No.	Microbes	Host Plant	Metals	References
1.	*Kluyvera ascorbata*	Tomato	Cd	[14]
2.	*Pseudomonas tolaasii*	Canola	Cu, Ni	[15]
3.	*Pseudomonas fluorescens*	Perennial grasses	Co	[16]
4.	*Variovorax paradoxus*	Indian mustard	Cr	[17]
5.	*Rhodococcus* sp.	Canola	Pb	[16]
6.	*Flavobacterium* sp.	Tomato	Zn	[15]
7.	*Pantoea* sp.	*Lolium perenne*	Diesel oil	[18]
8.	*Gordonia*	*Zea mays*	Diesel oil	[19]
9.	*Acinetobacter* sp.	*Lolium multiflorum*	Various PAHs	[20]
10.	*Rhizobium*	*Trifolium repens*	Chrysene	[21]

9.3 Rhizoremediation

Rhizoremediation is a type of bioremediation in which root-associated rhizobacterial strains play a positive role in the remediation of organic and inorganic environmental pollutants by trapping solar energy via plants to biostimulate pollutant reduction by the native soil microbial community [22, 23]. However, it is a plant-based *in-situ* phytorestoration technique, proven to be prudential, efficient, and easy to implement under field conditions [24]. Microorganisms living in the rhizosphere on plant-derived substrates (root exudates) play an inducible role in the degradation on transformation of xenobiotic compounds present in the soil [25]. The increased ability of root-associated microorganisms may be associated with the greater number of microorganisms or with the availability of growth-supporting substrates (root exudates) for c-metabolism.

Lorenz Hiltner (1904) [26] coined the term "rhizosphere" to describe a complex chemical matrix filled with diverse microbial species around the plant roots. The rhizosphere represents a micro environment that is totally different to nonrhizosphere soil. In this important zone the presence of plenty of microbes and their functions leads to cycling and availability of nutrients, improving soil quality indicators [27]. The enhanced microbial activities mainly in terms of carbon (microbial biomass carbon or soil carbon) and soil-nutrients-related enzymes like, dehydrogenase activity of numerous hydrolytic enzymes, phosphatases, sulfatases, proteases, amidases, and glucosidase are also well known for breaking down organically bound nutrients in soil [28]. The rhizosphere also influences the root zone of the plant and takes a significant part in regulating the decomposition of soil organic matter and nutrient cycling [29].

The maximum portion of plant rhizosphere is inhabited by fungi and bacterial community [30]. The effect of rhizosphere on fungal abundance is 10–20 times higher while bacterial abundance is influenced 2–20 times more in the bulk soil. Competition for nutrient sources between microbial communities is too high in the rhizospheric region [31]. Various tactics for nutrient sources of microorganisms give rise to a range of synergistic and antagonistic interaction between microbe and microbe as well as plant microbes [32]. Tremendous diversity of microorganisms and plants depends on variation in their interaction strategies [33]. It is critical to understand the various interaction tactics between plant and their adherent microorganisms to enhance the plant growth as well as for soil remediation practices.

The presence of beneficial rhizospheric microbes controls the decomposition rate of soil organic matter through the production of different metabolites [35] (See Figure 9.1). The food web of the soil ecosystem depends on root exudates, litter residues, and soil organic matter as a primary source of carbon. These carbon sources vary in their availability and accessibility to soil organisms; therefore,

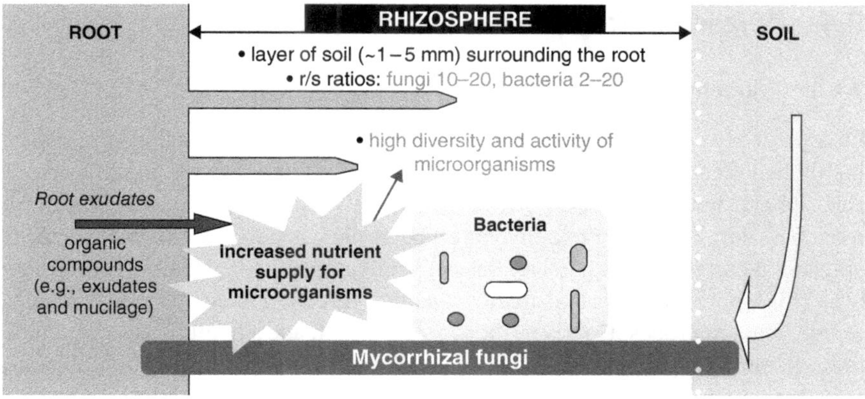

Figure 9.1 Rhizospheric microorganism link between plant and soil. (*Source:* Adapted from [34].)

it increases the carbon flow within the food web of soil. Processes that are shared both by roots and adhered microorganisms are mainly referred as rhizosphere processes. These processes are, exudation of soluble compounds, water uptake, nutrient mobilization, decomposition of soil organic matter, and the subsequent release of CO2 through respiration [28] (see Figure 9.1). Rhizosphere processes commonly serve as major gateways for nutrients and water to the plant from soil through different microbial activities [28]. Plant root exudates consist of carbohydrates, amino acids, organic acids, and a number of other complex compounds that are used by soil microorganisms as their nutrient source [28, 36]. Fifty percent of the energy fixed by photosynthesis in terrestrial ecosystems is utilized in processes associated with rhizospheric activity. It contributes approximately 50% of the total CO_2 emitted from terrestrial ecosystems, and mediates virtually all aspects of nutrient cycling [36] (Figure 9.1).

Hence, the interactions between plant roots and their microbial communities are at the center of soil ecosystem processes [37]. Rhizospheric-associated microorganisms are one of the major soil ecological agents that is involved in the colonization of soil microorganisms [38]. Root-associated microorganisms, in turn, provide nutrients to plant that favor plant health by stimulating root growth (regulators), enhancing water, and mineral uptake as well as inhibiting the growth of pathogenic or other, nonpathogenic, soil microbes [39, 40]. They also enhance remediation of organic and inorganic compounds in soils by volatilization or by increasing the humification of pollutants [41, 42]. For instance, the release of oxidoreductase enzymes (e.g., peroxidase) by microbes, as well as by plant roots, can catalyze the polymerization of a pollutant on root surfaces as well as the soil humic fraction [43].

9.4 Rhizospheric Microbial Community

9.4.1 Endophytes

Endophytes lives inside a plant tissue in a close mutual relationship that can be highly beneficial for plants. Endophytes are basically underlying the division of rhizospheric microbial region of roots, which also includes rhizoplane and root microbial flora [44, 45]. Endophytes can be easily categorized into three classes: obligate, facultative, and passive. Basically, obligate endophytes do not originate from rhizospheric regions and are not capable of growing in an open environment as free living microbes. However, they can survive inside a plant tissue and can be moved from one plant to other via seeds. Whereas, in comparison to obligates, facultative endophytes can be easily grown in soil and are free living in nature. They can reside in plants through coordinated infection breaks or openings. However, passive endophytes cannot easily colonize plant roots, they can inhabit inside a root with the help of root hairs via open wounds [46]. Bacterial endophytes, act as biocontrol agents, and help in plant protection against various phytopathogens. In fact, numerous reports have publicized that endophytes are highly beneficial for plant growth and that they are capable of controlling severe infections caused by insects, nematodes, and other phytopathogens [47, 48]. Endophytes are also reported to promote plant growth by enhancing seed germination and root and shoot growth activity under stress environments [49, 50].

Endophytes have a tremendous capacity for recalcitrant and xenobiotic degradation and sometimes act as vectors that can promote new degradative individuals for the protection of plants from toxic pollutants [51]. Various endophytes are able to grow and tolerate higher concentrations of toxic heavy metals, antibiotics, and organic and inorganic pollutants in contaminated environments. Endophytic bacteria can cope with such contaminated environments by developing various resistance mechanisms, which not only helps them to survive in such toxic contaminated environments but also promotes the degradation potential of endophytes helping bacteria promote rich and highly potential diversity of microorganism thus improving phytoremediation [52].

Endophytic bacteria are helpfully efficient in bioreduction and elimination of toxic heavy metals from contaminated environments by accelerating the phytoextraction ability of plants. They can also help in the degradation and detoxification of organic and chlorinated compounds such as polyaromatic hydrocarbons (PAH), chlorophenols, and pesticides. Endophytic microbe assisted phytoremediation of harmful soil pollutants can be a more pragmatic environmentally friendly approach that enhances degradation process via catalysis of atmospheric oxygen [53, 54]. Endophytic microbial communities as plant growth promoters are well reported [55]. But, the role of endophytic microbial communities in the biodegradation and transformation of organic and inorganic contaminants needs more advanced research.

9.4.2 Plant Growth Promoting Rhizobacteria

Plant growth-promoting rhizobacteria colonize the plant roots, comprise a large number of free-living bacteria that play an effective role in enhancing the growth and development of the host plant by inhibiting the harmful impacts of different contaminants present in polluted sites [56]. A number of plant growth promoting rhizobacterias (PGPRs) exist such as, *Azotobacter chroococcum* HKN-5, *Bacillus megaterium* HKP-1, *Bacillus mucilaginosus* HKK1, *Bacillus subtilis* SJ-101, *Brevundimonas* sp. KR013, *Pseudomonas fluorescens* CR3, *Rhizobium leguminosarumbv. Trifolii NZP561* and *Kluyvera ascorbata* SUD165. In a study it has have been reported that PGPR associated with plant growth in contaminated soils helps in bioremediation [57]. A large number of rhizobacteria like, *B. subtilis, Bacillus pumilus*, *Pseudomonas pseudoalcaligenes*, and *Brevibacterium halotolerans* together with hyper-accumulator plants is being utilized in rhizoremediation of various toxic metal present in the associated soil. [58].

PGPR play an important role in plant growth promotion by using two different modes, direct or indirect. In this way they benefited plants through root and shoot enlargement by producing a hormone (indole acetic acid) [59], enhancing essential nutrient supply [60, 61], and helping in solubilizing of insoluble phosphate [62], which is a macronutrient, stress mitigating 1-aminocyclopropane-1-carboxylate (ACC) deaminases enzyme [63]. Moreover, these PGPR also protect plants from environmental stress like salts, drought, nutrient deficiency, and toxic metals [64] (Figure 9.2).

Moreover, a study examining the functioning of PGPR found that rhizosphere inhabiting bacterial strains have been identified as a biocontrol mechanism against phytophathogens [65]. The achievement by PGPR in the rhizosphere is mainly due to their capability to compete with other microbes for nutrients, root exudates, sustain effective numbers, and colonize the root zone of growing plants [42].

Application of PGPR in problematic sites enhances the growth of plants by detoxify or preventing the harmful effects of heavy metal via their different mechanisms along with reprocessing of heavy metals and nutrients, maintenance of soil architecture, and control of pathogens and pests [67, 68].

PGPR mainly follows such strategies as, bioleaching, biomineralization, biotransformation, biosorption, and biodegradation for the detoxification or reduction in the toxic effects of metals in the environment. In this complex rhizosphere system, microbes consume the carbohydrates, alcohols, vitamins, proteins, and amino acids as a nutrient from the root exudates of plants for their growth and functioning. The presence of maximum amount of root exudates in

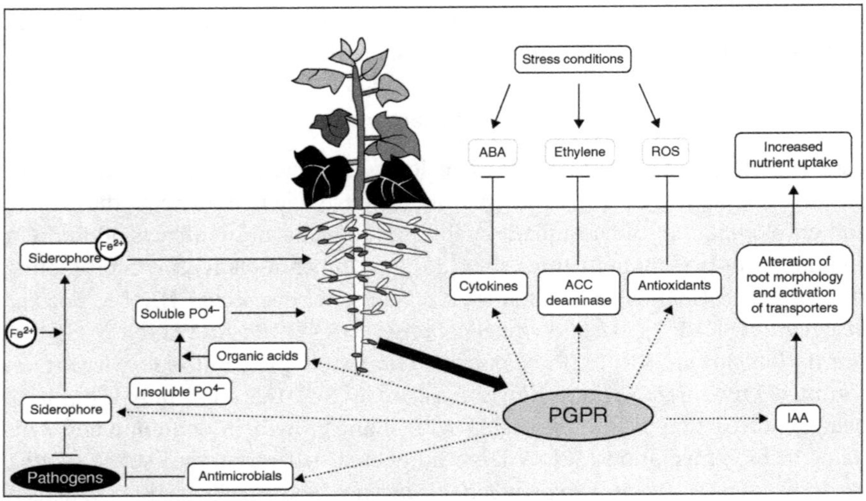

Figure 9.2 Plant growth promoting activities of rhizobacteria. (*Source:* Adapted from [66].)

the plant rhizosphere provides favorable conditions to the PGPR for more nutrient uptake than other microbes of the remaining soil [69]. In turn these benefitted rhizospheric microbes promote plant growth (Figure 9.2).

9.4.3 Mycorrhizal Association

Fungi are one of the major contributors involved in the conversion or degradation of toxic pollutants in association with plant community [70]. The symbiotic relationship between fungal cells and the roots of vascular plants termed as "Mycorrhiza" [71] [Figure 9.3]. Mycorrhizal fungi help in improving plant growth by increasing the root surface area of a plant, which enhances its nutrient and water uptake ability from soil. It increases the bioavailability of nutrients especially phosphorus and nitrogen, reduces environmental stress related to a low level of water, and increasing the resistance to pathogenic microorganisms by stimulating phytohormones [72–74]. Mycorrhizal Fungi degrades organic pollutants by virtue of organic acid production and by secreting several enzymes responsible for hydrolyzing compounds containing phosphorus and nitrogen [75]. Mycorrhizal fungi also help in the amelioration of heavy metal stress as well as diseases caused by soil-borne pathogens thereby increasing the fitness of plants. In addition, they promote plant growth by enhancing nutrient content using hyphal mantle in the fine roots of plants [75]. They can actually act as filters for plants that block toxic compounds with the help of their mycelium. Plants reciprocate by providing carbohydrates and other essential biomolecules required for proper fungal growth and

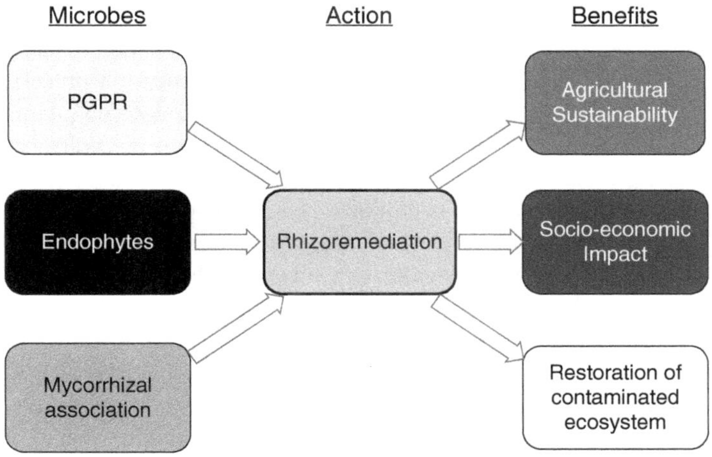

Figure 9.3 A graphical representation of microbe-mediated rhizoremediation and their benefits for environments.

their maintenance [76]. Mycorrhizae can only be found in plant roots hence they can also be called a modified form of rhizoremediation that explores the beneficial effects of mycorrhizal fungi. Sometimes is can also be related to phytoextraction and phytostabilization. However, it is totally different from phytoremediation in terms of the mechanism of removal of pollutants at a higher rate, depending on the area coverage by plants roots through fungal mycelium [77].

There are mainly two common forms of mycorrhizal associations found in plants: ectomycorrhizae (EM) and Arbuscular mycorrhizae (AM). Ectomycorrhizal fungi are commonly associated with woody plants like *Salicaceae*, *Pinaceae*, *Fagaceae*, *Betulaceae*, and *Dipterocarpaceae* [76].

Mycelium of EM fungi modifies the morphology of a plant root resulting in root bifurcation and formation of clusters thereby forming a thick mantle that increases surface area for better nutrient accumulation. Inward penetration of Hyphae of ectomycorrhizal fungi between the root cells form a netlike assembly known as the Hartig net [75], which establishes an interface facilitating photo assimilate exchange and nutrients immobilization between the fungal partner and respective host plant.

AM has been recently been proved to remove toxic pollutants and enhance the rate of rhizoremediation by improving soil fertility and tolerance of plants against toxic metals and organic and inorganic pollutants [78, 79]. It is generally found in association with herbaceous and graninaceous plants of semi-arid and temperate regions and in tropical and subtropical plants as well. The use of AM in rhizoremediation is becoming a promising approach for the degradation of various organic pollutants namely toxic metals, polycyclic aromatic hydrocarbons (PAHs), PAEs (phthalic acid esters), petroleum products, pesticides, chlorinated compounds by enzyme exudates [80] [Table 9.1]. AM can alleviate the root exudation process, which also helps to improve rhizospheric microbial activity and remove pollutants by plants uptake and accumulation [81]. Moreover, it modulates the physiology of the host plant making them more efficient in tolerating or resisting pathogenic attacks and a number of environmental conditions. The use of specific AM for the remediation of contaminated sites improved tolerance to toxic compounds by providing "glomalin" to the plants [82]. Glomalin is a substance that binds with soil aggregates and makes a stable soil structure [83]. The soil surrounded by mycorrhizal association near plant roots establishes a diverse bacterial population in comparison to the bulk soil [83]. The collective term for the rhizosphere in combination with the hyphosphere of mycorrhizal fungi is known as mycorrhizosphere. Inhabitants of the mycorrhizosphere comprise intrahyphal bacteria in association with EM fungi and bacterial intraspores in numerous arbuscular mycorrhizal fungi [84]. Earlier it has been reported that some bacterial species present in mycorrhizosphere can promote the formation of mycorrhizal association.

9.5 Practices that Improves Rhizoremediation

9.5.1 Phytoextraction

Phytoextraction is a subset of phytoremediation, which involves the use of PGPRs and hyper-accumulator plants for the remediation of polluted environments [6]. It is the best-suited technology for the reclamation of metal contaminated sites. Hyper-accumulator plants absorb metals pollutants from soil through root hairs, and rhizobacterial communities, which exist in the rhizospheric region of roots to help enhance uptake rate of toxic metals and other pollutants. Hyper-accumulator plants have the ability to accumulate higher concentrations (100–1000 times greater) of pollutants from contaminated soils with the help of rhizospheric microbial communities. Higher numbers of bacterial and fungal strains were reported in the rhizospheric region of hyper-accumulator plants. These microbes play an essential role in the accumulation, transportation, and detoxification mechanism of pollutants in soils and as well as in various parts of plants. Microbial biomass increases the bioremediation process as reported by [85] in arsenic contaminated soil. Management of arsenic contaminated soil with the use of higher biomass production of silverback fern (*Pityrogramma calomelanos*) helps in reduction of metal contamination. Rhizospheric microorganisms facilitate metal uptake availability for plants from higher to lower concentrations and increase biomass production under stressful environment.

9.5.2 Genetically Engineered Rhizobacteria

Rhizospheric microbial communities are a promising tool in the field of bioremediation. These communities provide a perfect microbial environment for the degradation of organic pollutants. Several microorganisms have the ability to adapt and survive in polluted environments but are unable to degrade or reduce the pollutants present in higher concentrations. Survival of microbes in higher concentrations of pollutants might be due to their resistance or tolerance, however survival and reduction are not necessarily interrelated and thus, not all resistant microorganisms can be utilized for developing remediation technology [86, 87]. With the advancement of molecular technology like proteomics, genomics, and transcriptomics, the application of pollutant degradative genes in genetically engineered rhizobacterial microorganisms gives a new aspect to the field of bioremediation. With the help of these advance technologies, specific degradative genes of rhizobacterial strains are developed, which are more competent and proficient rather than indigenous microbes. The use of genetically engineered rhizobacteria emerged as highly booming and successful practices in comparison to traditional treatment approaches for the reduction of environmental pollutants [88].

9.6 Future Perspective

The role of rhizospheric microorganisms for plant growth promotion is well known. However, the role such microbes in the field of bioremediation of contaminated environments are still unexplored. This technology exploits the relationship between plants and their rhizospheric microbial communities including endophytes, mycorrhiza, and PGPRs, which plays an important role in the degradation of organic pollutants like PAH, PCB (polychlorinated biphenyl), and PHC (petroleum hydrocarbon) with generating any kind of secondary pollutants. However, this technology needs further attention and research advancement for its proper application in commercial and industrial purposes. The most amazing aspect of this technology is the use of rhizobacterial communities, which not only hold a huge capacity to initiate plant growth but are also helpful in the decontamination of a polluted environment.

9.7 Advantages of Rhizoremediation

- Rhizoremediation is an appropriate effective reclamation technology for the degradation and detoxification of toxic metals, phenols, hydrocarbons, and other organic and inorganic pollutant contaminated soils.

- This technique involves a close association between microbes and plant root zone, which enhances the microbial activity in rhizosphere and helps in the greater breakdown of organic contaminants by the activation of degradative and catalytic genes.
- It is an ecofriendly, cost-effective, clean, and green approach for the removal of environmental pollutants.

9.8 Conclusion

Environmental pollution is a great challenge for present and future generations. To overcome environmental pollution problems rhizoremediation is emerging as an efficient environmentally friendly, esthetically pleasing, cleaner approach in the alleviation of environmental pollutants. This technology not only helps in the reclamation of contaminated environments, but also improves soil fertility, increases productivity, and increases the quality of crops. However, the proper application of this technology on a large scale is still not flourishing due to the selection of appropriate plants and microbes. Thus, to explore the potential of rhizoremediation technology for remediation of contaminated sites and plant growth promotion, extensive research is needed on the proper selection and compatibility between plants and microbes.

Acknowledgment

The study was supported by New Initiative (as a Cross Flow Technology project) "Development of microbial formulation to serve as stress buster and bio-fertilizers for improving crop improving crop productivity in stressed agriculture" (MLP 0022) from the Council of Scientific and Industrial Research (CSIR), New Delhi, India.

References

1 Mishra, S., Bharagava, R.N., More, N. et al. (2018). Heavy metal contamination: an alarming threat to environment and human health. In: *Environmental Biotechnology: For Sustainable Future* (eds. R.C. Sobti, N. Kumar Arora and R. Kothari), 103–125. Singapore: Springer Nature.

2 Sarwar, N., Imran, M., Shaheen, M.R. et al. (2017). Phytoremediation strategies for soils contaminated with heavy metals: modifications and future perspectives. *Chemosphere* 171: 710–721.

3 Mahar, A., Wang, P., Ali, A. et al. (2016). Challenges and opportunities in the phytoremediation of heavy metals contaminated soils: a review. *Ecotoxicology and Environmental Safety* 126: 111–121.

4 Mishra, S., Chowdhary, P., and Bharagava, R.N. (2019). Conventional methods for the removal of industrial pollutants, their merits and demerit. In: *Emerging and Eco-Friendly Approaches for Waste Management* (eds. R.M. Bharagava and P. Chowdhary), 1–31. Singapore: Springer Nature.

5 Dixit, R., Wasiullah, Malaviya, D. et al. (2015). Bioremediation of heavy metals from soil and aquatic environment: an overview of principles and criteria of fundamental processes. *Sustainability* 7: 2189–2212.

6 Ullah, A., Heng, S., Munis, M.F.H. et al. (2015). Phytoremediation of heavy metals assisted by plant growth promoting (PGP) bacteria: a review. *Environmental and Experimental Botany* 117: 28–40.

7 Lal, S., Ratna, S., Said, O.B. et al. (2018). Biosurfactant and exopolysaccharide-assisted rhizobacterial technique for the remediation of heavy metal contaminated soil: an advancement in metal phytoremediation technology. *Environmental Technology and Innovation* 10: 243–263.

8 Oberai, M. and Khanna, V. (2018). Rhizoremediation – plant microbe interactions in the removal of pollutants. *International Journal of Current Microbiology and Applied Sciences* 7 (1): 2280–2287.

9 Das, S. and Dash, H.R. (2014). Microbial bioremediation: a potential tool for restoration of contaminated areas. In: *Microbial Biodegradation and Bioremediation* (ed. S. Das), 1–21.

10 Pan, X., Achal, V., Zhao, C. et al. (2018). Microbial remediation of heavy metals and arsenic-contaminated environments in the arid zone of Northwest China. In: *Twenty Years of Research and Development on Soil Pollution and Remediation in China* (eds. Y. Luo and C. Tu), 477–486. Singapore: Springer Nature.

11 Jin, Y., Luan, Y., Ning, Y. et al. (2018). Effects and mechanisms of microbial remediation of heavy metals in soil: a critical review. *Applied Sciences* 8 (8): 1336. https://doi.org/10.3390/app8081336.

12 Mani, D. and Kumar, C. (2014). Biotechnological advances in bioremediation of heavy metals contaminated ecosystems: an overview with special reference to phytoremediation. *International Hournal of Environmental Science and Technology* 11: 843–872.

13 Megharaj, M., Ramakrishnan, B., Venkateswarlu, K. et al. (2011). Bioremediation approaches for organic pollutants: a critical perspective. *Environment International* 37: 1362–1375.

14 Burd, G.I., Dixon, D.G., and Glick, B.R. (1998). A plant growth-promoting bacterium that decreases nickel toxicity in seedlings. *Applied and Environmental Microbiology* 64 (3): 3663–3668.

15 Burd, G.I., Dixon, D.G., and Glick, B.R. (2000). Plant growth-promoting bacteria that decrease heavy metal toxicity in plants. *Canadian Journal of Microbiology* 46 (3): 237–245.

16 Dell'Amico, E., Cavalca, L., and Andreoni, V. (2005). Analysis of rhizobacterial communities in perennial *Graminaceae* from polluted water meadow soil, and screening of metal-resistant, potentially plant growth-promoting bacteria. *FEMS Microbiology Ecology* 52 (2): 153–162.

17 Belimov, A.A., Hontzeas, N., Safronova, V.I. et al. (2005). Cadmium-tolerant plant growth-promoting bacteria associated with the roots of Indian mustard (*Brassica juncea* L. Czern.). *Soil Biology and Biochemistry* 37 (2): 241–250.

18 Afzal, M., Yousaf, S., Reichenauer, T.G. et al. (2012). The inoculation method affects colonization and performance of bacterial inoculant strains in the phytoremediation of soil contaminated with diesel oil. *International Journal of Phytoremediation* 14: 35–47.

19 Hong, S.H., Ryu, H., Kim, J. et al. (2011). Rhizoremediation of diesel-contaminated soil using the plant growth promoting rhizobacterium *Gordonia* sp. S2RP-17. *Biodegradation* 22: 593–601.

20 Yu, X.Z., Wu, S.C., Wu, F.Y. et al. (2011). Enhanced dissipation of PAHs from soil using mycorrhizal ryegrass and PAH-degrading bacteria. *Journal of Hazardous Materials* 186: 1206–1217.

21 Johnson, D.L., Maguire, K.L., Anderson, D.R. et al. (2004). Enhanced dissipation of chrysene in planted soil: the impact of rhizobial inoculums. *Soil Biology and Biochemistry* 36: 33–38.

22 Habibi, P., Sa, M.F.G.D., Makhzoum, A. et al. (2017). Bioengineering hairy roots: phytoremediation, secondary metabolism, molecular pharming, plant-plant interactions and biofuels. In: *Sustainable Agriculture Reviews* (ed. E. Lichtfouse), 213–251. Switzerland: Springer.

23 Leigh, M.B., Prouzova, P., Mackova, M. et al. (2006). Polychlorinated biphenyl (PCB)-degrading bacteria associated with trees in a PCB-contaminated site. *Applied and Environmental Microbiology* 72: 2331–2342.

24 Kamaludeen, S.P.B. and Ramasamy, K. (2008). Rhizoremediation of metals: harnessing microbial communities. *Indian Journal of Microbiology* 48: 80–88.

25 Martin, B.C., George, S.J., Price, C.A. et al. (2014). The role of root exuded low molecular weight organic anions in facilitating petroleum hydrocarbon degradation: current knowledge and future directions. *Science of the Total Environment* 472: 642–653.

26 Hiltner, L. (1904). Uber neuere Erfahrungen und Probleme auf dem Gebiet der Bodenbakteriologie und unter besonderer Berucksichtigung der Grundungung and Brache. *Arbeiten der Deutschen Landwirtschaftlichen Gesellschaft* 98: 59–78.

27 Chaudhry, V., Rehman, A., Mishra, A. et al. (2012). Changes in bacterial community structure of agricultural land due to long-term organic and chemical amendments. *Microbial Ecology* 64: 450–460.

28 Bhaduri, D., Pal, S., Purakayastha, T.J. et al. (2015). Soil quality and plant-microbe interactions in the rhizosphere. In: *Sustainable Agriculture Reviews* (ed. E. Lichtfouse), 307–335. Switzerland: Springer Nature.

29 Dijkstra, F.A., Carrillo, Y., Pendall, E. et al. (2013). Rhizosphere priming: a nutrient perspective. *Frontiers in Microbiology* 4: 1–8.

30 Thavamani, P., Samkumar, R.A., Satheesh, V. et al. (2017). Microbes from mined sites: harnessing their potential for reclamation of derelict mine sites. *Environmental Pollution* 230: 495–505.

31 Morgan, J.A.W., Bending, G.D., and White, P.J. (2005). Biological costs and benefits to plant–microbe interactions in the rhizosphere. *Journal of Experimental Botany* 56: 1729–1739.

32 Rosier, A., Bishnoi, U., Lakshmanan, V. et al. (2016). A perspective on inter-kingdom signaling in plant-beneficial microbe interactions. *Plant Molecular Biology* 90 (6): 537–548.

33 Berg, G., Köberl, M., Rybakova, D. et al. (2017). Plant microbial diversity is suggested as the key to future biocontrol and health trends. *FEMS Microbiology Ecology* 93 (5) https://doi.org/10.1093/femsec/fix050.

34 Richardson, A.E., Barea, J.M., McNeill, A.M. et al. (2009). Acquisition of phosphorus and nitrogen in the rhizosphere and plant growth promotion by microorganisms. *Plant and Soil* 321: 305–339.

35 Dennis, P.G., Miller, A.J., and Hirsch, P.R. (2010). Are root exudates more important than other sources of rhizo deposits in structuring rhizosphere bacterial communities? *FEMS Microbiology Ecology* 72 (3): 313–327.

36 Rohrbacher, F. and Arnaud, M.S. (2016). Root exudation: the ecological driver of hydrocarbon rhizoremediation. *Agronomy* 6 (1): 19. https://doi.org/10.3390/agronomy6010019.

37 Pérez-Jaramillo, J.E., Mendes, R., and Raaijmakers, J.M. (2016). Impact of plant domestication on rhizosphere microbiome assembly and functions. *Plant Molecular Biology* 90 (6): 635–644.

38 Pandey, P. and Maheshwari, D.K. (2007). Two-species microbial consortium for growth promotion of *Cajanus cajan*. *Current Science* 92: 1137–1142.

39 Pilon-Smits, E. (2005). Phytoremediation. *Annual Review of Plant Biology* 56: 15–39.

40 Sharifi, R. and Ryu, C.M. (2017). Chatting with a tiny below-ground member of the holobiome: communication between plants and growth-promoting rhizobacteria. *Advances in Botanical Research* 82: 135–160.

41 Salt, D.E., Smith, R.D., and Raskin, I. (1998). Phytoremediation. *Annual Review of Plant Physiology and Plant Molecular Biology* 49: 643–668.

42 Bisht, S., Pandey, P., Bhargava, B. et al. (2015). Bioremediation of polyaromatic hydrocarbons (PAHs) using rhizosphere technology. *Brazilian Journal of Microbiology* 46 (1): 7–21.

43 Shahid, M., Shamshad, S., Rafiq, M. et al. (2017). Chromium speciation, bioavailability, uptake, toxicity and detoxification in soil-plant system: a review. *Chemosphere* 178: 513–533.

44 McKinnon, A.C., Saari, S., Moran-Diez, M.E. et al. (2017). *Beauveria bassiana* as an endophyte: a critical review on associated methodology and biocontrol potential. *BioControl* 62 (1): 1–17.

45 Sasse, J., Martinoia, E., and Northen, T. (2018). Feed your friends: do plant exudates shape the root microbiome? *Trends in Plant Science* 23 (1): 25–41.

46 Hardoim, P.R., Van Overbeek, L.S., and van Elsas, J.D. (2008). Properties of bacterial endophytes and their proposed role in plant growth. *Trends in Microbiology* 16: 463–471.

47 Etesami, H. and Alikhani, H.A. (2016). Rhizosphere and endorhiza of oil seed rape (*Brassica napus* L.) plant harbor bacteria with multifaceted beneficial effects. *Biological Control* 94: 11–24.

48 Azevedo, J.L., Maccheroni, W., Pereira, J.O. et al. (2000). Endophytic microorganisms: a review on insect control and recent advances on tropical plants. *Electronic Journal of Biotechnology* 3 (1) https://doi.org/10.2225/vol3-issue1-fulltext-4.

49 Chebotar, V.K., Malfanova, N.V., Shcherbakov, A.V. et al. (2015). Endophytic bacteria in microbial preparations that improve plant development (review). *Applied Biochemistry and Microbiology* 51 (3): 271–277.

50 Gouda, S., Kerry, R.G., Das, G. et al. (2018). Revitalization of plant growth promoting rhizobacteria for sustainable development in agriculture. *Microbiological Research* 206: 131–140.

51 Chakraborty, J. and Das, S. (2016). Molecular perspectives and recent advances in microbial remediation of persistent organic pollutants. *Environmental Science and Pollution Research* 23 (17): 16883–16903.

52 Ryan, R.P., Germaine, K., Franks, A. et al. (2008). Bacterial endophytes: recent developments and applications. *FEMS Microbiology Letters* 278: 1–9.

53 Olson, P.E. (2008). Effects of agronomic practices on phytoremediation of an aged PAH-contaminated soil. *Journal of Environmental Quality* 37 (4): 1439–1446.

54 Radwan, S. (2009). Phytoremediation for oily desert soils. In: *Advances in Applied Bioremediation* (eds. A. Singh, R.C. Kuhad and O.P. Ward), 279–298. Heidelberg: Springer Nature Berlin.

55 Thakore, Y. (2006). The biopesticide market for global agricultural use. *Industrial Biotechnology* 2: 194–208.

56 Etesami, H. and Maheshwari, D.K. (2018). Use of plant growth promoting rhizobacteria (PGPRs) with multiple plant growth promoting traits in stress agriculture: action mechanisms and future prospects. *Ecotoxicology and Environmental Safety* 156: 225–246.

57 Zhuang, X., Chen, J., Shim, H. et al. (2007). New advances in plant growth-promoting rhizobacteria for bioremediation. *Environment International* 33: 406–413.

58 Abou-Shanab, R.A., Ghanem, K., Ghanem, N. et al. (2007). The role of bacteria on heavy-metal extraction and uptake by plants growing on multi-metal-contaminated soils. *World Journal of Microbiology and Biotechnology* 24: 253–262.

59 Otieno, N., Lally, R.D., Kiwanuka, S. et al. (2015). Plant growth promotion induced by phosphate solubilizing endophytic *Pseudomonas* isolates. *Frontiers in Microbiology* 6: 745. https://doi.org/10.3389/fmicb.2015.00745.

60 Bano, A. and Fatima, M. (2009). Salt tolerance in *Zea mays* (L). Following inoculation with *Rhizobium* and *Pseudomonas*. *Biology and Fertility of Soils* 45 (4): 405–413.

61 Egamberdieva, D., Jabborova, D., and Hashem, A. (2015). *Pseudomonas* induces salinity tolerance in cotton (*Gossypium hirsutum*) and resistance to *Fusarium* root rot through the modulation of indole-3-acetic acid. *Saudi Journal of Biological Sciences* 22 (6): 773–779.

62 Glick, B.R., Patten, C.L., Holguin, G. et al. (1999). *Biochemical and Genetic Mechanisms Used by Plant Growth Promoting Bacteria*. London, UK: Imperial College Press.

63 Glick, B.R. (1995). The enhancement of plant growth by free-living bacteria. *Canadian Journal of Microbiology* 41 (2): 109–117.

64 Zhang, H., Kim, M.S., Sun, Y. et al. (2008). Soil bacteria confer plant salt tolerance by tissue-specific regulation of the sodium transporter HKT1. *Molecular Plant-Microbe Interactions* 21 (6): 737–744.

65 Barea, J.M., Pozo, M.J., Azcón, R. et al. (2005). Microbial co-operation in the rhizosphere. *Journal of Experimental Botany* 56: 1761–1778.

66 Ganesan, V. (2012). Rhizoremediation: a pragmatic approach for remediation of heavy metal-contaminated soil. In: *Toxicity of Heavy Metals to Legumes and Bioremediation* (ed. A. Zaidi). Wien: Springer-Verlag.

67 Rashid, M.H., Krehenbrink, M., and Akhtar, M.S. (2015). Nitrogen-fixing plant-microbe symbioses. In: *Sustainable Agriculture Reviews* (ed. E. Lichtfouse), 193–234. Cham: Springer International Publishing.

68 Karthik, C., Elangovan, N., Kumar, T.S. et al. (2017). Characterization of multifarious plant growth promoting traits of rhizobacterial strain AR6 under Chromium (VI) stress. *Microbiological Research* 204: 65–71.

69 Babalola, O.O. (2010). Beneficial bacteria of agricultural importance. *Biotechnology Letters* 32: 1559–1570.

70 Deng, Z. and Cao, L. (2017). Fungal endophytes and their interactions with plants in phytoremediation: a review. *Chemosphere* 168: 1100–1106.

71 Viollet, A., Pivato, B., Mougel, C. et al. (2017). *Pseudomonas fluorescens* C7R12 type III secretion system impacts mycorrhization of *Medicago truncatula* and associated microbial communities. *Mycorrhiza* 27 (1): 23–33.

72 Merkl, N., Schultze Kraft, R., and Infante, C. (2005). Phytoremediation in the tropics-influence of heavy crude oil on root morphological characteristics of graminoids. *Environmental Pollution* 138: 86–91.

73 Kohler, J., Caravaca, F., Carrasco, L. et al. (2007). Interactions between a plant growth-promoting rhizobacterium, an AM fungus and a phosphate-solubilizing fungus in the rhizosphere of *Lactuca sativa*. *Applied Soil Ecology* 35: 480–487.

74 Berruti, A., Lumini, E., Balestrini, R. et al. (2016). Arbuscular mycorrhizal fungi as natural biofertilizers: let's benefit from past successes. *Frontiers in Microbiology* 6: 1559. https://doi.org/10.3389/fmicb.2015.01559.

75 Hrynkiewicz, K. and Baum, C. (2012). The potential of rhizosphere microorganisms to promote the plant growth in disturbed soils. In: *Environmental Protection Strategies for Sustainable Development, Strategies for Sustainability* (eds. A. Malik and E. Grohmann), 35–64. Freiburg: Springer Science + Business Media B.V.

76 Smith, S.E. and Read, D.J. (1997). *Mycorrhizal Symbiosis*. Cambridge, UK: Academic Press.

77 Gao, Y., Cheng, Z., Ling, W. et al. (2010). Arbuscular mycorrhizal fungal hyphae contribute to the uptake of polycyclic aromatic hydrocarbons by plant roots. *Bioresource Technology* 101: 6895–6901.

78 Malachowska-Jutsz, A. and Kalka, J. (2010). Influence of mycorrhizal fungi on remediation of soil contaminated by petroleum hydrocarbons. *Fresenius Environmental Bulletin* 19: 3217–3223.

79 Chibuike, G.U. (2013). Use of mycorrhiza in soil remediation: a review. *Scientific Research and Essays* 8: 1679–1687.

80 Yadav, K.K., Gupta, N., Kumar, A. et al. (2018). Mechanistic understanding and holistic approach of phytoremediation: a review on application and future prospects. *Ecological Engineering* 120: 274–298.

81 Li, Q., Ling, W., Gao, Y. et al. (2006). Arbuscular mycorrhizal bioremediation and its mechanisms of organic pollutants-contaminated soils. *The Journal of Applied Ecology* 17 (11): 2217–2221.

82 Lenoir, I., Lounes-Hadj Sahraoui, A., and Fontaine, J. (2016). Arbuscular mycorrhizal fungal-assisted phytoremediation of soil contaminated with persistent organic pollutants: a review. *European Journal of Soil Science* 67 (5): 624–640.

83 Wright, S.F., Green, V.S., and Cavigelli, M.A. (2007). Glomalin in aggregate size classes from three different farming systems. *Soil and Tillage Research* 94: 546–549.

84 Bertaux, J., Schmid, M., Prevost-Boure, N.C. et al. (2003). In situ identification of intracellular bacteria related to *Paenibacillus* spp. in the mycelium of the ectomycorrhizal fungus *Laccaria bicolor* S238N. *Applied and Environmental Microbiology* 69 (7): 4243–4248.

85 Jankong, P., Visoottiviseth, P., and Khokiattiwong, S. (2007). Enhanced phytoremediation of arsenic contaminated land. *Chemosphere* 68 (10): 1906–1912.

86 Singh, R., Lal, S., and Dixit, V.K. (2015). Rhizoremediation approaches: a sustainable perspective for remediation of PAHs compound assisted with PGPR. *Journal of Chemical, Biological and Physical Sciences* 5 (3): 3067–3082.

87 Vergani, L., Mapelli, F., Zanardini, E. et al. (2017). Phyto-rhizoremediation of polychlorinated biphenyl contaminated soils: an outlook on plant-microbe beneficial interactions. *Science of the Total Environment* 575: 1395–1406.

88 Furukawa, K. (2003). Super bugs for bioremediation. *Trends in Biotechnology* 21: 187–190.

10

Bioremediation Potential of Rhizobacteria associated with Plants Under Abiotic Metal Stress

Shrvan Kumar[1], Saroj Belbase[1], Asha Sinha[2], Mukesh Kumar Singh[1], Brajesh Kumar Mishra[1], and Ravindra Kumar[3]

[1] *Rajiv Gandhi South Campus, IAS, Banaras Hindu University, Mirzapur, Uttar Pradesh, India*
[2] *Mycology and Plant Pathology, IAS, Banaras Hindu University, Varanasi, Uttar Pradesh, India*
[3] *Indian Agricultural Research Institute, Regional Station, Karnal, Haryana, India*

10.1 Introduction

Agricultural productivity, water quality, and climate change are greatly influenced by the exchange of nutrients, energy, and carbon between soil organic matter, soil environment, aquatic ecosystem, and atmosphere [1]. According to the FAO (Food and Agriculture Organization of the United Nations) report, 38.47% of the world's land area is covered by agricultural land, and although 28.43% of the land is arable, only 3.13% is permanently used for crop production. The situation is further deteriorated as 20–25% of land worldwide is being degraded every year and another 5–10 Mha will be degraded each year [2]. Nowadays extensive industrialization imparts detrimental effects on soil as well as on crop productivity by accumulating heavy metals [3]. Damage to soil texture, i.e., pH of soil, presence of different elements, and accumulation of heavy metals cause direct and/or indirect reduction of plant growth by adversely affecting various physiological, prophological, and molecular activities of plants [4, 5]. Heavy metals such as Zn, Cu, Mo, Mn, Co, and Ni are essential for crucial biological processes and developmental pathways [3, 6].

Nonetheless, these metals along with four other highly toxic heavy metals including, arsenic (As), mercury (Hg), lead (Pb), cadmium (Cd), chromium (Cr), zinc (Zn), uranium (U), selenium (Se), silver (Ag), gold (Au), and nickel (Ni), which are not useful to plants, are capable of reducing plant growth due to reduced photosynthetic activities, plant mineral nutrition, and reduced activity of essential

Soil Bioremediation: An Approach Towards Sustainable Technology, First Edition.
Edited by Javid A. Parray, Abeer Hashem Abd Elkhalek Mahmoud, and Riyaz Sayyed.

enzymes [7–10]. These toxic elements cause morphological abnormalities, and metabolic disorders that lead to yield reduction in plants [11].

Heavy metals are cytotoxic at low concentrations and could lead to cancer in humans [12]. This leads to base damage through formation of reactive oxygen species (ROS), which includes oxygen radicals [13]. Whereas nonradical derivatives of molecular oxygen (O_2) such as hydrogen peroxide (H_2O_2) which is responsible for breaking into the DNA molecule of the cell [14, 15]. Heavy metal toxicity increases the production of ROS thereby decreasing the antioxidant systems (glutathione, superoxide dismutase [SOD], etc.), which protect cells. If this condition continues, the normal functioning of the organism is affected and may invariably lead to cell death [3, 16–18].

To endure heavy metal stress and metal toxicity, plants have evolved numerous defense mechanisms viz., reduced heavy metal uptake, sequestration of metal into vacuoles, binding to phytochelatins/metallothioneins (MT), and activation of various antioxidants [3]. In combination with different functional genomic approaches, the transcriptomics, proteomics, and metabolomics approaches help to develop improved varieties with enhanced abiotic stress tolerance [19]. Several quantitative trait loci (QTLs) and candidate genes have been identified for Zn, Fe, and Cd tolerance in various plant species, which can be utilized for crop improvement by marker-assisted selection or QTL pyramiding [20–22]. Numerous plant-associated microbes namely, bacteria, and fungi are known to exhibit plant-growth promoting traits under heavy metal stress. These microbes impart favorable effects on plants via several direct and indirect mechanisms such as biofilm formation, siderophores, exopolysaccharide (EPS), and phytohormones production [23, 24].

Over the past decade, green, and sustainable remediation is gaining importance as a beneficial approach to optimize all phases of remediation. Bioremediation mediated by biological agents such as the species *Pseudomonas, Bacillus, Enterobacter, Klebsiella, Azobacter, Variovorax Azosprillum*, and *Serratia* [25] or plants is considered a cost-effective, green, and sustainable approach for restoring the contaminated sites. However, bioremediation has its limitations for field-scale application as an efficient remediation technology. The available remediation technologies including bioremediation for both organic and inorganic contaminants have been critically reviewed [26–29]. This chapter presents an overview of bioremediation technologies in the context of rhizobacteria sustainability, their applicability, and limitations for reclamation of contaminated sites. Also, bioremediation using advanced molecular techniques and genetically engineered organisms in combination with other technologies where bioremediation alone is not efficient and is discussed with some examples.

The rhizosphere is a supporting area for important and intensive interactions between the plant, soil, and soil microfauna. In fact, biochemical interactions and exchanges of signal molecules between plants and rhizobacteria take place in the rhizosphere [30, 31]. Such interactions can significantly influence plant growth and yields. Rhizobacteria are rhizosphere competent bacteria that aggressively colonize plant roots; they are able to multiply and colonize all the ecological niches found on the roots at all stages of plant growth, presence of such rhizobia in the rhizosphere can have beneficial, detrimental, or neutral effects on plants [32]. It harbors an extremely complex microbial community that includes saprophytes, endophytes, epiphytes, and pathogens as well as many useful microorganisms [33] like bacteria, fungi, nematodes, protozoa, and algae. Yadav et al. [34] reported that 1200×10^6 bacteria/g dry soil are present in the rhizosphere, which is very high as compared to fungi (12×10^5 fungi/g dry soil), algae (5×10^5 algae/g dry soil), and actinomycetes (46×10^6 actinomycetes/g dry soil). The rhizosphere effect is calculated in terms of the rhizosphere ratio, i.e., R : S, which is obtained by dividing the total number of microorganisms in the rhizosphere (R) by the corresponding number in the bulk soil (S) [35]. R : S is the measure of degree of microbial activity, the higher the R : S ratio the higher is the activity in rhizosphere. The R : S ratio higher is also higher in bacteria 23.0 as compared to fungi (12.0), algae (0.2), and actinomycetes (7.0) [34]. Rhizosphere bacteria (Rhizobacteria), which have the capacity to influence the root in a positive way are called plant growth promoting rhizobacteria (PGPR), these rhizobacteria exert a beneficial effect on plant growth [36–45]. PGPR are the microorganism present in the rhizosphere, which includes the bacterial species including *Alcaligenes, Azospirillum, Arthrobacter, Acinetobacter, Azoarcus*, *Bradyrhizobium, Bacillus, Burkholderia, Enterobacter, Erwinia, Flavobacterium, Pseudomonas, Pantoea, Paenibacillus,* Rhizobium *Azorhizobium, Bradyrhizobium, Allorhizobium, Sinorhizobium, Methylobacterium, Frankia,* and *Mesorhizobium* [36, 41–44, 46, 47].

Among all gram-negative soil bacteria *Pseudomonas* spp. is the most abundant genus in the rhizosphere, known for many years for its PGPR activities, resulting in a broad knowledge of the mechanisms involved [48, 49]. The name *Pseudomonas* (Pseudes^G "false" and monas^G "a single unit" or "false unit") comprises one of the most diverse and ecologically fit groups of bacteria on this planet. The most effective strains of *Pseudomonas* have been Fluorescent *Pseudomonas* they includes *Pseudomonas aerogenosa*, *Pseudomonas fluorescence*, *Pseudomonas putida*, and *Pseudomonas syringae* [50]. Kloepper and Schroth [51] introduced the term PGPR for these beneficial microbes in 1978. PGPR plays an important role in enhancing plant growth through a wide variety of mechanisms. The mode of action of PGPR that promotes plant growth includes (i) abiotic stress tolerance in plants; (ii) nutrient fixation for easy uptake by plant; (iii) plant growth regulators; (iv) the production of siderophores; (v) the production of volatile organic compounds; and (vi) the production of protection enzymes such as chitinase, glucanase, and

ACC-deaminase for the prevention of plant diseases [52, 53]. However, the mode of action of different PGPR varies depending on the type of host plants [54].

10.2 Mode of Action Rhizobacteria

10.2.1 Role of PGPR in Plant Growth Under Abiotic Stress

Most rhizospheric bacteria naturally tolerate environmental contaminants; hence, they are used in phytoremediation to remove organic pollutants and heavy metal contaminants in food crops. Rhizospheric bacteria make nutrients available in soils for plant growth, produce phytohormones such as Indole three-acetic acid, and also protect the plant against pathogens as well as remediating the soil of contaminants [55–59]. These PGPR live in the surrounding of the rhizosphere of the host plant where they boost plant growth and development by direct or indirect mechanisms [60]. Direct mechanisms include siderophore production, phosphate solubilization, and 1-aminocyclopropane-1-carboxylate deaminase synthesis, which enable the plant to withstand abiotic stress conditions by reducing ethylene levels, and enhances plant growth hormone production [56, 60, 61]; while the indirect mechanism of growth promotion involves the PGPR acting as biocontrol agents (BCAs) and detoxifying noxious substances such as heavy metals and pesticides [25, 62]. The PGPRs associated with the rhizosphere could be extracellular plant growth promoting rhizobacteria (ePGPR) or intracelluar plant growth promoting rhizobacteria (iPGPR) depending on the level of interaction with the host plant root cells [63, 64]. Most bacterial genera are ePGPR and include *Erwinia, Flavobacterium, Micrococcus, Pseudomonas, Serratia, Chromobacterium, Caulobacter, Azospirillium, Azotobacter,* and *Agrobacterium* [60]. These PGPR produce various substances that enhance plant growth under abiotic stress.

10.2.2 Mechanism Action of PGPR Species

- *Pseudomonas* sp. is particularly suited as BCAs because they can use diverse root exudates as nutrient source.
- They are abundantly present in natural soils, especially in the rhizosphere.
- They directly promote plant growth and have the ability to control diseases by a variety of mechanisms.
- Besides plant growth promotion and disease resistance these strains are also involved in bioremediation, pesticide degradation, and stress management of plants [65].
- Sequestration of toxic metals by cell wall components or by intracellular metal binding proteins and peptides such as MT and phytochelatins along with

compounds such as bacterial siderophores, which are mostly catecholates, compared to fungi that produce hydroxamate siderophores.
- Alteration of biochemical pathways to block metal uptake.
- Conversion of metals to innocuous forms by enzymes.
- Reduction of intracellular concentration of metals using precise efflux systems [66].

Bacteria produce iron-chelating substances called siderophores, which enhances mobility and reduces bioavailability of metals and its subsequent removal from soil. Sulphate-reducing bacteria such as *Desulfovibrio desulfuricans* have the ability to convert sulphate to hydrogen sulphate, which then reacts with heavy metals such as Cd and Zn to form insoluble forms of these metal sulphides [13]. Mustapha and Halimoon [67] obtained biosorption efficiency of 15.3–84.6% using algae, which is high compared to other microbial biosorbents. Microbes make use of heavy metals and trace elements as terminal electron acceptors from which they acquire the needed energy to detoxify metals via enzymatic and nonenzymatic processes [12]. The most essential constituent in such bacterial cells having ion sequestration capability is EPS. EPS protects the bacteria against environmental stresses such as heavy metal toxicity, drought, and salinity. Microorganisms such as *Agrobacterium* spp., *Alcaligenes faecalis, Xanthomonas campestris, Bacillus* spp., *Zygomonas mobilis, Leuconostoc, Pseudomonas* spp. *and Acetobacter xylinum,* have been identified as genera of EPS-producing microorganisms [61]. Some commercial bacterial EPS with the required anionicity include: alginate (*Pseudomonas aeruginosa, Azotobacter vinelandii*), gellan (*Sphingomonas paucimobilis*), hyaluronan (*P. aeruginosa, Pasteurella multocida, Streptococci* attenuated strains), xanthan (*X. campestris*), galactopol (*Pseudomonas oleovorans*), and fucopol (*Enterobacter* A47) [68]. EPS production is associated with processes such as biofilm production, which is essential in the biosorption and biomineralization of metal ions [69]. EPS can be modified chemically by acetylation, carboxymethylation, methylation, phosphorylation, and sulphonylation, which modify the biological activities of EPS, thereby enhancing the applicability of the polymer [68]. More EPS-producing microorganisms need to be investigated for metal-ion-chelation abilities as they might come up with potent polysaccharide rich in anionic groups, which will enhance the clean-up of the environment from toxic metals [70].

10.3 Bioremediation

Biological remediation is considered as the most effective method of toxic metal removal as these are natural, environmentally friendly, low-cost, and high societal acceptance technologies [112]. One such technology is the use of plant growth promoting microbes for bioremediation of heavy metal polluted soil and is quite important in the context of global climate change and excessive fertilizer use in agricultural soils [23, 113]. This is feasible because microorganisms have developed many resistance mechanisms for survival in the presence of toxic heavy metals in their environment [67, 114]. Microbes also enhance bioavailability of metals from soil by chelation, acidification, and precipitation as, for example, organic acids released by microbes and plant roots lower the soil pH and helps in sequestration of metal ions [115]. The role of phytohormones ABA, SA, JA, and ethylene have also been elucidated in PGPR inoculated plants under stressed conditions [23, 24]. The induction of these genes, which are central to heavy metal stress signaling, in the presence of plant-associated microbes as well indicate the complex cross-talk between plant, microbes, and heavy metals in stress response and tolerance. Therefore, an understanding of the intricate metal stress signaling pathways and the existing cross-talk among the networks of plant–microbe–metal interaction is extremely important to elucidate the stress-responsive networks in plants.

The effectiveness of bioremediation depends on several factors such as the nature of the organisms utilized, the prevailing environmental factors at the contaminated site, as well as the degree of the pollutants in that environment [116]. Bioremediation can be achieved with the use of microorganisms (microbial bioremediation), which depends on the metabolic potential of the microorganisms to degrade environmental pollutants and change them to innocuous forms through redox processes [66]. Utilization of natural microorganisms in the environment for *in-situ* bioremediation is affected by the nonavailability of suitable nutrient levels and/or environmental setting at the polluted location [117, 118]. *Ex-situ* bioremediation involves taking the contaminated media from its original site to a different location for treatment based on the cost of treatment, deepness of contamination, pollutant type, and the extent of pollution, geographical locality, and geology of the contaminated site [116]. The bioremediation of heavy metal by some rhizobacterial species associated with plants is outlined in Table 10.1.

10.4 Genetically Modified Plant-Associated Microbes for Heavy Metal Stress Tolerance

Genetically engineered microorganisms (GEMs) contain foreign genes inserted into their genome from another organism of the same or different species using recombinant DNA technology. These engineered microbes have been used to obtain competent strains for bioremediation of contaminated environment by

Table 10.1 Bioremediation of heavy metal by rhizobacterial strains associated with plants.

Microorganisms	Heavy Metals	Plants	Reference
Bacillus cereus, Pseudomonas moraviensis	Cu, Cr, Co, Cd, Ni, Mn, Pb	*Triticum aestivum*	[4]
Microbacterium sp. CE3R2, *Curtobacterium* sp. NM1R1	Zn, Pb, Cu, As	*Brassica nigra*	[71]
Bacteroidetes bacterium, Pseudomonas fluorescens	Cd, Cu, Pb, Zn	*Brassica napus*	[72]
Kocuria sp. CRB15	Cu	*Brassica nigra*	[73]
Klebsiella pneumoniae	Cd	*Oryza sativa*	[74]
Enterobacter ludwigii, Klebsiella pneumoniae	Hg	*Triticum aestivum*	[75]
Azospirillum	Pb, Cd	*Panicum virgatum*	[76]
Enterobacter, Leifsonia, Klebsiella, Bacillus	Cd	*Zea mays*	[77]
Pseudomonas putida	Cd	*Eruca sativa*	[78]
Rhodococcus erythropolis, Achromobacter sp., *Microbacterium* sp.	Zn, Cd	*Trifolium repens*	[79]
Variovorax paradoxus, Rhodococcus sp., *Flavobacterium* sp.	Cd	*Brassica juncea*	[80]
Bacillus pumilus E2S2, *Bacillus* sp. E1S2	Cd, Zn	*Sedum plumbizincicola*	[81]
Rhizobium leguminosarum	Zn	*Brassica juncea*	[82]
Glomus versiforme	Cd	*Solanum nigrum*	[83]
Rhizophagus clarus	Pb	*Cymbopogon citratus*	[84]
Pseudomonas sp. LK9	Cd, Zn, Cu	*Solanum nigrum*	[85]
Bacillus licheniformis	Ni	*Oryza sativa*	[86]
Rahnella sp. JN6	Cd, Pb, Zn	*Brassica napus*	[87]
Bacillus thuringiensis GDB-1	As, Cu	*Alnus firma*	[88]
Ralstonia eutropha, Chryseobacterium humi	Zn, Cd	*Helianthus annuus*	[89]
Staphylococcus arlettae	As	*Brassica juncea*	[90]
Ochrobactrum sp., *Bacillus* sp.	Cd, Pb, As	*Oryza sativa*	[91]
Paenibacillus macerans, Bacillus endophyticus, Bacillus pumilus	Cu, Ni, Zn	*Brassica juncea*	[92]
Bacillus sp. MN3–4	Pb	*Alnus firma*	[93]

(*Continued*)

Table 10.1 (Continued)

Microorganisms	Heavy Metals	Plants	Reference
Psychrobacter sp. SRS8	Ni	*Helianthus annuus, Ricinus communis*	[94]
Bacillus sp. SLS18	Cd	*Solanum nigrum*	[95]
Glomus mosseae	Cd, Pb	*Cajanus cajan*	[96]
Bacillus cereus, Candida parapsilosis	Fe, Mn, Zn, Cd	*Trifolium repens*	[97]
Paecilomyces lilacinus NH1	Cd	*Solanum nigrum*	[98]
Bradyrhizobium sp. 750, *Pseudomonas* sp., *Ochrobactrum cytisi*	Cu, Cd, Pb	*Lupinus luteus*	[99]
Enterobacter sp. B16, *Stenotrophomonas* sp. B19	Se	*Triticum aestivum* L. cv. Fritz, Inoculation Root	[100]
Bacillus sp. R8, *Enterobacter* sp. B16, *Pseudomonas* sp. R12, *Stenotrophomonas* sp. B19	Se	*Triticum aestivum* L. cv. Puelche, Inoculation Root	[101]
Bacillus sp. YAM2	Se	*Triticum aestivum* L. Seed palletization	[102]
Acinetobacter sp. E6. 1, *Bacillus* sp. R8, *Klebsiella* sp. E2	Se	*Triticum aestivum* L. cv., Seed palletization	[103]
Azotobacter chroococcum	Pb, Zn	*Brassica juncea*	[104]
Bacillus subtilis	Ni	*Brassica juncea*	[105]
Kluyvera ascorbata	Ni, Zn, Pb	*Indian mustard*	[106]
Mesorhizobium huakuii	Cd	*Astragalus sinicus*	Sriprang et al. [107]
Bacillus subtilis	Cr	*Zea mays* And *Sorghum bicolor*	[108]
Bacillus pumilus	Zn, Pb, Cr, Cu	*Zea mays*	[108]
Brevibacterium halotolerans	Cu, Zn, Pb, Cr	*Zea mays* and *S. bicolor*	[108]
Microbacterium arabinogalactanolyticum	Ni	*Alyssum murale*	[109, 110]
Nitrobacteria irancium	Cr, Zn	*Water hyacinth*	[111]

possessing enhanced ability to breakdown a variety of contaminants [12]. Previous studies reported Cd resistant *Ochrobactrum* sp. and Pb and As resistant *Bacillus* spp. have several PGPR traits that help in bioremediation and growth promotion of a rice cultivar [91]. Recently elimination of toxic metals through a novel phytobacterial strategy, i.e., via the use of genetically transformed PGPR has been suggested [119]. Genetically transformed bacteria possess one or more genes to increase remediation of heavy metals. In this context, genes for metal chelators, metal homeostasis, transporters, biodegradative enzymes, metal uptake regulators, and biotic and abiotic stress tolerance are important candidates for making recombinant bacteria [120]. Reports have been made on the use of genetically engineered *Eschericia coli* strain M109 and *P. putida* containing the merA gene to effectively eradicate Hg from contaminated soils and sediments [121–123]. Genes that have been widely used include the merA gene for Hg uptake, phenol catabolic genes (pheA, pheB, pheC, pheD, and pheR) [124] and the ArsM gene for the removal of As from contaminated soils [125]. The *Cupriavidus metallidurans* strain MSR33 genetically engineered with a pTP6 plasmid that provided genes (organomercurial lyase protein [merB] and mercuric reductase [merG]) regulating Hg biodegradation along with the synthesis of merB and merA also gave the microorganism the ability to reduce Hg contamination from polluted sites [126, 127]. *Pseudomonas* strains have also been made resistant to Hg due to the introduction of novel genes into the strain using pMR68 plasmid [128]. The n-alkane-degrading microorganisms possess specific genes such as alkB, alkB1, alkB2, alkM, aromatic hydrocarbons: xylE, and polycyclic aromatic hydrocarbons (PAHs): ndoB, nidA, most often located on plasmids to enable horizontal gene transfer and used as markers for the identification of microbial biodegradation [129]. Advancements of modern techniques in genetics and omics such as genomics, proteomics, and metabolomics have enabled scientists to study catabolism of organic pollutants by microorganisms, which has made it possible to understand the physiology, ecology, and biochemistry of PAH-degrading microorganisms [130]. DNA microarrays, a high-throughput technique that identifies several genes in a single test, has overcome the shortcoming of other culture-independent approaches. The GeoChip array is the most widespread gene array technique used for studying the function of genes. It targets numerous genes that participate in various geochemical cycles of carbon, nitrogen, phosphorus, and sulfur, as well as metal resistance and reduction and degradation of pollutants [131]. Recombinant DNA technology is essential for the bioremediation process as it enables researchers to analyze, monitor, and assess the implementation of the process [132]. However, it should be used with caution and in accordance with biosafety regulations. The list of genetically modified plant-associated microbes for heavy metal stress tolerance is documented in Table 10.2.

Table 10.2 List of genetically modified plant-associated microbes for heavy metal stress tolerance.

Genetically engineered microbe	Modified gene expression	Associated plant	Heavy metal(s)	References
Pseudomonas putida	Phytochelatin synthase	*Triticum aestivum*	Cd, Hg, Ag	[133]
Mesorhizobium huakuii	Metallothionein, phytochelatin synthase	*Astragalus sinicus*	Cd, Cu, Zn, As	[134–136] [107]
Pseudomonas putida	Expression of metal binding peptide	*Helianthus annuus*	Cd	[137]
Enterobacter cloacae	EC 4.1.99.4	*Brassica napus*	As	[138]
Ralstonia eutropha	Metallothionein	*Nicotiana benthamiana*	Cd	[139]

(*Source:* Adapted from [119].)

10.5 Highly Toxic Metals

The nonbiodegradability of heavy metals makes it hard to remove them from contaminated biological tissues and this is a major concern for global health because of their lethal nature [140]. Heavy metals such as cobalt (Co), copper (Cu), iron (Fe), manganese (Mn), and molybdenum (Mo) are required in small quantities for the survival of living organisms [141] but, at higher concentrations they could become detrimental. The heavy metals Hg, Cr, As, Zn, Cd, Pb, Ur, Se, Ag, Au, and Ni are hazardous heavy metals that contaminate the environment and adversely affect the quality of the soil, crop production as well as public health [68, 140, 142–145], if their concentration exceeds the maximum permissible concentration in water given by the US Comprehensive Environmental Response Compensation and Liability Act: Ar (0.01/mgl), Cd (0.05/mgl), Cr (0.01/mgl), Pb (0.015/mgl), Hg (0.002/mgl), and Ag (0.05/mgl) respectively [141, 146]. These pollutants are major sources of life-threatening degenerative diseases affecting humans such as cancer, Alzheimer's disease, atherosclerosis, and Parkinson's disease [147]. Among organisms greatly affected by heavy metal toxicity are plants as their normal physiological activities are severely hampered. For example, the processes of respiration, photosynthesis, electron transport chain, and cell division are negatively affected by elevated levels of heavy metals as documented by laboratory experiments [18, 148]. Moreover, high metal toxicity inhibits cytoplasmic enzymes in plant cells and causes damage to cell structures due to oxidative stress [13, 149], which consequently affects plant growth and metabolism. It accumulates in the cells of microbes in aquatic bodies where it gets converted to methyl mercury in the microbes and becomes detrimental to aquatic lives. Consumption of fish and other aquatic animals by man can cause the transfer of toxic methyl mercury to man. Due to the detrimental effects of these heavy metals, concerted efforts need to be made to effectively eradicate them from the environment and stabilize the ecosystem.

10.6 Effects of Highly Toxic Metals on Plant Growth

The heavy metals Hg, Cr, As, Zn, Cd, Pb, Ur, Se, Ag, Au, and Ni are hazardous. Among them, As, Pb, Cd, and Hg are considered as the most toxic metals by the Agency for Toxic Substances and Disease Registry [150], based on their toxicity, frequency of occurrence, and most importantly, their exposure potential to flora and fauna.

10.6.1 Arsenic ($^{33}As_{74.922}$)

Arsenic is typically originated via volcanic action, erosion of rocks, and by human activities such as applications of pesticides and wood preservatives, mining, and smelting operations [151–153]. The contamination of As in groundwater used for irrigation and drinking is a worldwide problem as it not only affects crop productivity, but also accumulates in different plant tissues including grains and contaminates the food chain [154]. The physiological and molecular mechanisms of As are toxicity, accumulation, detoxification, and tolerance in various plants including rice, lettuce, spinach, and carrot [155]. It interrupts biological functions in plants via a different manner as, for example, it binds to proteins with sulfhydryl groups, interfering with their functions [154]. It also generates ROS, inhibiting respiration by binding to vicinal thiols in pyruvate dehydrogenase and 2-oxo-glutarate dehydrogenase, and acts indirectly as a mutagen by inducing intrachromosomal homologous recombination [156]. In plants, As (V) interferes with oxidative phosphorylation and ATP synthesis during energy metabolism [154, 157].

10.6.2 Lead ($^{82}Pb_{207.2}$)

Lead contamination of the environment has resulted from mining and smelting activities, Pb containing paints, gasoline, and explosives as well as from the disposal of Pb-rich municipal sewage sludges [158]. Despite regulatory measures adopted in many countries to limit Pb input in the environment, it continues to be one of the most serious global environmental and human hazards. As many of the Pb pollutants are indispensable for modern human life, soil contamination with Pb is not likely to decrease in the near future [159]. The content of Pb in various plant organs tends to decrease in the following order, roots>leaves>stem>inflore scence>seeds [160]. Lead changes cell membrane permeability by reacting with active groups of different metabolic enzymes, with the phosphate groups of ADP or ATP, and by replacing essential ions, thus causing phytotoxicity [18]. Pb toxicity leads to inhibition of ATP production, induces lipid peroxidation, and DNA damage by over production of ROS [161].

10.6.3 Cadmium($^{48}Cd_{112.41}$)

Cadmium is highly soluble in water, it is easily taken up by plants representing the main entry pathway into the food chain causing serious human health hazards [162]. Cd has been classified as a potent human carcinogen by The International Agency for Research on Cancer [163–165]. It is commonly released into the arable soil from industrial processes and farming practices [166]. Even at low concentrations Cd can severely alter several enzyme activities including those involved in the Calvin cycle, carbohydrate and phosphorus metabolism, and CO_2 fixation [164, 167–169] ultimately resulting in stunted growth, chlorosis, leaf epinasty, alterations in chloroplast ultrastructure, inhibition of photosynthesis and pollen germination and tube growth, induction of lipid peroxidation, and alterations in nitrogen (N) and sulfur (S) metabolism and disruption of antioxidant machinery [164].

10.6.4 Chromium ($^{24}Cr_{51.9961}$)

Chromium is the seventh most abundant metal in the earth's crust [170] and an important environmental contaminant into the atmosphere due to its huge industrial use [171]. In nature it exists in two different stable oxidation states; trivalent (Cr^{III}) and hexavalent (Cr^{IV}). Cr^{IV} is more soluble in water than Cr^{III}. Cr phytotoxicity can result in inhibition of seed germination, degrade pigment status, nutrient balance, antioxidant enzyme, and induce oxidative stress in plants [172–176]. Besides these effects, Cr can alter chloroplast and membrane ultrastructure in plants [177, 178]. It can affect growth, water balance, pigment content, and initiate lipid peroxidation causing oxidative damage to plants [172, 174, 176, 179], reduces the number of palisade and spongy parenchyma cells in leaves, and increases the number of vacuoles and electron dense material along the walls of xylem and phloem [180]. However, in many plants an increase in DNA content has been observed under Cr and the amount of the DNA increased with the increase in concentration of Cr [181–183]. During seed germination, hydrolysis of protein and starch takes place, providing amino acids and sugars. Under Cr treatment, a decrease in both alpha and beta-amylase has been reported, which is one of the important factors for germination inhibition in many plants in view of the impaired supply of sugar to developing embryo axes [183]. Lipid peroxidation, which is considered an indication of oxidative stress in plants, can be induced via free radicals or ROS that are generated as a result of heavy metal toxicity in plants. Lipid peroxidation can degrade biological membranes making them susceptible to oxidative damage [184]. Under Cr, lipid peroxidation can be initiated resulting in oxidative stress. ROS that are common consequences of most biotic and abiotic stresses are also formed as a result Cr toxicity [175, 178, 185].

10.6.5 Cobalt ($^{27}As_{58.933}$)

Toxic concentrations inhibit active ion transport. In higher plants, absorption of Co^{2+} by roots involves active transport. Transport through the cortical cells is operated by both passive diffusion and active process. In the xylem, the metal is mainly transported by the transpirational flow. Distribution through the sieve tubes is acropetal by complexing with organic compounds. The lower mobility of Co^{2+} in plants restricts its transport to leaves from stems. Increased amounts of exogenously added cobalt caused morphological damage in proplastids of interphasic meristematic cells in root tips of *Vicia faba* cv. *zborovicky*. It also altered the division of proplastids, changed physiological activity, and inhibited starch grain differentiation [186]. Alterations are induced in the number and structure of plastids and their chlorophyll contents with increasing concentration of cobalt. 0.01 and 0.005 mg/l of cobalt increased the cell size, height of palisade and spongy parenchyma and longitudinal diameter of cell, and number and size of chloroplasts in parenchyma cells of the cucumber [187] and barley [188]. But the number of chloroplasts per cell and leaf unit increased with a reduction in Co concentration [189]. Cobalt may affect growth processes by increasing the active form of auxin [190], cobalt was shown to exert a positive influence on Hill reaction, with a simultaneous decrease in the quantity of chlorophyll and an increase in the number of chloroplasts per unit leaf surface. Increase in activity may be related to the redistribution of pigments (increase in quantity of chlorophyll b and decrease of chlorophyll a) or its direct participation in one of the light or dark stages of Hill reaction [191]. Co^{2+} destroys chloroplast function by primarily damaging the chloroplast membrane structure and integrity and perhaps the reaction center complex of PSII [192]. At the ultrastructural level, cobalt caused an enlargement of the cell vacuole, accumulation of lipids, and overload of plastids with starch. The nucleus became amoebiform accompanied by stickiness of the membrane. Vacuoles contained electron-dense bodies in both algae and higher plants [193–195]. A Co^{2+} range 0.001–0.2% caused despiralization of prophase chromosomes in meristematic cells of *V. faba* [194].

10.6.6 Copper ($^{29}Cu_{63.546}$)

Copper is a naturally occurring element in the soil and it can be found as a metal or in a variety of ores. It is a constituent of many manmade alloys and is used in wire and some coins. At concentrations above those required for optimal growth Cu was shown to inhibit growth and to interfere with important cellular processes such as photosynthesis and respiration [196, 197]. Plants grown in the presence of high levels of Cu normally show reduced biomass and chlorotic symptoms. A lower content of chlorophyll and altercations of chloroplast structure and thylakoid membrane composition was found in the leaves under such conditions [198–201]. In particular, degradation of grana stacking and stroma lamellae, an increase in the number and size of plastoglobuli, and the appearance of intrathylakoidal inclusions were observed. It was proposed that Cu interferes with the biosynthesis of the photosynthetic machinery modifying the pigment and protein composition of photosynthetic membranes [199, 202]. Furthermore, lipid peroxidations, decrease of lipid content, and changes in fatty acid composition of thylakoid membranes was observed [202–204]. It is well known that transition metals like Cu catalyze the formation of hydroxyl radicals (OH^-) from the nonenzymatic chemical reaction between superoxide (O_2^-) and H_2O_2 [205]. Hence, the presence of excess Cu can cause oxidative stress in plants and subsequently increase the antioxidant responses due to increased production of highly toxic oxygen free radicals.

10.6.7 Gold ($^{79}Au_{196.97}$)

Gold may exist in soil from natural sources, through the anthropogenic placement of mine tailings in the ecosystem, and more recently from the escalating use of nanoparticles in commercial products [206]. Although gold inhibits plant growth, the exact mechanisms for this inhibition are not well understood. Gold has been reported to covalently bind to cysteine residues and sulfhydryl groups of aquaporins [207, 208]. The inhibition of aquaporins could reduce the water permeability of root cells, decreasing the ability of the plant to take up water. Gold inhibition of proteins via the disruption of disulfide bonds has also been reported and it is possible that, in addition to disrupting protein structure, gold may displace other essential metals from within proteins inhibiting protein function and leading to metal deficiency symptoms [209, 210].

10.6.8 Iron ($^{26}Fe_{55.845}$)

Iron (Fe) toxicity leads to increased polyphenol oxidase activity, leading to the production of oxidized polyphenols. It also causes leaf bronzing and reduced root oxidation power [211]. Oxidative stress due to excessive production of ROS viz., superoxide radical ($O_2^{\cdot -}$), hydrogen peroxide (H_2O_2) and hydroxyl radical ($OH^{\cdot}$) is a common consequence of most abiotic and biotic stresses [212]. The toxicity of ROS depends on the presence of a Fenton catalyst such as iron or copper ions, giving rise to an extremely reactive $OH^{\cdot}$-radical in the presence of H_2O_2 and $O_2^{\cdot -}$ [205, 213]. These reactive toxic oxygen species causes damage to DNA, proteins, lipids, chlorophyll, and almost every other organic constituent of the living cell [214–220]. Plants protect cell and subcellular systems from the cytotoxic effects of these ROS with antioxidant enzymes such as superoxide dismutase (SOD), ascorbate peroxidase (APOX), glutathione reductase (GR), mono-dehydro ascorbate reductase (MDAR), dehydro ascorbate reductase (DAR), catalase (CAT), and metabolites such as glutathione, ascorbic acid, α-tocopherol and carotenoids [219, 221–225].

10.6.9 Manganese ($^{25}Mn_{54.938}$)

Manganese is ubiquitous in the environment. It comprises about 0.1% of the Earth's crust [226, 227]. Manganese does not occur naturally as a base metal but is a component of more than 100 minerals, including various sulfides, oxides, carbonates, silicates, phosphates, and borates [227]. The most commonly occurring manganese bearing minerals include pyrolusite (manganese dioxide), rhodocrosite (manganese carbonate [$MnCO_3$]), rhodonite (manganese silicate), and hausmannite (manganese tetroxide [Mn_3O_4]) [227–230]. Crustal rock is a major source of manganese found in the atmosphere. Ocean spray, forest fires, vegetation, and volcanic activity are other major natural sources of manganese in the atmosphere [231]. Stokes et al. [232] estimated that two-thirds of manganese air emissions were from natural sources. Atmospheric particulate matter collected in the Antarctic indicated that manganese was derived from either crustal weathering or the ocean [233].

The toxicity limit of Mn, as well as tolerance to excesses of this metal, is dependent on the plant species, in addition to varieties or genotypes [234, 235]. However, its toxicity is favored in acidic soils [236]. Excessive concentrations of Mn in plant tissues may alter various processes, such as enzymatic activity, uptake, redistribution, and the use of other nutrients (Ca, Fe, Mg, N, and P), in addition to the productive responses of agricultural crops [237]. The literature has reported that toxicity by metals, e.g., Mn, induces oxidative stress [238, 239]. However, the antioxidative system response to this stress presents variable and controversial results [16, 240, 241]. There are several factors responsible for this lack of consensus: heavy metals that exert different stress induction mechanisms; compartmentalization of antioxidants, leading to differences between responses among the organelles, cells, and tissues; and the detoxification and complexation of heavy metals reducing its stimulating effects on oxidative stress [240]. The toxicity caused by Mn results in the degradation of lipids, proteins, carbohydrates, and nucleic acids, damaging the cell metabolism and causing cell death in some cases [242]. Plants signal Mn toxicity by means of receptors that trigger molecular signal cascades that transmit the signals to the regulating systems via ion channels, signaling proteins, and second messenger compounds [243–245].

10.6.10 Molybdenum ($^{42}Mo_{95.94}$)

Molybdenum is present as many different complexes depending on the chemical speciation of the soil zone. Mineral forms of molybdenum found in rocks include molybdenite (MoS_2), wulfenite ($PbMoO_4$), and ferrimolybdenite [$Fe_2(MoO_4)$] [246]. Release of molybdenum from solid mineral forms is through weathering, a process involving continual solution and oxidation reactions [247, 248]. The total molybdenum content of soils usually is about 1–3 ppm. Soils containing 75 ppm of molybdenum have been found in Hawaii. Molybdenum differs from most of the essential trace elements in that its availability is highest at soil pH values that approach neutrality. Most of the other trace elements are most available in acid soils.

Molybdenum (Mo), the last of the required micronutrients, is needed in the smallest quantities by plants. The normal range for most plant tissue is between 0.3 and 1.5 ppm and in the growing medium between 0.01 and 0.20 ppm Molybdenum is an essential component in two enzymes that convert nitrate into nitrite (a toxic form of nitrogen) and then into ammonia before it is used to synthesize amino acids within the plant. It also needed by symbiotic nitrogen fixing bacteria in legumes to fix atmospheric nitrogen. Plants also use molybdenum to convert inorganic phosphorus into organic forms in the plant [249].

Severe Mo toxicity occurs as marginal leaf scorch and abscission in the plant, and yellowing or browning of leaves and depressed tillering. Excess levels in plants are more of a concern to animal life, especially ruminants. Over-consumption of natural forages containing 20–100 ppm Mo can lead to a condition called Molybdenosis [250]. This toxicity causes a condition called "teart" in which cattle develop scours. Scours can be mild to debilitating, resulting in permanent injury to the digestive tract and sometimes death. This condition can be corrected by adding a copper-sulfate supplement to the cattle diet [251].

10.6.11 Mercury ($^{80}Hg_{200.59}$)

The large input of Hg onto arable lands has resulted in the widespread occurrence of Hg-contamination in the entire food chain. In the environment several forms of Hg exist such as elemental (Hg^0), inorganic (Hg^{2+}), associated with ions (HgS, $ClHg_2$, Hg_2Cl_2), and organic (CH_3–Hg) but in agricultural soils the ionic form is predominant (Hg^{2+}) [252, 253]. Increasing evidence has shown that Hg^{2+} can readily accumulate in higher plants [254, 255]. Binding of Hg^{2+} to water channel proteins leads to leaf stomata closure and physical impediment of water flow in plants [253, 256]. Additionally, it has also been reported to interfere with mitochondrial activity [253]. Mercuric ions are further reported to induce oxidative stress by stimulating generation of ROS in plants leading to disruption of biomembrane lipids and cellular metabolism, as well as increased activities of antioxidant enzymes like SOD, POD, or Ascorbate peroxides (APX) indicating the degree of stress [257, 258].

10.6.12 Nickel ($^{28}Ni_{58.693}$)

Nickel (Ni) and nickel compounds have many industrial and commercial uses and the progress of industrialization has led to an increased emission of pollutants into the environment. Ni is an essential micronutrient for plant growth and it is also a component of the enzyme urease, which is required for nitrogen metabolism in higher plants. Ni is absorbed and redistributed in plants via cation and/or metal–ligand complex transport system. Ni concentration in the majority of plant species is very low (0.05–10 mg/kg dry weight). Toxic effects of high concentrations of Ni in plants have been frequently reported; for example, inhibition of mitotic activity of pigeon pea (*Cajanus cajan* L.), reductions in plant growth of cabbage (*Brassica oleraceae*), and adverse effects on fruit yield and quality of wheat (*Triticum aestivum* L.). Extremely high in-soil Ni concentrations have left some farmland unsuitable for growing crops, fruits, and vegetables [259]. The toxicity of Ni^{2+} reduced the area of transpiring surface (leaf blades) of plants [260]. The 40% reduction in leaf area of the *Cajanu scajan* plant was observed under the Ni stress of 1 mM in nutrient solution. When 4-day old plants in sand culture were treated with 10 mM Ni added in nutrient solution transpiration rate, stomatal, conductance, leaf water potential, and total moisture content decreased [261]. A high uptake of Ni induces a decline in water content of dicot and monocot plant species. The decrease in water can act as an indicator for Ni toxicity in plants. Ni is associated with protein inhibition germination and chlorophyll synthesis. In several plants, Ni induces changes in activity of antioxidant enzymes like SOD, APX, and catalayse (CAT) at higher concentration. The most common symptoms of nickel toxicity in plants are the inhibition of growth, induction of chlorosis, necrosis, and wilting. Nickel strongly influences metabolic reaction in plants and has the ability to generate ROSs, which may cause oxidative stress [262].

10.6.13 Selenium ($^{34}Se_{78.96}$)

There are some factors that help identify the susceptibility of a certain plant to toxicity, such as Se concentrations, levels of sulphate in the soil, the stage of growth, and the chemical form of the concentrated Se. It has been observed that both selenite (SeO_3^{2-}) and selenate (SeO_4^{2-}) are major forms of toxic to nonaccumulators because they are readily absorbed and assimilated by plants [263]. It has been reported that toxic Se concentrations in nonaccumulator plants result in a 10% decrease of yield, without any visible symptoms (Se contents between of 2 and 330 mg/kg in rice and white clover resp.). In accumulator plants, Se concentration may reach 4000 mg/kg, without any negative effects [264].

Toxic levels of Se in plant tissues are generally above 5 mg/kg [265], but among agricultural crops, the phytotoxicity of Se is variable [266]. Various toxic symptoms shown by plants include stunting of growth, chlorosis, withering, and drying of leaves, premature death of plant as well as decreased protein synthesis [267]. In nontolerant plant species, a surplus of Se may destroy germination and growth and cause chlorosis and black spots on leaves [268]. Tolerance mechanisms include processes of exclusion of active Se amino acids, thus preventing their incorporation into proteins and the damaging effects on plant functions [269].

The deprivation of Se from proteins in accumulator plants is the basis for their tolerance to Se. Generally, food crops have a low Se tolerance; however, most other plants may accumulate amounts of Se that are toxic for humans and animals.

Selenium toxicity might be due to the substitution of Se for sulfur in proteins or may be due to inhibition of methylation [270], owing to its similarity with sulfur. This similarity of Se-amino acids to their S analogue of cysteine and methionine can disrupt the normal biological reactions and enzyme functions within the cell. Se accumulator and nonaccumulator plant species differ in their Se absorption to show the symptoms of toxicity. Also, the distribution of Se in various parts of the plant differs according to species, stage of development, and its physiological condition [271]. In the case of Se accumulators, Se is accumulated in young leaves during the early vegetative stage of growth, but during the reproductive stage, high levels of Se are found in seeds while the Se content in leaves is drastically reduced [272]. Shrift [273] observed that the Se accumulators *Morinda reticulata* and *Neptunia amplexicaulis* may accumulate Se up to 4000 mg Se/kg without showing toxic symptoms. There are many plants where selenium has reduced growth and yield and hence productivity, however, the toxic levels of Se for plants varies. In the case of rice, 10% reduction in yield at 2 mg/kg of soil was observed. White clover (*Trifolium repens*) plants showed reduction at 330 mg Se/kg (dry weight) present in the shoot tissue [266]. Ryegrass (*Lolium perenne*) plants showed oxidative stress at 10 mg/kg selenium in soil, causing loss of yield [274]

10.6.14 Silver ($^{47}Ag_{107.87}$)

Oat, lettuce, and radish plants were cultivated in soil contaminated with particles of bentonite organomodified with silver (Ag^+ bentonite), silver phosphate glass (Ag^+ phosphate), and silver nanoparticles adsorbed on fumed silica (AgNp silica). Plant development and microorganisms' abundance was evaluated. To some degree, Ag^+ bentonite impacted plants development and AgNp silica caused an adverse effect on microbial abundance. The impact on plants and microorganisms was contradictory and varied according to the soil and particles' physico-chemical characteristics [275, 276]. Silver is an important component in the area of nanomaterials (NMs), with over 450 metric tons of silver nanoparticles (nAg) produced in 2010 [277]. Even though some studies have reported the effects of nAg in the environment [278, 279], their toxicity on crop plants is not yet well understood. Previous investigations indicate that the effects of nAg on seed germination vary with NP characteristics and plant species. Kumari et al. [280] reported that nAg decrease the mitotic index and cause multiple chromosomal breaks and cell disintegration in onion (*Allium cepa*). Stampoulis et al. [281] found that nAg at 500 and 100 mg/l reduced plant biomass and transpiration in zucchini (*Cucurbita pepo*) by 57 and 41%, respectively. In addition, Pokhrel and Dubey [282] found that coated nAg promoted histological changes in maize (*Zea mays* L.) by inducing elongation of root cells, and reduced root growth in cabbage (*Brassica oleracea var. capitata* L.) by 24%. Among macro-elements, only Ca and Mg were significantly reduced by the highest concentrations of nAg; 20 and 33% of Mg reduced calcium by 10 and 19% at 250 and 500 mg/l, respectively, compared with the control in radish.

10.6.15 Uranium ($^{92}U_{238.03}$)

Uranium is a radioactive material that is very reactive. As a result, it cannot be found in the environment in its elemental form. Uranium compounds that have consisted during reactions of uranium with other elements and substances dissolve in water to their own extent. The water-solubility of a uranium compound determines its mobility in the environment, as well as its toxicity. While uranium itself is not particularly dangerous, some of its decay products do pose a threat, especially radon, which can build up in confined spaces such as basements. Uranium in air exists as dust that will fall into surface water, on plants or on soils through settling or rainfall. It will then sink to the sediment in water or to the lower soil layers, where it will mix with uranium that is already present. Water containing low amounts of uranium is usually safe to drink. Because of its nature, uranium is not likely to accumulate in fish or vegetables and uranium that is absorbed will be eliminated quickly through urine and feces. The compounds in the soil will combine with other compounds, which can stay in the soil for years without moving toward the groundwater. Uranium concentrations are often higher in phosphate-rich soil, but this does not have to be a problem, because concentrations often do not exceed normal ranges for uncontaminated soil. Plants absorb uranium through their roots and store it there. Root vegetables such as radishes may contain higher than usual concentrations of uranium as a result. When the vegetables are washed the uranium will be removed [283].

10.6.16 Zinc ($^{30}Zn_{65.38}$)

Zinc acts as a plant nutrient [284, 285]; but at higher concentrations it is toxic. Zinc toxicity in plants is clearly visible with the inhibition of growth and decrease in biomass production; severe toxicity can also be fatal. Zn toxicity might be the result of complex interactions of the major toxic ions, i.e., Cd^{+2}, Cu^{+2}, Pb^{+2} with Ca, Mg, Fe and P, and other environmental factors. Zinc toxicity was almost certainly involved with metabolism through competition for uptake, inactivation of enzymes, and displacement of essential elements from functional sites. Generally, Zn toxicity caused chlorosis and inhibited Fe translocation in some cases [286]. The physiology and biochemistry of zinc toxicity have been less studied in intact plants.

The cytotoxic effects of zinc on plants are elucidated. A major change was seen in the nucleus of the root tip cells due to zinc toxicity. The chromatin material was highly condensed and some of the cortical cells showed disruption and dilation of nuclear membrane in presence of 7.5 mM zinc. The cytoplasm became structureless, disintegration of cell organelles and the development of vacuoles were also observed. The number of nucleoli also increased in response to zinc resulting in the synthesis of new protein involved in heavy metal tolerance.

10.7 Conclusion

Recent developments in sequencing technologies and software now permit research into the community engineering of *in-situ* microbial communities to augment the phytoextraction of heavy metals in soil. Using *in-situ* microbial communities to assist phytoextraction could further improve the technology by utilizing microorganisms optimally adapted to contaminated soil and rhizosphere conditions. Considering the good impact of PGPR in terms of biofertilization, biocontrol, and bioremediation, all of which exert a positive influence on crop productivity and ecosystem functioning, encouragement should be given to its implementation in agriculture. Hoping for the betterment of technology in developing successful research and development, PGPR use will surely become a reality and will be instrumental to crucial processes that ensure the stability and productivity of agro-ecosystems, thus leading us toward an ideal agricultural system. PGPR have been successfully used to reduce plant stress in metal-contaminated soils, rhizospheric microbes establish a synergistic relationship with plant roots, which enhances nutrient absorption and improves plant performance, as well as the quality of soils. Nanotechnology inclusion in the agricultural sector should be intensified to reduce the damage to the ecosystem and meet global crop demand. Over the past decades, promising results and applications have already been developed in the area of delivery of fertilizers, pesticides, and genetic material for plant transformation. Based on that, gigantic effort is needed to develop the aspect of nanotechnology with plant-growth promoting bacteria. Environmental factors such as temperature and low nutrient concentrations and other factors, which are not easily controlled, can hamper their utilization and the effectiveness of the bioremediation process. Moreover, more research is required to fully understand the metabolic pathways of transgenic plants and microbes used in bioremediation so as to ascertain their effectiveness and possible side effects if used for bioremediation.

References

1 Lehmann, J. and Kleber, M. (2015). The contentious nature of soil organic matter. *Nature* 528: 60–68.

2 Abhilash, P.C., Tripathi, V., Edrisi, S.A. et al. (2016). Sustainability of crop production from polluted lands. *Energy, Ecology and Environment* 1 (1): 54–65.

3 Shahid, M., Khalid, S., Abbas, G. et al. (2015). *Heavy metal stress and crop productivity*. In: *Crop Production and Global Environmental Issues* (ed. K.R. Hakeem), 1–25. Cham: Springer International Publishing.

4 Hassan, T.U., Bano, A., and Naz, I. (2017). Alleviation of heavy metals toxicity by the application of plant growth promoting *rhizobacteria* and effects on wheat grown in saline sodic field. *International Journal of Phytoremediation* 19: 522–529. https://doi.org/10.1080/15226514.2016.1267696.

5 Panuccio, M.R., Sorgona, A., Rizzo, M. et al. (2009). Cadmium adsorption on vermiculite, zeolite and pumice: batch experiment studies. *Journal of Environmental Management* 90: 364–374. https://doi.org/10.1016/j.jenvman.2007.10.005.

6 Salla, V., Hardaway, C.J., and Sneddon, J. (2011). Preliminary investigation of Spartina alterniflora for phytoextraction of selected heavy metals in soils from Southwest Louisiana. *Microchemical Journal* 97: 207–212. https://doi.org/10.1016/j.microc.2010.09.005.

7 Kabata, P.A. (2010). *Trace Elements in Soils and Plants*. New York;, NY, USA: CRC Press.

8 Nematian, M.A. and Kazemeini, F. (2013). Accumulation of Pb, Zn, C and Fe in plants and hyperaccumulator choice in galali iron mine area. *International Journal of Agriculture and Crop Sciences* 5: 426.

9 Pierart, A., Shahid, M., Séjalon-Delmas, N. et al. (2015). Antimony bioavailability: knowledge and research perspectives for sustainable agricultures. *Journal of Hazardous Materials* 289: 219–234. https://doi.org/10.1016/j.jhazmat.2015.02.011.

10 Xiong, T., Leveque, T., Shahid, M. et al. (2014). Lead and cadmium phytoavailability and human bioaccessibility for vegetables exposed to soil or atmospheric pollution by process ultrafine particles. *Journal of Environmental Quality* 43: 1593–1600. https://doi.org/10.2134/jeq2013.11.0469.

11 Amari, T., Ghnaya, T., and Abdelly, C. (2017). Nickel, cadmium and lead phytotoxicity and potential of halophytic plants in heavy metal extraction. *South African Journal of Botany* 111: 99–110. https://doi.org/10.1016/j.sajb.2017.03.011.

12 Dixit, R., Malaviya, D., Pandiyan, K. et al. (2015). Bioremediation of heavy metals from soil and aquatic environment: an overview of principles and criteria of fundamental processes. *Sustainability* 7: 2189–2212.

13 Chibuike, G. and Obiora, S. (2014). Heavy metal polluted soils: effect on plants and bioremediation methods. *Applied Environmental Soil Science* 2014 (752708): 1–12. https://doi.org/10.1155/2014/752708.

14 Chandra, K., Salman, A.S., Mohd, A. et al. (2015). Protection against fca induced oxidative stress induced DNA damage as a model of arthritis and in vitro; anti-arthritic potential of costus speciosus rhizome extract. *International Journal of Pharmaceutical and Phytopharmacological Research.* 7: 383–389.

15 Mani, S. (2015). Production of reactive oxygen species and its implication in human diseases. In: *Free Radicals in Human Health and Disease* (eds. R. Vibha and C.S.Y. Umesh), 3–15. New Delhi, India: Springer.

16 Gill, S.S. and Tuteja, N. (2010). Reactive oxygen species and antioxidant machinery in abiotic stress tolerance in crop plants. *Plant Physiology and Biochemistry* 48: 909–930. https://doi.org/10.1016/j.plaphy.2010.08.016.

17 Ibrahim, M.R., Khan, P.R., Hegazy, S.S. et al. (2015). Improving the phytoextraction capacity of plants to scavenge heavy-metal infested sites. *Environmental Reviews* 23: 1–22.

18 Pourrut, B., Jean, S., Silvestre, J. et al. (2011). Lead-induced DNA damage in Vicia faba root cells: potential involvement of oxidative stress. *Mutation Research* 726: 123–128. https://doi.org/10.1016/j.mrgentox.2011.09.001.

19 Mosa, K.A., Ismail, A., and Helmy, M. (2017). Functional genomics combined with other omics approaches for better understanding abiotic stress tolerance in plants. In: *Plant Stress Tolerance* (ed. R. Sunkar), 55–73. Cham: Springer International Publishing.

20 Courbot, M., Willems, G., Motte, P. et al. (2007). A major quantitative trait locus for cadmium tolerance in Arabidopsis halleri colocalizes with HMA4, a gene encoding a heavy metal ATPase. *Physiologia Plantarum* 144 (s): 1052–1065. https://doi.org/10.1104/pp.106.095133.

21 Meyer, C.L., Pauwels, M., Briset, L. et al. (2016). Potential preadaptation to anthropogenic pollution: evidence from a common quantitative trait locus for zinc and cadmium tolerance in metallicolous and nonmetallicolous accessions of Arabidopsis halleri. *New Phytologist* 212: 934–943. https://doi.org/10.1111/nph.14093.

22 Zhang, J., Xu, J., Pasuquin, J. et al. (2017). QTL mapping and candidate gene analysis of ferrous iron and zinc toxicity tolerance at seedling stage in rice by genome-wide association study. *BMC Genomics* 18: 828–830. https://doi.org/10.1186/s12864-017-4221-5.

23 Tiwari, S., Lata, C., Chauhan, P.S. et al. (2016). Pseudomonas putida attunes morphophysiological, biochemical and molecular responses in *Cicer arietinum* L. during drought stress and recovery. *Plant Physiology and Biochemistry* 99: 108–117. https://doi.org/10.1016/j.plaphy.2015.11.001.

24 Tiwari, S., Prasad, V., Chauhan, P.S. et al. (2017b). *Bacillus amyloliquefaciens* confers tolerance to various abiotic stresses and modulates plant response to phytohormones through osmoprotection and gene expression regulation in rice. *Frontiers in Plant Science* 8: 1510. https://doi.org/10.3389/fpls.2017.01510.

25 Glick, B.R. (2012). *Plant Growth-Promoting Bacteria: Mechanisms and Applications*. Waterloo, Canada: Hindawi Publishing Corporation, Scientifica.

26 Chen, M., Xu, P., Zeng, G. et al. (2015). Bioremediation of soils contaminated with polycyclic aromatic hydrocarbons, petroleum, pesticides, chlorophenols and heavy metals by composting: applications, microbes and future research needs. *Biotechnology Advances* 33: 745–755.

27 Kuppusamy, S., Palanisami, T., Megharaj, M. et al. (2016a). Ex situ remediation technologies for environmental pollutants: a critical perspective. *Reviews of Environmental Contamination and Toxicology* 236: 117–192.

28 Kuppusamy, S., Palanisami, T., Megharaj, M. et al. (2016b). In situ remediation approaches for the management of contaminated sites: a comprehensive overview. *Reviews of Environmental Contamination and Toxicology* 236: 1–115.

29 Megharaj, M., Ramakrishnan, B., Venkateswarlu, K. et al. (2011). Bioremediation approaches for organic pollutants: a critical perspective. *Environmental International* 37: 1362–1375.

30 Weller, D.M. (1983). Colonization of wheat roots by a fluorescent pseudomonad suppressive to take all. *Phytopathology* 73: 1548–1553.

31 Werner, D. (2001). *Organic signals between plants and microorganisms*. In: *The Rhizosphere. Biochemistry and Organic Substances at the Soil-Plant Interface* (eds. R. Pinton, Z. Varanini and P. Nannipieri), 197–222. New York;: Marcel Dekker, Inc.

32 Antoun, H. and Prévost, D. (2005). Ecology of plant growth promoting rhizobacteria. In: *PGPR: Biocontrol and Biofertilization* (ed. Z.A. Siddiqui), 1–38. Amsterdam: Springer.

33 Avis, T.J., Gravel, V., Antoun, H. et al. (2008). Multifaceted beneficial effects of rhizosphere microorganisms on plant health and productivity. *Soil Biology and Biochemistry* 40: 1733–1740.

34 Yadav, B.K., Akhtar, M.S., and Panwar, J. (2015). *Rhizospheric plant-microbe interactions: key factors to soil fertility and plant nutrition*. In: *Plant Microbes Symbiosis: Applied Facets* (ed. N.K. Arora), 127–145. India: Springer.

35 Barbhaiya, H.B. and Roa, K.K. (1985). Production of Pyoverdin, the fluorescent pigments of Pseudomonas aeruginosa PAO1. *FEMS Microbiology Letters* 27: 233–235.

36 Bashan, Y., de-Bashan, L.E., and Prabhu, S.R. (2016). Superior polymeric formulations and emerging innovative products of bacterial inoculants for sustainable agriculture and the environment. In: *Agriculturally Important Microorganisms* (eds. H.B. Singh, B.K. Sarma and C. Keswani), 15–46. Singapore: Springer.

37 Hayat, R., Ahmed, I., and Sheirdil, R.A. (2012). An overview of plant growth promoting Rhizobacteria (PGPR). for sustainable agriculture. In: *Crop Production for Agricultural Improvement* (eds. M. Ashraf, M. Öztürk, M.S.A. Ahmad, et al.), 557–579. Dordrecht: Springer Science Business Media B.V.

38 Hayat, R., Ali, S., Amara, U. et al. (2010). Soil beneficial bacteria and their role in plant growth promotion: a review. *Annals of Microbiology* 60: 579–598.

39 Liu, K., Newman, M., McInroy, J.A. et al. (2017). Selection and assessment of Plant Growth-Promoting Rhizobacteria (PGPR). for biological control of multiple plant diseases. *Phytopathology* 107: 928–936.

40 Rubin, R.L., Groenigen, K.J., and Hungate, B.A. (2017). Plant growth promoting rhizobacteria are more effective under drought: a meta-analysis. *Plant and Soil* 416: 1–15.

41 Shaikh, S.S. and Sayyed, R.Z. (2015). Role of plant growth promoting Rhizobacteria and their formulation in biocontrol of plant diseases. In: *Plant Microbes Symbiosis: Applied Facets* (ed. N.K. Arora), 337–351. New Delhi: Springer.

42 Shaikh, S.S., Patel, P.R., Patel, S.S. et al. (2014). Production of biocontrol traits by Banana field fluorescent pseudomonads and their comparison with chemical fungicides. *Indian Journal of Expermental Biology* 52: 917–920.

43 Shaikh, S.S., Wani, S.J., and Sayyed, R.Z. (2016). Statistical based optimization of siderophore production and scale-up on bioreactor. *Biotechniques* 6: 69–72.

44 Sharma, P., Kumawat, K.C., and Kaur, S. (2016). *Plant growth promoting Rhizobacteria in nutrient enrichment: current perspectives.* In: *Biofortification of Food Crops* (eds. U. Singh, C.S. Praharaj, S.S. Singh, et al.), 263–289. New Delhi: Springer.

45 Vessey, J.K. (2003). Plant growth promoting rhizobacteria as biofertilizers. *Plant and Soil* 255: 571–586.

46 Chauhan, H., Bagyaraj, D.J., Selvakumar, G. et al. (2015). Novel plant growth promoting rhizobacteria-prospects and potential. *Applied Soil Ecology* 95: 38–53.

47 Lal, S., Chiarini, L., and Tabacchioni, S. (2016). New insights in plant-associated Paenibacillus species: biocontrol and plant growth-promoting activity. In: *Bacilli and Agro Biotechnology* (eds. M.T. Islam, M.M. Rahman, P. Pandey, et al.), 237–279. London: Springer International Publishing.

48 Gutiérrez, F.J., Probanza, A., Ramos, B. et al. (2003). Effects of culture filtrates of Rhizobacteria isolated from wild lupine on germination, growth, and biological nitrogen fixation of lupine seedlings. *Journal of Plant Nutrition* 26: 1101–1115.

49 Patten, C.L. and Glick, B.R. (2002). Role of *Pseudomonas putida* Indole acetic acid in development of the host plant root system. *Applied Environmental Microbiology* 68: 3795–3801.

50 Haas, D. and Defago, G. (2005). Biological control of soil-borne pathogens by fluorescent pseudomonads. *Nature Reviews Microbiology* 3: 307–319.

51 Kloepper, J.W. and Schroth, M.N. (1978). Plant growth-promoting rhizobacteria on radishes. In: Station de Pathologie, Proceedings of the 4th International Conference on "Plant Pathogenic Bacteria", Tours, France, 27 August–2 September 1978. Végétale et Phyto-Bactériologie. 879–882.

52 Choudhary, D.K., Sharma, K.P., and Gaur, R. (2011). Biotechnological perspectives of microbes in agro-ecosystems. *Biotechnology Letters* 33 (10): 1905–1910.

53 García, P., Menéndez, E., and Rivas, R. (2015). Role of bacterial biofertilisers in agriculture and forestry. *AIMS Bioengineering* 2: 183–205.

54 Dey, R., Pal, K.K., Bhatt, D.M. et al. (2004). Growth promotion and yield enhancement of peanut (*Arachis hypogeal* L.). by application of plant growth-promoting *rhizobacteria*. *Microbiology Resrearch* 159: 371–394.

55 Ahemad, M. (2012). Implications of bacterial resistance against heavy metals in bioremediation: a review. *Journal of the Institute of Integrative Omics and Applied Biotechnolgy* 3 (3): 12–25.

56 Ahemad, M. and Kibret, M. (2014). Mechanisms and applications of plant growth promoting Rhizobacteria. *Journal of King Saud University - Science* 26: 1–20.

57 Nehra, V. and Choudhary, M.A. (2015). Review on plant growth promoting rhizobacteria acting as bioinoculants and their biological approach towards the production of sustainable agriculture. *Journal of Applied Natural Science* 7: 540–556.

58 Paul, D. and Lade, H. (2014). Plant-growth-promoting rhizobacteria to improve crop growth in saline soils: a review. *Agronomy for Sustainable Development* 34: 737–752.

59 Rajkumar, M., Ae, N., Prasad, M.N.V. et al. (2010). Potential of siderophore-producing bacteria for improving heavy metal phytoextraction. *Trends in Biotechnology* 28: 142–149.

60 Bhattacharyya, P. and Jha, D. (2012). Plant growth-promoting rhizobacteria (PGPR): emergence in agriculture. *World Journal of Microbiology and Biotechnology* 28: 1327–1350.

61 Donot, F., Fontana, A., Baccou, J. et al. (2012). Microbial exopolysaccharides: main examples of synthesis, excretion, genetics and extraction. *Carbohydrate Polymers* 87: 951–962.

62 Choudhary, D.K. and Varma, A. (2016). *Microbial-Mediated Induced Systemic Resistance in Plants*. New York;, NY: Springer.

63 Martínez, O., Jorquera, M., Crowley, D. et al. (2010). Mechanisms and practical considerations involved in plant growth promotion by rhizobacteria. *Journal of Soil Science and Plant Nutrition* 10: 293–319.

64 Ramadan, E.M., AbdelHafez, A.A., Hassan, E.A. et al. (2016). Plant growth promoting rhizobacteria and their potential for biocontrol of phytopathogens. *African Journal of Microbiology Research* 10: 486–504.

65 Selvakumar, G., Panneerselvam, P., Bindu, G.H. et al. (2015). Pseudomonads: plant growth promotion and beyond. In: *Plant Microbes Symbiosis: Applied Facets* (ed. N.K. Arora), 193–208. India: Springer.

66 Jan, A.T., Azam, M., Ali, A. et al. (2014). Prospects for exploiting bacteria for bioremediation of metal pollution. *Critical revies in Environmental Science and Technology* 44: 519–560.

67 Mustapha, M.U. and Halimoon, N. (2015). Screening and isolation of heavy metal tolerant bacteria in industrial effluent. *Procedia Environmental Sciences* 30: 33–37. https://doi.org/10.1016/j.proenv.2015.10.006.

68 Gupta, A., Joia, J., Sood, A. et al. (2016). Microbes as potential tool for remediation of heavy metals: a review. *Journal of Microbology, Biochemistry and Technology* 8: 364–372.

69 François, F., Lombard, C., Guigner, J.M. et al. (2012). Isolation and characterization of environmental bacteria capable of extracellular biosorption of mercury. *Applied and Environmental Biology* 78: 1097–1106.

70 Shaikh, S.S., Wani, S.J., and Sayyed, R.Z. (2018). Impact of interactions between Rhizosphere and Rhizobacteria: a review. *Journal of Bacteriology and Mycology* 5 (1): 1058.

71 Román, B., Reza-vázquez, D.M., Gutiérrez-paredes, S. et al. (2017). Plant growth-promoting traits in rhizobacteria of heavy metal-resistant plants and their effects on Brassica nigra seed germination. *Pedosphere* 27: 511–526. https://doi.org/10.1016/S1002-0160(17)60347-3.

72 Dabrowska, G., Hrynkiewicz, K., Trejgell, A. et al. (2017). The effect of plant growth-promoting rhizobacteria on the phytoextraction of Cd and Zn by *Brassica napus* L. *International Journal of Phytoremediation* 19: 597–604. https://doi.org/10.1080/15226514.2016.1244157.

73 Hansda, A. and Kumar, V. (2017). Cu-resistant Kocuria sp. CRB15: a potential PGPR isolated from the dry tailing of Rakhà copper mine. *Biotechniques* 7: 132. https://doi.org/10.1007/s13205-017-0757-y.

74 Pramanik, K., Mitra, S., Sarkar, A. et al. (2017). Characterization of cadmium-resistant Klebsiella pneumoniae MCC 3091 promoted rice seedling growth by alleviating phytotoxicity of cadmium. *Environmental Science and Pollution Research* 24: 24419–24437. https://doi.org/10.1007/s11356-017-0033-z.

75 Gontia, I., Sapre, S., Sharma, A. et al. (2016). Alleviation of mercury toxicity in wheat by the interaction of mercury-tolerant plant growth promoting *rhizobacteria*. *Journal of Plant Growth Regulation* 35: 1000–1012. https://doi.org/10.1007/s00344-016-9598-x.

76 Arora, K., Sharma, S., and Monti, A. (2016). Bio-remediation of Pb and cd polluted soils by switchgrass: a case study in India. *International Journal of Phytoremediation* 18: 704–709.

77 Ahmad, I., Akhtar, M.J., Asghar, H.N. et al. (2016). Differential effects of plant growth-promoting Rhizobacteria on maize growth and cadmium uptake. *Journal of Plant Growth Regulation* 35: 303–315. https://doi.org/10.1007/s00344-015-9534-5.

78 Kamran, M.A., Syed, J.H., Eqani, S.A. et al. (2015). Effect of plant growth-promoting rhizobacteria inoculation on cadmium (Cd). uptake by *Eruca sativa*. *Environmental Science and Pollution Research* 22: 9275–9283. https://doi.org/10.1007/s11356-015-4074-x.

79 Pereira, S.I.A., Barbosa, L., and Castro, P.M.L. (2015). Rhizobacteria isolated from a metal polluted area enhance plant growth in zinc and cadmium contaminated soil. *International Journal of Environmental Science and Technology* 12: 2127–2142. https://doi.org/10.1007/s13762-014-0614-z.

80 Belimov, A.A., Puhalsky, I.V., Safronova, V.I. et al. (2015). Role of plant genotype and soil conditions in symbiotic plant-microbe interactions for adaptation of plants to cadmium polluted soils. *Water Air Soil Pollution* 226: 264. https://doi.org/10.1007/s11270-015-2537-9.

81 Ma, Y., Oliviera, R.S., Nai, F. et al. (2015). The hyperaccumulator sedum plumbizincicola harbors metal-resistant endophytic bacteria that improve its phytoextraction capacity in multi-metal contaminated soil. *Journal of Environmental Management* 156: 62–69. https://doi.org/10.1016/j.jenvman.2015.03.024.

82 Adediran, G.A., Ngwenya, B.T., Mosselmans, J.F.W. et al. (2015). Mechanism behind bacteria induced plant growth promotion and Zn accumulation in Brassica juncea. *Journal of Hazardous Materials* 283: 490–499. https://doi.org/10.1016/j.jhazmat.2014.09.064.

83 Liu, H., Yuan, M., Tan, S.Y. et al. (2015). Enhancement of arbuscular mycorrhizal fungus (Glomus versiforme). on the growth and Cd uptake by Cd-hyperaccumulator Solanum nigrum. *Applied Soil Ecology* 89: 44–49. https://doi.org/10.1016/j.apsoil.2015.01.006.

84 Lermen, C., Morelli, F., Gazim, Z.C. et al. (2015). Essential oil content and chemical composition of Cymbopogon citratus inoculated with arbuscular mycorrhizal fungi under different levels of lead. *Industrial Crops and Products* 76: 734–738. https://doi.org/10.1016/j.indcrop.2015.07.009.

85 Chen, L., Luo, S., Li, X. et al. (2014). Interaction of cd hyperaccumulator *Solanum nigrum* L. and functional endophyte *Pseudomonas* sp. Lk9 on soil heavy metals uptake. *Soil Biology and Biochemistry* 68: 300–308. https://doi.org/10.1016/j.soilbio.2013.10.021.

86 Jamil, M., Zeb, S., Anees, M. et al. (2014). Role of Bacillus licheniformis in phytoremediation of nickel contaminated soil cultivated with rice. *International Journal of Phytoremediation* 16: 554–571. https://doi.org/10.1080/15226514.2013.798621.

87 He, H., Ye, Z., Yang, D. et al. (2013). Characterization of endophytic Rahnella sp. JN6 from Polygonum pubescens and its potential in promoting growth and Cd, Pb, Zn uptake by Brassica napus. *Chemosphere* 90 (6): 1960–1965.

88 Babu, A.G., Kim, J.D., and Oh, B.T. (2013). Enhancement of heavy metal phytoremediation by Alnus firma with endophytic *Bacillus thuringiensis* GDB-1. *Journal of Hazardous Materials* 250: 477–483. https://doi.org/10.1016/j.jhazmat.

89 Marques, A.P., Moreira, H., Franco, A.R. et al. (2013). Inoculating Helianthus annuus (sunflower). grown in zinc and cadmium contaminated soils with plant growth promoting bacteria-effects on phytoremediation strategies. *Chemosphere* 92: 74–83. https://doi.org/10.1016/j.chemosphere.2013.02.055.

90 Srivastava, S., Verma, P.C., Chaudhary, V. et al. (2013). Inoculation of arsenic-resistant Staphylococcus arlettae on growth and arsenic uptake in *Brassica juncea* (L.). Czern. Var. R-46. *Journal of Hazardous Materials* 262: 1039–1047. https://doi.org/10.1016/j.jhazmat.2012.08.019.

91 Pandey, S., Ghosh, P.K., Ghosh, S. et al. (2013). Role of heavy metal resistant Ochrobactrum sp. and Bacillus spp. strains in bioremediation of a rice cultivar and their PGPR like activities. *Journal of Microbiology* 51: 11–17. https://doi.org/10.1007/s12275-013-2330-7.

92 Tiwari, S., Singh, S.N., and Garg, S.K. (2012). Stimulated phytoextraction of metals from fly ash by microbial interventions. *Environmental Technology* 33: 2405–2413. https://doi.org/10.1080/09593330.2012.670269.

93 Shin, M., Shim, J., You, Y. et al. (2012). Characterization of lead resistant endophytic Bacillus sp. MN3-4 and its potential for promoting lead accumulation in metal hyperaccumulator Alnus firma. *Journal of Hazardous Materials* 19: 314–320. https://doi.org/10.1016/j.jhazmat.2011.11.010.

94 Ma, Y., Prasad, M.N.V., Rajkumar, M. et al. (2011). Plant growth promoting rhizobacteria and endophytes accelerate phytoremediation of metalliferous soils. *Biotechnolog Advances* 29: 248–258. https://doi.org/10.1016/j.biotechadv.2010.12.001.

95 Luo, S.L., Chen, L., Chen, J.I. et al. (2011). Analysis and characterization of cultivable heavy metal-resistant bacterial endophytes isolated from Cd hyperaccumulator *Solanum nigrum* L. and their potential use for phytoremediation. *Chemosphere* 85: 1130–1138. https://doi.org/10.1016/j.chemosphere.2011.07.053.

96 Garg, N. and Aggarwal, N. (2011). Effects of interactions between cadmium and lead on growth, nitrogen fixation, phytochelatin, and glutathione production in mycorrhizal *Cajanus cajan* (L.). Millsp. *Journal of Plant Growth Regulation* 30: 286–300. https://doi.org/10.1007/s00344-010-9191-7.

97 Azcón, R., Perálvarez, M.D.C., Roldán, A. et al. (2010). Arbuscular mycorrhizal fungi, Bacillus cereus, and *Candida parapsilosis* from a multicontaminated soil alleviate metal toxicity in plants. *Microbial Ecology* 59: 668–677. https://doi.org/10.1007/s00248-009-9618-5.

98 Gao, Y., Miao, C., Mao, L. et al. (2010). Improvement of phytoextraction and antioxidative defense in *Solanum nigrum* L. under cadmium stress by application of cadmium-resistant strain and citric acid. *Journal of Hazardous Materials* 181: 771–777. https://doi.org/10.1016/j.jhazmat.2010.05.080.

99 Dary, M., Chamber-Pérez, M.A., Palomares, A.J. et al. (2010). "In situ" phytostabilisation of heavy metal polluted soils using Lupinus luteus inoculated with metal resistant plant-growth promoting *rhizobacteria*. *Journal of Hazardous Materials*. 177: 323–330. https://doi.org/10.1016/j.jhazmat.2009.12.035.

100 Acuna, J.J., Jorquera, M.A., Barra, P.J. et al. (2013). Selenobacteria selected from the rhizosphere as a potential tool for Se biofortification of wheat crops. *Biology and Fertility of Soils* 49 (2): 175–185.

101 Duran, P., Acuña, J.J., Jorquera, M.A. et al. (2013). Enhanced selenium content in wheat grain by co-inoculation of selenobacteria and arbuscular mycorrhizal fungi: a preliminary study as a potential Se biofortification strategy. *Journal of Cereal Science* 57 (3): 275–280.

102 Yasin, M., El-Mehdawi, A.F., Anwar, A. et al. (2015). Microbial-enhanced selenium and iron biofortification of wheat (*Triticum aestivum* L.)-applications in phytoremediation and biofortification. *International Journal of Phytoremediation* 17 (4): 341–347.

103 Duran, P., Acuña, J.J., Gianfreda, L. et al. (2015). Endophytic selenobacteria as new inocula for selenium biofortification. *Applied Soil Ecology* 96: 319–326.

104 Wu, C.H., Wood, T.K., Mulchandani, A. et al. (2006). Engineering plant-microbe symbiosis for rhizoremediation of heavy metals. *Applied and Environmental Microbiology* 72: 1129–1134. https://doi.org/10.1128/AEM.72.2.1129-1134.2006.

105 Zaidi, S., Usmani, S., Singh, B.R. et al. (2006). Significance of *Bacillus subtilis* strain SJ-101 as a bioinoculant for concurrent plant growth promotion and nickel accumulation in *Brassica juncea*. *Chemosphere* 64 (6): 991–997.

106 Burd, G.I., Dixon, D.G., and Glick, B.R. (2000). Plant growth-promoting bacteria that decrease heavy metal toxicity in plants. *Canadian Journal of Microbiology* 46 (3): 237–245.

107 Sriprang, R., Hayashi, M., Ono, H. et al. (2003). Enhanced accumulation of Cd2C by a Mesorhizobium sp. transformed with a gene from Arabidopsis thaliana coding for phytochelatin synthase. *Applied Environmental Microbiology* 69: 1791–1796. https://doi.org/10.1128/AEM.69.3.1791-1796.2003.

108 Abou-Shanab, R.A., Ghanem, K., Ghanem, N. et al. (2008). The role of bacteria on heavy-metal extraction and uptake by plants growing on multi-metal-contaminated soils. *World Journal of Microbiology and Biotechnology* 24 (2): 253–262.

109 Abou-Shanab, R.A., Angle, J.S., Delorme, T.A. et al. (2003). Rhizobacterial effects on nickel extraction from soil and uptake by *Alyssum murale*. *New Phytologist* 158 (1): 219–224.

110 Abou-Shanab, R.A.I., Angle, J.S., and Chaney, R.L. (2006). Bacterial inoculants affecting nickel uptake by *Alyssum murale* from low, moderate and high Ni soils. *Soil Biology and Biochemistry* 38 (9): 2882–2889.

111 Abou-Shanab, R.A.I., Van Berkum, P., and Angle, J.S. (2007). Heavy metal resistance and genotypic analysis of metal resistance genes in gram-positive and gram-negative bacteria present in Ni-rich serpentine soil and in the rhizosphere of *Alyssum murale*. *Chemosphere* 68 (2): 360–367.

112 Doble, M. and Kumar, A. (2005). *Biotreatment of Industrial Effluents*, 19–38. Oxford: Butterworth-Heinemann https://doi.org/10.1016/B978-075067838-4/50004-X.

113 Nautiyal, C.S., Srivastava, S., Chauhan, P.S. et al. (2013). Plant growth-promoting bacteria Bacillus amylo-liquefaciens NBRISN13 modulates gene expression profile of leaf and rhizosphere community in rice during salt stress. *Plant Physiology and Biochemistry* 66: 1–9. https://doi.org/10.1016/j.plaphy.2013.01.020.

114 Thassitou, P.K. and Arvanitoyannis, I. (2001). This is feasible because microorganisms have developed many resistance mechanisms for survival in the presence of toxic heavy metals in their environment. *Trends in Food Science and Technology* 12 (5–6): 185–196. https://doi.org/10.1016/S0924-2244(01)00081-4.

115 Mishra, J., Singh, R., and Arora, N.K. (2017). Alleviation of heavy metal stress in plants and remediation of soil by rhizosphere microorganisms. *Frontiers in Microbiology* 8: 1706. https://doi.org/10.3389/fmicb.2017.01706.

116 Azubuike, C.C., Chikere, C.B., and Okpokwasili, G.C. (2016). Bioremediation techniques-classification based on site of application: principles, advantages, limitations and prospects. *World Journal of Microbiology and Biotechnology* 32: 180–194.

117 Lu, L., Huggins, T., Jin, S. et al. (2014). Microbial metabolism and community structure in response to bio-electrochemically enhanced remediation of petroleum hydrocarbon-contaminated soil. *Environmental Science and Technology* 48: 4021–4029.

118 Smith, E., Thavamani, P., Ramadass, K. et al. (2015). Remediation trials for hydrocarbon-contaminated soils in arid environments: evaluation of bioslurry and biopiling techniques. *International Biodeterioration and Biodegradation* 101: 56–65.

119 Ullah, A., Heng, S., Munis, M.F.H. et al. (2015). Phytoremediation of heavy metals assisted by plant growth promoting (PGP). bacteria: a review. *Environmental and Experimental Botany* 117: 28–40. https://doi.org/10.1016/j.envexpbot.2015.05.001.

120 Singh, J.S., Abhilash, P.C., Singh, H.B. et al. (2011). Genetically engineered bacteria: an emerging tool for environmental remediation and future research perspectives. *Gene* 480: 1–9. https://doi.org/10.1016/j.gene.2011.03.001.

121 Barkay, T., Miller, S.M., and Summers, A.O. (2013). Bacterial mercury resistance from atoms to ecosystems. *FEMS Microbiology Review* 27: 355–384.

122 Chen, S. and Wilson, D.B. (1997). Genetic engineering of bacteria and their potential for Hg^{2+} bioremediation. *Biodegradation* 8: 97–103.

123 Deckwer, W.D., Becker, F., Ledakowicz, S. et al. (2004). Microbial removal of ionic mercury in a three-phase fluidized bed reactor. *Environmental Science and Technology* 38: 1858–1865. [CrossRef] [PubMed].

124 Marconi, A.M., Kieboom, J., and Bont, J.A. (1997). Improving the catabolic functions in the toluene-resistant strain pseudomonas putidas. *Biotechnology Letterrs* 19: 603–606.

125 Liu, S., Zhang, F., Chen, J. et al. (2011). Arsenic removal from contaminated soil via biovolatilization by genetically engineered bacteria under laboratory conditions. *Journal of Environmental Science* 23: 1544–1550.

126 Rojas, L.A., Yáñez, C., González, M. et al. (2011). Characterization of the metabolically modified heavy metal-resistant cupriavidus metallidurans strain MSR33 generated for mercury bioremediation. *PLoS One* 6: e17555.

127 Tak, H.I., Ahmad, F., and Babalola, O.O. (2013). *Advances in the application of plant growth-promoting rhizobacteria in phytoremediation of heavy metals.* In: *Reviews of Environmental Contamination and Toxicology*, vol. 223 (ed. D.M. Whitacre), 33–52. New York;, NY: Springer.

128 Sone, Y., Mochizuki, Y., Koizawa, K. et al. (2013). Mercurial-resistance determinants in pseudomonas strain k-62 plasmid pmr68. *AMB Express* 3: 41.

129 Wolejko, E., Wydro, U., and Loboda, T. (2016). The ways to increase efficiency of soil bioremediation. *Ecological Chemistry and Engineering* 23: 155–156. https://doi.org/10.1515/eces-2016-0011.

130 Ghosal, D., Ghosh, S., Dutta, T.K. et al. (2016). Current state of knowledge in microbial degradation of polycyclic aromatic hydrocarbons (PAHS): a review. *Frontiers in Microbiology* 7: 1369–1379.

131 Van Nostrand, J.D., He, Z. et al. (2012). Use of functional gene arrays for elucidating in situ biodegradation. *Frontiers in Microbiology* 3: 339–342.

132 Ebbs, S.D. and Kochian, L.V. (1997). Toxicity of zinc and copper to brassica species: implications for phytoremediation. *Journal of Environmental Quality* 26: 776–781.

133 Yong, X., Chen, Y., Liu, W. et al. (2014). Enhanced cadmium resistance and accumulation in Pseudomonas putida KT2440 expressing the phytochelatin synthase gene of *Schizosaccharomyces pombe*. *Letters in Applied Microbiology* 58: 255–261. https://doi.org/10.1111/lam.12185.

134 Ike, A., Sriprang, R., Ono, H. et al. (2007). Bioremediation of cadmium contaminated soil using symbiosis between leguminous plant and recombinant rhizobia with the MTL4 and the PCS genes. *Chemosphere* 66: 1670–1676. https://doi.org/10.1016/j.chemosphere.2006.07.058.

135 Ike, A., Sriprang, R., Ono, H. et al. (2008). Promotion of metal accumulation in nodule of Astragalus sinicus by the expression of the iron-regulated transporter gene in Mesorhizobium huakuii subsp. *Journal of Bioscience and Bioengineering* 105: 642–648. https://doi.org/10.1263/jbb.105.642.

136 Sriprang, R., Hayashi, M., Yamashita, M. et al. (2002). A novel bioremediation system for heavy metals using the symbiosis between leguminous plant and genetically engineered rhizobia. *Journal of Biotechnology* 99: 279–293. https://doi.org/10.1016/S0168-1656(02)00219-5.

137 Wu, S.C., Cheung, K.C., Luo, Y.M. et al. (2006). Effects of inoculation of plant growth-promoting rhizobacteria on metal uptake by *Brassica juncea*. *Environmental Pollution* 140: 124–135.

138 Nie, L., Shah, S., Rashid, A. et al. (2002). Phytoremediation of arsenate contaminated soil by transgenic canola and the plant growth-promoting bacterium Enterobacter cloacae CAL2. *Plant Physiology and Biochemistry* 40: 355–361. https://doi.org/10.1016/S0981-9428(02)01375-X.

139 Valls, M., Atrian, S., Lorenzo, V. et al. (2000). Engineeringa mouse metallothionein on the cell surface of Ralstonia eutropha CH34 for immobilization of heavy metals in soil. *Nature Biotechnology* 18: 661–665. https://doi.org/10.1038/76516.

140 Ayangbenro, A.S. and Babalola, O.O. (2017). A new strategy for heavy metal polluted environments: a review of microbial biosorbents. *International Journal of Environmental Research and Public Health* 14: 81–94.

141 Kushwaha, A., Rani, R., Kumar, S. et al. (2015). Heavy metal detoxification and tolerance mechanisms in plants: implications for phytoremediation. *Environmental Reviews* 24: 39–51.

142 Chaturvedi, A.D., Pal, D., Penta, S. et al. (2015). Ecotoxic heavy metals transformation by bacteria and fungi in aquatic ecosystem. *World Journal of Microbiology and Biotechnology* 31: 1595–1603.

143 Ndeddy Aka, R.J. and Babalola, O.O. (2016). Effect of bacterial inoculation of strains of pseudomonas aeruginosa, alcaligenes feacalis and Bacillus subtilis on germination, growth and heavy metal (Cd, Cr, and Ni). uptake of *Brassica juncea*. *International Journal of Phytoremediation* 18: 200–209.

144 Skórzynska, P.E., Dra̜z˙kiewicz, M., and Krupa, Z. (2010). Lipid peroxidation and antioxidative response in arabidopsis thaliana exposed to cadmium and copper. *Acta Physiologiae Plantarum* 32: 169–172.

145 Upadhyay, N., Vishwakarma, K., Singh, J. et al. (2017). Tolerance and reduction of chromium (VI). by *Bacillus* sp. Mnu16 isolated from contaminated coal mining soil. *Frontiers in Plant Science* 8: 778–781.

146 Jaishankar, M., Tseten, T., Anbalagan, N. et al. (2014). Toxicity, mechanism and health effects of some heavy metals. *Interdisciplinary Toxicology* 7: 60–72.

147 Muszynska, E. and Hanus-Fajerska, E. (2015). Why are heavy metal hyperaccumulating plants so amazing? *Journal of Biotechnology, Computational Biology and Bionanotechnology* 96: 265–271.

148 Jadia, C.D. and Fulekar, M. (2009). Phytoremediation of heavy metals: recent techniques. *African Journal of Biotechnology* 8: 921–928.

149 Gaur, N., Flora, G., Yadav, M. et al. (2014). Review with recent advancements on bioremediation-based abolition of heavy metals. *Environmental Science: Processes and Impacts* 16: 180–193.

150 ATSDR. (2003). Agency for Toxic Substances and Disease Registry () http://www.atsdr.cdc.gov.

151 Neumann, R.B., Ashfaque, K.N., Badruzzaman, A.B.M. et al. (2010). Anthropogenic influences on groundwater arsenic concentrations in Bangladesh. *Nature Geoscience* 3: 46–52. https://doi.org/10.1038/ngeo685.

152 Tripathi, R.D., Srivastava, S., Mishra, S. et al. (2007). Arsenic hazards: strategies for tolerance and remediation by plants. *Trends in Biotechnology* 25: 158–165. https://doi.org/10.1016/j.tibtech.2007.02.003.

153 Wang, S. and Mulligan, C.N. (2006). Occurrence of arsenic contamination in Canada: sources, behavior and distribution. *Science of the Total Environment* 366: 701–721. https://doi.org/10.1016/j.scitotenv.2005.09.005.

154 Verma, S., Verma, P.K., Meher, A.K. et al. (2016). A novel arsenic methyltransferase gene of Westerdykella aurantiaca isolated from arsenic contaminated soil: phylogenetic, physiological, and biochemical studies and its role in arsenic bioremediation. *Metallomics* 8: 344–353. https://doi.org/10.1039/c5mt00277j.

155 Kumar, S., Dubey, R.S., Tripathi, R.D. et al. (2015). Omics and biotechnology of arsenic stress and detoxification in plants: current updates and prospective. *Environmental International* 74: 221–230. https://doi.org/10.1016/j.envint.2014.10.019.

156 Helleday, T., Nilsson, R., and Jenssen, D. (2000). Arsenic (III). and heavy metal ions induce intrachromosomal homologous recombination in the hprt gene of V79 Chinese hamster cells. *Environment and Molecular Mutagenesis* 35: 114–122. https://doi.org/10.1002/(SICI).1098-2280(2000)35:2<114::AID-EM6>3.0.CO;2-Q.

157 Carbonell, A.A., Aarabi, M.A., Delaune, R.D. et al. (1998). Arsenic in wetland vegetation: availability, phytotoxicity, uptake and effects on plants growth and nutrition. *Science of the Total Environment* 217: 189–199. https://doi.org/10.1016/S0048-9697(98)00195-8.

158 Chaney, R.L. and Ryan, J.A. (1994). *Risk Based Standards for Arsenic Lead and Cadmium in Urban Soils*. Frankfurt, Germany: Dechema.

159 Yang, Y.Y., Jung, J.Y., Song, W.Y. et al. (2000). Identification of rice varieties with high tolerance or sensitivity to lead and characterization of the mechanism of tolerance. *Plant Physiology.* 124: 1019–1026.

160 Sharma, P. and Dubey, R.S. (2005). Lead toxicity in plants. *Brazilian Journal of Plant Physiology* 17 (1): 35–52.

161 Kumar, B., Smita, K., and Flores, L.C. (2017). Plant mediated detoxification of mercury and lead. *Arabian Journal of Chemistry* 10: S2335–S2342. https://doi.org/10.1093/pcp/pcu117.

162 Buchet, J.P., Lauwerys, R., Roels, H. et al. (1990). Renal effects of cadmium body burden of the general population. *Lancet* 336: 699–702. https://doi.org/10.1016/0140-6736(90)92201-R.

163 Gianazza, E., Wait, R., Sozzi, A. et al. (2007). Growth and protein profile changes in *Lepidium sativum* L. plantlets exposed to cadmium. *Environmental and Experimental Botany* 59: 179–187. https://doi.org/10.1016/j.envexpbot.2005.12.005.

164 Gill, S.S. and Tuteja, N. (2011). Cadmium stress tolerance in crop plants: probing the role of sulfur. *Plant Signaling and Behaviour* 6: 215–222. https://doi.org/10.4161/psb.6.2.14880.

165 IARC (1993). *IARC Monographs on the Evaluation of Carcinogenic Risk to Humans. Beryllium, Cadmium, Mercury and Exposures in the Glass Manufacturing Industry*, vol. 58, 41–117. Lyon: International Agency for Research on Cancer Publication.

166 Wagner, G.J. (1993). Accumulation of cadmium in crop plants and its consequences to human health. *Advances in Agronomy* 51: 173–212. https://doi.org/10.1016/S0065-2113(08)60593-3.

167 Sandalio, L.M., Dalurzo, H.C., Gomez, M. et al. (2001). Cadmium-induced changes in the growth and oxidative metabolism of pea plant. *Journal of Experimental Botany* 52: 2115–2126. https://doi.org/10.1093/jexbot/52.364.2115.

168 Sharma, P. and Dubey, R.S. (2006). Cadmium uptake and its toxicity in higher plants. In: *Cadmium Toxicity and Tolerance in Plants* (eds. N.A. Khan and Y. Samiullah), 63–86. New Delhi: Narosa Publishing House.

169 Verma, S. and Dubey, R.S. (2001). Effect of Cadmium on soluble sugars and enzymes of their metabolism in rice. *Biolgia Plantarum* 44: 117–123. https://doi.org/10.1023/A:1017938809311.

170 Katz, S.A. and Salem, H. (1994). *The Biological and Environmental Chemistry of Chromium*. Weinheim, pappelallae 3, Postfach: Verlagsgesellschaft mbh.

171 Nriagu, J.O. and Pacyna, J.M. (1988). Quantitative assessment of worldwide contamination of air, water and soils by trace metals. *Nature* 333: 134–138.

172 Barcelo, J. and Poschenrieder, C. (1997). Chromium in plants. In: *Chromium Environmental Issue* (eds. S. Carati, F. Tottarelli and P. Seqmi), 101–129. Milan: Francotangati Press.

173 Panda, S.K. (2003). Heavy metal phytotoxicity induces oxidative stress in *Taxithelium* species. *Current Science* 84: 631–633.

174 Panda, S.K. and Patra, H.K. (2000). Does Cr(III). produces oxidative damage in excised wheat leaves. *Journal of Plant Biology* 27 (2): 105–110.

175 Panda, S.K., Chaudhury, I., and Khan, M.H. (2003). Heavy metals induce lipid peroxidation and affects antioxidants in wheat leaves. *Biologia Plantarum* 46: 289–294.

176 Poschenrieder, C., Vazquez, M.D., Bonet, A. et al. (1991). Chromium III iron interaction in iron sufficient and iron deficient bean plants. II Ultrastructural aspects. *Journal of Plant Nutrition* 14: 415–428.

177 Bassi, M., Corradi, M.G., and Realini, M. (1990). Effects of chromium (VI). on two freshwater plants, *Lemna minor* and *Pistia stratiotes*, morphological observations. *Cytobios* 62: 27–38.

178 Choudhury, S. and Panda, S.K. (2004). Induction of oxidative stress and ultrastructural changes in moss Taxithelium nepalense (Schwaegr.). broth under lead (Pb). and arsenic (As). phytotoxicity. *Current Science* 87: 342–348.

179 Bonet, A., Poschenriede, C., and Barcelo, J. (1991). Chromium 111–iron interaction in Fe deficient and Fe-sufficient bean plants. I. Growth and nutrient content. *Journal of Plant Nutrition* 14: 403–414.

180 Han, F.X., Sridhar, B.M., Monts, D.L. et al. (2004). Phytoavailability and toxicity of trivalent and hexavalent chromium to *Brassica juncea*. *New Phytologist* 162 (2): 489–499.

181 Bishnoi, N.R., Anita, D., Gupta, V.K. et al. (1993). Effect of chromium on seed germination seedling growth and yield of peas. *Agriculture, Ecosystems abd Environment* 47: 47–57.

182 Sheoran, V., Sheoran, A.S., and Poonia, P. (2009). Phytomining: a review. *Mineral Engineering* 22: 1007–1020.

183 Zeid, I.M. (2001). Responses of *Phaseolus vulgaris* to chromium and cobalt treatments. *Biologia Plantarum* 44: 111–115.

184 Panda, S.K. (2002). *The biology of oxidative stress in green cells: a review*. In: *Advances of Stress Physiology of Plants* (ed. S.K. Panda), 1–13. Jhodpur, India: Scientific Publishers.

185 Dixit, V., Pandey, V., and Shayam, R. (2002). Chromium ions inactivate electron transport and enhance superoxide generation in vivo; in pea (*Pisum sativum*). root mitochondria. *Plant, Cell and Environment* 25: 687–693.

186 Herich, R. and Bobak, M. (1977). Influence of exogenously added cobalt on the submicroscopical structure of the proplastids. *Facultatis Rerum Naturalium Universitatis Comenianae, P1 Physiologica* . 13: 3–8.

187 Lipskaya, G.A. (1972). Accumulation of chlorophyll in chloroplasts of cucumber leaves under the effect of cobalt and manganese applied separately and together. *Biologicheskie Nauki* 15: 90–94.

188 Lipskaya, G.A. (1990). Development of the photosynthetic apparatus in barley grown from seeds varying in cobalt content under sterile and nonsterile conditions. *Fiziologiia Rastenii* 37: 668–674.

189 Lipskaya, G.A. (1970). Anatomical-cytological features of cucumber leaves in the presence of cobalt and manganese in the nutrient mixture. *Fiziologiia Rastenii* 17: 475–981.

190 Lipskaya, G.A. (1974). Effect of cobalt and heteroauxin on the morphology and structure of barley leaf. *Biyalagichnykh Navuk* 5: 121–123.

191 Lipskaya, G.A., Matvyeyentsava, V.S., and Charkaskaya, S.K. (1973). Effect of various combination of cobalt with other trace elements on the change of activity of the Hill reaction. *Biyalagichnykh Navuk* 2: 32–37.

192 Tripathy, B.C. and Mohanty, P. (1981). Stabilization by glutaraldehyde fixation by chloroplast membrane structure and function against heavy metal ion induced damage. *P1. Science Letters* 42: 253–261.

193 Blankenship, M.L. and Wilbur, K.M. (1975). Cobalt effects on cell division and calcium uptake in the coccolithophoroid Cricosphaera carterae (Haptophyceae). *Journal of Physiology* 11: 211–219.

194 Bobak, M. (1974). Influence of exogenous added cobalt upon the submicroscopic structure and the chromosomes of meristematic cells of the horse bean (*Viciafaba L.C.V. Zborovicky*). *Acta Facultatis Rerum Naturalium Universitatis Comenianae Physiologica* 8: 17–24.

195 Burca, S., Cachita-Cosma, D., and Trifu, M. (1978). Morphological changes induced by four microelements, Mn, Zn, co and cd in tomato seedlings (*Solanum lycopersicum*). *Studia Universitatis Babes-Bolyai: Biologia* 2: 11–18.

196 Marschner, M. (1995). *Mineral Nutrition of Higher Plants*, 2nde, 200–255. London: Academic Press.

197 Prasad, M.N.V. and Strzalka, K. (1999). *Impact of heavy metals on photosynthesis.* In: *Heavy Metal Stress in Plants* (eds. M.N.V. Prasad and J. Hagemeyer), 117–138. Berlin: Springer Publishers.

198 Ciscato, M., Valcke, R., Van Loven, K. et al. (1997). Effects of in vivo; copper treatment on the photosynthetic apparatus of two *Triticum durum* cultivars with different stress sensitivity. *Physiologia Plantarum* 100 (4): 901–908.

199 Lidon, F.C. and Henriques, F.S. (1991). Limiting step in photosynthesis of rice plants treated with varying copper levels. *Journal of Plant Physiology* 138: 115–118.

200 Pätsikkä, E., Aro, E.M., and Tyystjärvi, E. (1998). Increase in the quantum yield of photoinhibition contributes to copper toxicity in vivo;. *Plant Physiology* 117: 619–627.

201 Quartacci, M.F., Pinzino, C., Sgherri, C.L.M. et al. (2000). Growth in excess copper induces changes in the lipid composition and fluidity of PSII-enriched membranes in wheat. *Physiologia Plantarum* 108: 87–93.

202 Maksymiec, W., Russa, R., and Baszyñski, U.S.T. (1994). Effect of excess Cu on the photosynthetic apparatus of runner bean leaves treated at two different growth stages. *Physiologia Plantarum* 91: 715–721.

203 Luna, C.M., Gonzalez, C.A., and Trippi, V.S. (1994). Oxidative damage caused by an excess of copper in oat leaves. *Plant Cell Physiology* 35: 11–15.

204 Sandmann, G. and Böger, P. (1980). Copper-mediated lipid peroxidation processes in photosynthetic membranes. *Plant Physiology* 66: 797–800.

205 Halliwell, B.J. and Gutteridge, M.C. (1984). Oxygen toxicity, oxygen radicals, transition metals and disease. *Biochemical Journal* 219: 1–14.

206 Best, S. and Sadler, P. (1996). Gold drugs: mechanism of action and toxicity. *Gold Bulletin* 29: 87–93.

207 Binder, B.M., Rodriguez, F.I., Bleecker, A.B. et al. (2007). The effects of group 11 transition metals, including gold, on ethylene binding to the ETR1 receptor and growth of Arabidopsis thaliana. *FEBS Letters* 581: 5105–5109.

208 Starnes, D.L., Jain, A., and Sahi, S.V. (2010). In planta engineering of gold nanoparticles of desirable geometries by modulating growth conditions: an environment-friendly approach. *Environmental Science and Technology* 44: 7110–7115.

209 Niemietz, C.M. and Tyerman, S.D. (2002). New potent inhibitors of aquaporins: silver and gold compounds inhibit aquaporins of plant and human origin. *FEBS Letters* 531: 443–447.

210 Rodriguez, E., Parsons, J.G., Peralta-Videa, J.R. et al. (2007). Potential of Chilopsis linearis for gold phytomining: using XAS to determine gold reduction and nanoparticle formation within plant tissues. *International Journal of Photoremediation* 9: 133–147.

211 Dobermann, A. and Fairhurst, T.H. (2000). *Nutrient Disorders and Nutrient Management*, 191. Manila, Philippines: The International Rice Research Institute.

212 Polle, A. and Rennenberg, H. (1993). *Significance of antioxidants in plant adaptation to environmental stress.* In: *Plant Adaptation to Environmental Stress* (eds. T. Mansfield, L. Fowden and F. Stoddard), 263–273. London: Chapman and Hall.

213 Chen, S.X. and Schopfer, P. (1999). Hydroxyl-radical production in physiological reactions: a novel function of peroxidase. *European Journal of Biochemistry* 260: 726–735.

214 Becana, M., Moran, J.F., and Iturbe-Ormaetxe, I. (1998). Iron dependent oxygen free radical generation in plants subjected to environmental stresses: toxicity and antioxidants. *Plant and Soil* 201: 137–147.

215 Davies, K.J.A. (1987). Protein damage and degradation by oxygen radicals. I. General aspects. *Journal of Biological Chemistry* 262: 9895–9901.

216 Fridovich, I. (1986). Biological effects of superoxide radical. *Archives of Biochemistry and Biophysics* 247: 1–11.

217 Halliwell, B. (1987). Oxidative damage, lipid peroxidation and antioxidant protection in chloroplasts. *Chemistry and Physics of Lipids* 44: 327–340.

218 Imlay, J.A. and Linn, S. (1988). DNA damage and oxygen radical toxicity. *Science* 240: 1302–1309.

219 Liebler, D.C., Kling, D.S., and Reed, D.J. (1986). Antioxidant protection of phospholipid bilayers by a-tocopherol. Control of a-tocopherol status and lipid peroxidation by ascorbic acid and glutathione. *Journal of Biology andChemistry* 261: 12114–12119.

220 Wise, R.R. and Naylor, A.W. (1987). Chilling enhanced peroxidation. The peroxidative distruction of lipids during chilling injury to photosynthesis and ultra-structure. *Plant Physiology* 83: 272–277.

221 Alscher, R.G., Donahue, J.L., and Cramer, C.L. (1997). Reactive oxygen species and antioxidants: relationship in green cells. *Physiologia Plantarum* 100: 224–233.

222 Bowler, C., Montagu, M.V., and Inze, D. (1992). Superoxide dismutase and stress tolerance. *Annual Review of Plant Physiology and Plant Molecular Biology* 43: 83–116.

223 Elstner, E.F. (1986). Metabolism of activated oxygen species. In: *The Biochemistry of Plants: Biochemistry of Metabolism*, 2e (ed. D.D. Davies), 253–315. San Diego: Academic Press.

224 Menconi, M., Sgherri, C.L.M., Pinzino, C. et al. (1995). Activated oxygen production and detoxification in wheat plants subjected to a water deficit programme. *Journal of Experimental Botany* 46: 1123–1130.

225 Moran, J.F., Becana, M., Iturbe-Ormaetxe, I. et al. (1994). Drought induces oxidative stress in pea plants. *Planta* 194: 346–352.

226 Graedel, T.E. (1978). Inorganic elements, hydrides, oxides, and carbonates. In: *Chemical Compounds in the Atmosphere* (ed. T.E. Graedel), 35–41. New York;, NY: Academic Press and 44–49.

227 NAS (1973). *Medical and Biological Effects of Environmental Pollutants: Manganese*. Washington, DC: National Academy of Sciences, National Academy Press.

228 HSDB (1998). *Hazardous Substances Data Bank*. Bethesda, MD: National Institutes of Health, National Library of Medicine.

229 US EPA (1984). Health assessment document for manganese. Final draft. Cincinnati, OH, US Environmental Protection Agency, Office of Research and Development (Report No. EPA/600/8–83/013F).

230 Windholz, M. (ed.) (1983). *The Merck Index: An Encyclopedia of Chemicals, Drugs and Biologicals*. Rahway, NJ: Merck and Company.

231 Schroeder, W.H., Dobson, M., and Kane, D.M. (1987). Toxic trace elements associated with airborne particulate matter: a review. *Journal of the Air Pollution Control Association* 37: 1267–1285.

232 Stokes, P.M., Campbell, P.G.C., Schroeder, W.H. et al. (1988). *Manganese in the Canadian environment*. Ottawa, Ontario: National Research Council of Canada, Associate Committee on Scientific Criteria for Environmental Quality NRCC No. 26193.

233 Zoller, W.H., Gladney, E.S., and Duce, R.A. (1974). Atmospheric concentrations and sources of trace metals at the South Pole. *Science* 183: 198–200.

234 Ducic, T. and Polle, A. (2005). Transport and detoxification of manganese and copper in plants. *Brazilian Journal of Plant Physiology* 17: 103–112.

235 Kochian, L.V., Hoekenga, O.A., and Pineros, M.A. (2004). How do crop plants tolerate acid soils? Mechanisms of aluminum tolerance and phosphorous efficiency. *Annual Review of Plant Biology* 55: 459–493.

236 Fageria, N.K. and Stone, L.F. (2008). Micronutrient deficiency problems in South America. In: *Micronutrient Deficiencies in Global Crop Production* (ed. B.J. Alloway), 245e266. Dordrecht: Springer.

237 Lavres, J.J., Reis, A.R., Rossi, M.L. et al. (2010). Changes in the ultrastructure of soybean cultivars in response to manganese supply in solution culture. *Science and Agriculture* 67: 287–294.

238 St. Clair, S.B. and Lynch, J.P. (2005). Element accumulation patterns of deciduous and evergreen tree seedlings on acid soils: implications for sensitivity to manganese toxicity. *Tree Physiology* 25: 85–92.

239 St. Clair, S.B., Carlson, J.E., and Lynch, J.P. (2005). Evidence for oxidative stress in sugar maple stands growing on acidic, nutrient imbalanced forest soils. *Oecologia* 145: 257–268.

240 Gratao, P.L., Polle, A., Lea, P.J. et al. (2005). Making the life of heavy metal stressed plants a little easier. *Functional Plant Biology* 32: 481–494.

241 Schutzendubel, A. and Polle, A. (2002). Plant responses to abiotic stresses: heavy metalinduced oxidative stress and protection by mycorrhization. *Journal of Experimental Botany* 53: 1351–1365.

242 Fernando, D.R., Marshall, A.T., Forster, P.I. et al. (2013). Multiple metal accumulation within a manganese-specific genus. *American Journal of Biology* 100: 690–700.

243 Choudhary, D.K., Varma, A., and Tuteja, N. (2017). *Plant-Microbe Interaction: An Approach to Sustainable Agriculture*. New Delhi, India: Springer.

244 Li, Y., Yang, X.X., Ren, B.B. et al. (2012). Why nitrogen use efficiency decreases under high nitrogen supply in rice (Oryza sativa L.). seedlings. *Journal of Plant Growth Regulation* 31: 47–52. https://doi.org/10.1007/s00344-011-9218-8.

245 Schmutz, J., Cannon, S.B., and Schlueter, J. (2010). Genome sequence of the palaeopolyploid soybean. *Nature* 463: 178–183.

246 Reddy, K.J., Munn, L.C., and Wang, L. (1997). *Chemistry and mineralogy of molybdenum in soils.* In: *Molybdenum in agriculture* (ed. U.C. Gupta), 4–21. Cambridge: Cambridge University Press.

247 Gupta, U.C. (1997a). Soil and plant factors affecting molybdenum uptake by plants. In: *Molybdenum in Agriculture*, 71–91. Cambridge: Cambridge University Press.

248 Lindsay, W.L. (1979). *Chemical Equilibria in Soils.* New York;: Wiley.

249 Promix (2018). Role of molybdenum in plant culture. (accessed September 29, 2020) https://www.pthorticulture.com/en/training-center/role-of-molybdenum-in-plant-culture.

250 Spectrum Analytic (n.d.). Molybdenum MoO_4^-. (accessed September 29, 2020) https://www.spectrumanalytic.com/support/library/ff/Mo_Basics.htm.

251 Mahler, R.L. (n.d.) Molybdenum in Idaho. () http://cals.uidaho.edu/edcomm/pdf/CIS/CIS1087.pdf.

252 Azevedo, R. and Rodriguez, E. (2012). Phytotoxicity of mercury in plants: a review. Zhoudoi: https://doi.org/10.1155/2012/848614.

253 Zhou, Z.S., Wang, S.J., and Yang, Z.M. (2008). Biological detection and analysis of mercury toxicity to alfalfa (*Medicago sativa*). plants. *Chemosphere* 70: 1500–1509. https://doi.org/10.1016/j.chemosphere.2007.08.028.

254 Israr, M., Sahi, S., Datta, R. et al. (2006). Bioaccumulation and physiological effects of mercury in Sesbania drummondii. *Chemosphere* 65: 591–598. https://doi.org/10.1016/j.chemosphere.2006.02.016.

255 Yadav, S.K. (2010). Heavy metals toxicity in plants: an overview on the role of glutathione and phytochelatins in heavy metal stress tolerance of plants. *South African Journal of Botany* 76: 167–179. https://doi.org/10.1016/j.sajb.2009.10.007.

256 Zhang, W.H. and Tyerman, S.D. (1999). Inhibition of water channels by HgCl2 in intact wheat root cells. *Plant Physiology* 120: 849–858. https://doi.org/10.1104/pp.120.3.849.

257 Cargnelutti, D., Tabaldi, L.A., Spanevello, R.M. et al. (2006). Mercury toxicity induces oxidative stress in growing cucumber seedlings. *Chemosphere* 65: 999–1006. https://doi.org/10.1016/j.chemosphere.

258 Zhou, Z.S., Huang, S.Q., Guo, K. et al. (2007). Metabolic adaptations to mercury-induced oxidative stress in roots of *Medicago sativa* L. *Journal of Inorganic Biochemistry* 101: 1–9. https://doi.org/10.1016/j.jinorgbio.2006.05.011.

259 Duarte, B., Delgado, M., and Caador, I. (2017). The role of citric acid in cadmium and nickel uptake and translocation in halimioneportulacoides. *Chemosphere* 69: 836–840.

260 Chen, C., Chen, T.H., Lo, K.F. et al. (2004). Effects of proline on copper transport in rice seedlings under excess copper stress. *Plant Science* 166: 103–111.

261 Bishnoi, N.R., Sheoran, I.S., and Singh, R. (1993). Influence of cadmium and nickel on photosynthesis and water relations in wheat leaves of different insertion level. *Photosynthetica (Praha)* 28 (3): 473–479.

262 Bishnoi, N.R., Sheoran, I.S., and Singh, R. (1993). Influence of cadmium and nickel on photosynthesis and water relations in wheat leaves of different insertion level. *Photosynthetica* 28: 473–479.

263 Wu, L., Huang, Z.Z., and Burau, R.G. (1988). Selenium accumulation and selenium-salt cotolerance in five grass species. *Crop Science* 28 (3): 517–522.

264 Kabata, P.E. (2011). *Trace Elements in Soils and Plants*, 4e. Taylor & Francis, Boca Raton: CRC Press.

265 Reilly, C. (1996). Selenium. In: *Selenium in Food and Health* (eds. E. Chapman and W. Hall), 1–24. London: Blackie Academic & Professional.

266 Mikkelsen, R.L., Haghnia, G.H., and Page, A.L. (1989). Factors affecting selenium accumulation by crop plants. In: *Selenium in Agricultural and the Environment* (ed. L.W. Jacobs), 65–93. Madison, WI: American Society of Agronomy and Soil Science Society of America.

267 Mengel, K. and Kirby, E.A. (1987). *Principles of Plant Nutrition*, 687. Bern: International Potash Institute.

268 Garousi, F. (2017). Toxicity of selenium, application of selenium in fertilizers, selenium treatment of seeds, and selenium in edible parts of plants. *Alimentaria - Acta Universitatis Sapientiae* 10 (1): 61–74. https://doi.org/10.1515/ausal-2017-0004.

269 Terry, N., Zayed, A.M., de Souza, M.P. et al. (2000). Selenium in higher plants. *Annual Review of Plant Physiology and Plant Molecular Biology* 51 (1): 401–432. https://doi.org/10.1146/annurev.arplant.51.1.401.

270 Spallholz, J.E. and Hoffman, D. (2002). Selenium toxicity: cause and effects in aquatic birds. *Aquatic Toxicology* 57: 27–37. https://doi.org/10.1016/S0166-445X(01)00268-5.

271 Trelease, S.F. and Beath, O.A. (1949). *Selenium: Its Geological Occurrence and Its Biological Effects in Relation to Botany, Chemistry, Agriculture, Nutrition and Medicine*, 292. New York;, NY: Trelease and Beath.

272 Ernst, W. (1982). Selenpflanzen (Selenophyten). In: *Pflanzenökologie und Mineralstoffwechsel* (ed. H. Kinzel), 511–519. Stuttgart: Verlag Eugen Ulmer.

273 Shrift, A. (1969). Aspects of selenium metabolism in higher plants. *Annual Review of Plant Physiology* 20: 475–494. https://doi.org/10.1146/annurev.pp.20.060169.002355.

274 Hartikainen, H., Xue, T., and Piironen, V. (2000). Selenium as an antioxidant and pro-oxidant in ryegrass. *Plant and Soil* 225: 193–200. https://doi.org/10.1023/A:1026512921026.

275 Shah, V. and Belozerova, I. (2009). Influence of metal nanoparticles on the soil microbial community and germination of lettuce seeds. *Water Air Soil Pollution* 197: 143–148.

276 Tomacheski, D., Pittol, M., Simões, D.N. et al. (2017). Impact of silver ions and silver nanoparticles on the plant growth and soil microorganisms. *Global Journal of Environmental Science and Management* 3 (4): 341–350.

277 Keller, A.A., Mcferran, S., Lazareva, A. et al. (2013). Global life cycle releases of engineered nanomaterials. *Journal of Nanoparticle Research* 15: 1–17. https://doi.org/10.1007/s11051-013-1692-4.

278 Beer, C., Foldbjerg, R., Hayashi, Y. et al. (2012). Toxicity of silver nanoparticles—nanoparticle or silverion? *Toxicology Letters* 208: 286–292. https://doi.org/10.1016/j.toxlet.

279 Daniel, S.C.G.K., Tharmaraj, V., Sironmani, T.A. et al. (2010). Toxicity and immunological activity of silver nanoparticles. *Applied Clay Science* 48: 547–551. https://doi.org/10.1016/j.clay.2010.03.001.

280 Kumari, M., Mukherjee, A., and Chandrasekaran, N. (2009). Genotoxicity of silver nanoparticles in *Allium cepa*. *Science of the Total Environment* 407: 5243–5246.

281 Stampoulis, D., Sinha, S.K., and White, J.C. (2009). Assay-dependent phytotoxicity of nanoparticles to plants. *Environmental Science and Technology* 43: 9473–9479. https://doi.org/10.1021/es901695c.

282 Pokhrel, L.R. and Dubey, B. (2013). Evaluation of developmental responses of two crop plants exposed to silver and zinc oxide nanoparticles. *Science of the Total Environment* 453: 321–332.

283 Lenntech (n.d.) Uranium – U. () https://www.lenntech.com/periodic/elements/u.htm.

284 Shen, Z.G., Zhao, F.J., and McGrath, S.P. (1997). Uptake and transport of zinc in the hyperaccumulator Thlaspi caerulescens and the nonhyperaccumulator Thlaspi ochroleucum. *Plant Cell and Environment* 20: 898–906.

285 Watanabe, F.S., Lindsay, W.L., and Olsen, S.K. (1965). Nutrient balance involving P, Fe and Zn. *Soil Science Society of America Journal* 29: 562–565.

286 Ambler, J.E., Brown, J.C., and Gauch, H.G. (1970). Effect of zinc on translocation of iron in soybean plants. *Plant Physiology* 46: 320–323.

11

Molecular and Enzymatic Mechanism Pathways of Degradation of Pesticides Pollutants

Rangasamy Kirubakaran[1], Athiappan Murugan[2], and Javid A. Parray[3]*

[1] *Department of Biotechnology, Vysya College, Salem, Tamil Nadu, India*
[2] *Department of Microbiology, Periyar University, Salem, Tamil Nadu, India*
[3] *Department of Environmental Science, Govt Degree College, Edigah, Srinagar, Jammu and Kashmir, India*
**Corresponding author: rangasamykirubakaran@gmail.com*

11.1 Introduction

The most abundant soil pollutant in India is organophosphorus (OP) [1]. The modern agriculture sector uses pesticides of over four million tons annually [2]. OP compounds like chlorpyrifos, coumaphos, diazinon, fenamiphos, glyphosate, methyl parathion, monocrotophos, parathion, and phorate have been extensively used since 1952 [2–5]. The organophosphate pesticide constitutes heterogeneous compounds of monocrotophos and phosphoric acid that share a phosphoric acid derivative chemical structure [6–8]. Several pesticides serve as a food resource for several microfloras [9–11]. Livingstone [10] identified that pesticide contamination resulting from direct or indirect introduction can cause changes at all levels of biological systems. In the case of selecting pesticides for farming, it is very important to consider the toxicity to plant pathogens and also to farm animals (which should be low or nil) [12–14]. Several studies reported that the application of a specific or broad spectrum of pesticides in agricultural sites leads to an accumulation of poisonous pesticides in agricultural fields [8, 15] The essence of pesticides implicates toxicity only to the target organisms, whereas its impact on nontarget organisms should be negative leading to the pesticide's degradation [16–20]. Bioaccumulation of pesticides inhibits growth of aquatic invertebrates and causes allergic skin reactions and eye irritation in humans, acting as a possible carcinogen [18, 21]. The poisoning effect of pesticides on agricultural products leads to ailments such as asthma, allergies, and hypersensitivity.

Soil Bioremediation: An Approach Towards Sustainable Technology, First Edition.
Edited by Javid A. Parray, Abeer Hashem Abd Elkhalek Mahmoud, and Riyaz Sayyed.

The pesticide poisoned food intake is also linked to cancer and hormone disruption in reproduction and fetal problems [9, 22–24].

Food and agricultural organizations (FAOs) reported that more than 700 pest species were resistant to pesticides [25–27]. All inorganic pesticide compounds tend to inhibit the activity of microorganisms, living creatures, plants, and animals [28, 29] with the exception of some organisms that escape as they are genetically adapted and altered [30]. Development of antibiotic and pesticide resistance is often presented as a modern example of evolution by mutations, as a clear evidence for Darwinism [31]. The risks associated with use of certain pesticides and antibiotics have direct and indirect effects on human health [32, 33]. The multiple antibiotic resistances in disease-causing bacteria were developed by herbicides 2,4-dichlorophenoxy acetic acids and dicamba are the best example for the proposed hypothesis that pesticides can induce multidrug resistance [34]. Meta-analysis reviews critically discuss the drug-resistant property of bacteria imparted from pesticide-resistant bacteria on polluted agricultural land. The utilization of pesticides in the agricultural sector induces selective resistance in microorganisms *Verticillium* sp., *Bacillus* sp., *Providencia stuartii*, *Pseudomonas putida Lipomyces kononenkoae*, *Stenotrophomonas* sp., *Bacillus cereus*, and *Bacillus* sp., *Pseudomonas* sp. [35–37].

11.2 Effect of Pesticides on Soil Enzymes

Soil enzymes include free enzymes, bounds forms of extracellular enzymes, and enzymes within microbial cells [35, 38]. They are the indicators of biological equilibrium [39–41], fertility [42–44], quality [45, 46], and changes in the biological status of soil [47, 48]. The role of soil enzymes and their activities is defined by their relationships with soil and other environmental factors (e.g., acid rain, heavy metals, pesticides, and other industrial chemicals) that affect their activities [49, 50]. Previous reports have proven that pesticides negatively influence soil enzymes e.g., hydrolases, oxidoreductases, and dehydrogenase activities [44, 51–56]. Soil enzymes capable of ATP can transform pesticides into energy. Various researchers [57–60] have attributed such differences to the dual behavior of pesticides (both harmful and beneficial for soil enzymes), diversity, and various stages of the processes taking place in soil that frequently overlapped. Enzyme activity in soil reflects its state and fertility [61–63], which are also influenced by organic matter [64, 65], moisture [66, 67], and temperature [68]. Soil enzymes are responsible for multidimensional behaviors in a complex soil medium and the greater complexity of soil microbial and biochemical interactions [43, 62, 69, 70] researchers are facing difficulties in understanding the effects of pesticides on extracellular enzyme activities in soil. Further, it has been revealed that acetamiprid increased dehydrogenase activity up to 22% after the first pesticide application. Chen, Edwards, and Subler [71] studied three fungicides and all three fungicides had different effects on DHA in the amended soils. Metalaxyl (fungicide) application initially increased and then decreased the DHA in fungicide-treated soil [72–74]. Hydrolases are of particular importance on account of their role in soil nitrogen, phosphorus, carbon, and sulfur cycles. Urease is an enzyme that catalyzes the hydrolysis of urea into CO_2 and NH_3 and is a key component in the nitrogen cycle in soils [57]. Urease activity is found in a large number of soil bacteria and fungi [75]. It is also reported that there is unchanged, increased, or decreased urease activity following various pesticide applications [42, 71, 76, 77]. Urease produce NH_4 ions from urea, which is taken up by plants. Nitrification of an NH_4 ion into NO_3 ion in soil is rapid and NO_3 can leach into water. Singh et al. [78] identified that the chlorimuron ethyl and furadan enhanced urease activity up to 14–18% and 13–21%, respectively. Contrarily, acetamiprid reduced urease activity up to 35% in soil at 43 days following crop sowing. Phosphatase is an exocellular enzyme produced by many soil microorganisms and is responsible for the hydrolysis of organic phosphate compounds to inorganic phosphate [54]. Acid and alkaline phosphatases are mostly found in microorganisms and animals [40, 41]. Klose et al. [79] reported that soil pesticides reduced the activity and the total oxidative potential of microorganisms. They also indicated that soil pesticides could change microbial communities and important biochemical reactions involved in the cycling of elements.

11.3 Introduction of Plasmids in Soil Bacteria

Emergence of pesticide resistance is a global concern and it has been well documented that pesticide resistant bacteria are present in a wide number of agricultural products throughout the world [21, 80, 81]. Soil bacteria adapt rapidly to new environmental conditions and the presence or development of novel antimicrobial molecules have consequently increased parallel resistance [82]. Such selective processes leading to the emergence and maintenance of bacteria resistant to pesticide are mainly due to the abuse of pesticides [83]. Many bacteria living in contaminated environments develop multiple resistant properties [7]. Bacteria and insects often acquire resistance through a single, continuous, or spontaneous mutation. Changes in common genes allow the overproduction of an enzyme that breaks down pesticide compounds, antibiotics, and other poisons as a means of resistance mechanism [84].

In some cases, horizontal or vertical transfer of resistant genes may facilitate the development of resistance. Gene transfer seems to be of particular importance in environmental conditions, as resistant genes are often found on plasmids that are mobile [30, 85]. The emergence of numerous genes conferring resistance has recently increased due to use of diverse pesticides as suggested by Bernier and Surette [86]. Pesticide resistant genes are widely spread by the horizontal transfer of genetic material. These genes may be transferred by different mechanisms of conjugation, transformation, or transduction. Resistant genes can be incorporated into recipient chromosomes by means of a recombination mechanism, which may contain point mutations acquiring more sequence changes in a single gene [84, 87–90]. Pesticide resistance in most bacterial species, due to the acquisition of new genes is often associated with plasmids or transposable elements that are often conjugative. Among gram-negative and gram-positive bacteria, conjugative transposons are predictable as important mediators of genetic exchange on large R-plasmids [91–93].

The conjugative element naturally raises the risk of spreading resistance genes to other soil bacteria. Mutated pesticide resistance genes that have vertical and horizontal transfer mechanisms for the dissemination of multiple resistance genes are found as gene clusters [94–97]. These mutated genes are enabled by the existence of specific DNA structures called integrons, which are DNA elements with the ability to capture genes and by site-specific recombination with the element located on either bacterial chromosomes or plasmid DNA [93, 98, 99]. Integrons promote the capture of one or more gene cassettes within the same attachments thereby forming clusters of antibiotic and pesticide resistance genes [100–103], which lead to different types of resistance gene cassettes. Integron movements allow the transfer of the gene cluster-associated resistance gene from one DNA replicon to another [92] thus horizontal transfer of resistance genes is also achieved when an integron is incorporated into a broad host range of plasmids [104]. Mechanisms of drug resistance are found to be different with normal development of drug resistance among bacteria [105]. Drug resistant bacteria may over produce an enzyme that breaks down pesticide and other poisons [106]. Experimental proof of formaldehyde resistance in *Enterobacteriaceae* has shown high formaldehyde dehydrogenase activity that is responsible for multidrug resistance [25].

In the majority of the drug-resistance cases, the removal of the selective agent does not necessarily cause the organism to revert to a susceptible state [15]. The use of antibiotics and pesticides has increased extensively in recent times due to the increasing number of infectious diseases to human and plants. The bacterial plasmid DNA, in turn, strives to develop resistance to these xenobiotics gradually by induced mutations or through gene-regulation mechanisms [36]. Plasmid DNA molecules are capable of autonomous replication and transferable between different bacterial species and clones [107]. This leads to multidrug-resistant and pesticide-resistant bacterial strains in natural flora [108]. There are a number of ways through which the catabolic genes will be transferred from one bacteria to the other: transformation, transduction, and conjugation [92, 109].

11.4 Microbial Degradation of Pesticides

Biodegradation of pesticides is generally carried out using different biological systems, as microorganisms. It has been reported that a fraction of soil microbiota develops the ability to degrade certain pesticides, when they are continuously applied to the soil. These chemicals are found to provide an adequate carbon source and electron donors for certain soil microorganisms [110], creating a way for the treatment of pesticide-contaminated sites [111, 112]. However, the transformations of such compounds depend not only on the presence of microorganisms with appropriate degrading enzymes, but also on a wide range of environmental parameters [113]. Physiological, ecological, biochemical, and molecular aspects play an important role in the microbial transformation of pollutants [114, 115]. There are diverse sources of microorganisms, which have the ability to degrade pesticides. Because pesticides are mainly applied to agricultural crops, soil is the medium that mostly receives these chemicals, alongside pesticide industry's effluent, sewage sludge, activated sludge, wastewater, natural waters, sediments, areas surrounding the manufacture of pesticides, and even some live organisms. In general, microorganisms that have been identified as pesticide degraders can be isolated from a wide variety of sites contaminated with some kind of pesticide. At present, in different laboratories around the world, microorganisms are characterized by their identification, growth, and degradation of pesticides. The isolation and characterization of such microorganisms that are able to degrade pesticides has given the possibility for the restoration of polluted environments and to treat wastes before disposal [116].

New analytical and molecular tools have expanded our insights into the mechanisms of microbial degradation, their identity, and the occurrence of active players that effect biodegradation of environmental pollutants like pesticides [117]. Biodegradation that involves the capabilities of microorganisms in the removal of pollutants is the most promising, relatively efficient, and cost-effective technology. Biodegradation involves the complete separation of an organic compound in its inorganic constituents. Microbial transformation may be driven by energy needs or a need to detoxify the pollutants or may be fortuitous in nature [114–119].

Microbial populations of soil or aquatic environments are composed of diverse, synergistic, or antagonistic communities rather than a single strain. In natural environments, biodegradation involves transferring the substrates and products within a well-coordinated microbial community, a process referred to as metabolic cooperation [120]. Amer and Aly [121] reported that *pseudomonas*, is claimed to be the most efficient bacterial genus for the degradation of toxic compounds. The ability of these bacteria to degrade these compounds, is related to the contact time with the compound, the environmental conditions in which they develop, and their physiological versatility. Vasquez and Reyes [122], stated that *Pseudomonas* species are capable of biodegradation of herbicides. Moreover, *Botrytis cinereal* was found to eliminate the linuron and metroburon herbicides almost completely and other 31 isolated species were observed to eliminate metroburon. Senthilkumar et al. [123] reported that *Rhodococcus* sp. degrades triazines to nitrate. Fournier et al. [124] conducted a test to study the atrazine herbicide conversions resulting from microbial decomposition. It was found that after microbial action this compound was transformed into nitrite (30%), nitrous oxide (3.2%), ammonia (10%), and formaldehyde (27%). These microorganisms were reported to have enzymes involved in the hydrolysis of P—O, P—F, P—S, and P—C bonds, which are generally found in a wide variety of OP pesticides [125]. There are also studies reporting bacteria isolated from the soil are capable of degrading pesticides as ethyl-parathion and methyl-parathion (MP) [57]. Extracellular enzymes that act on a broad array of organic compounds such as lignin peroxidase, manganese peroxidase, laccase, and oxidases are involved in lignin degradation [35, 126, 127] have been produced. Several bacterial enzymes that help to degrade pesticides have been isolated and the list is expanding rapidly. The three main enzyme families implicated in degradation are esterases, glutathione S-transferases (GSTs), and cytochrome P450 [126]. Applying enzymes to degrade pesticides is currently an innovative treatment technique for removal of these chemicals from polluted environments. Enzyme-catalyzed degradation of a pesticide may be more effective than existing chemical methods [35].

Enzymes are also involved both in the target organism, through intrinsic detoxification mechanisms and evolved metabolic resistance, and in the wider environment, via biodegradation by soil and water microorganism [35]. Scott et al. [128] suggested that (i) the fundamental metabolism of the universal biodegradation networks involves transferases, isomerases, hydrolases, and ligases; (ii) linear pathways converging on particular intermediates are found to form a funnel topology; (iii) novel reactions occur in the exterior part of the network; and (iv) the possible pathway between compounds and the central metabolism can be arrived at by considering all the enzymes in a given organism as well as the intermediate compounds present [127, 128]. Several enzymes catalyze metabolic reactions including hydrolysis, oxidation, addition of an oxygen to a double bound, oxidation of an amino group (NH_2) to a nitro group, addition of a hydroxyl group to a benzene ring, dehalogenation, reduction of a nitro group (NO_2) to an amino group, replacement of a sulfur with an oxygen, metabolism of side chains, and ring cleavage. The process of biodegradation depends on the metabolic potential of microorganisms to detoxify or transform the pollutant molecule, which is dependent on both accessibility and bioavailability [39].

11.5 Enzymatic Degradation

Enzymatic metabolism of pesticides can be mediated through oxidation, reduction, hydrolysis, peroxidation, and dehalogenation mechanisms [17, 39, 129]. Cytochrome P450 enzymes can metabolize chlorpyrifos, these enzymes can cause chlorpyrifos to undergo dearylation (oxidative ester cleavage) forming 3,5,6-trichloro-2-pyridinol (TCP) and diethylthiophosphate. Chlorpyrifos degradation leads mainly to 3,5,6-trichloro-2-pyridinol, which will be degraded by oxidation and hydrolytic reactions of bacterial enzymes [130]. Sequential dechlorination of TCP by one oxidative and two hydrolytic steps forms 3,6-dihydroxypyridine-2,5-dione [39, 131, 132]. Degradation of 2–4 Dichlorophenoxy acetic acid was proved to have two different pathways, in one pathway the sixth carbon is oxidized by the addition of a OH group yielding 6-OH – 2,4-D, this is reaction is followed by the removal of acetate resulting in formation of 3,5 – Dichlorocatechol, whereas in another pathway two carbon side chains are removed resulting in glyoxylate and 2,4-DCP. These two degradation pathways were mediated by the isolates of *Pseudomonas* sp. and *Alcaligenes* sp. [133]. The oxygenases synthesized by *Pseudomonas* sp. have the ability to degrade the tetrachlorobenzene to trichlorocatechol by eliminating HCL (hydrochloric acid) from the compound [133–135] stated that mono and dioxygenases are actively involved in dehalogenation mediated degradation of halogen-based pesticides. Peroxidases synthesized by the fungus and bacteria are known to have a biodegradation role in pesticides and derived products, there are number of studies describing the mechanism of peroxidase mediated degradation. The 3,4-dichloroaniline is converted to 4,4-tetrachloroazobenzene by the peroxidases produced by soil microbes [136]. The peroxidases secreted by *Phanerochaete chrysosponum* added one chlorine to 2,4,-di, tri, 2,4,6-tri, and pentachlorophenol at the para position to form p-benzoquinone, as reported by Hammel and Tardone [137]. These peroxides were found to be mineralize 2,4,5 TCP rapidly.

11.6 Organophosphorus Hydrolase and Organophosphorus Dehydrogenase (OPH and OPD)

Among bacterial enzymes, OPH from *Pseudomonas diminuta* has received most attention due to its wide range of substrate specificity. OPH is a dimer of two identical subunits containing 336 amino acid residues [138] that folds into a (ab)8-barrel motif with a molecular weight of 72 kDa. Each subunit comprises binuclear zinc situated at the C-terminal portion. The two zinc atoms are separated by about 3.4 A° and are linked to the protein through the side chain of His 55, His 57, His 201, His 230, Asp 301, and a carboxylated Lys 169. Both the Lys 169 and the water molecule (or hydroxide ion) act to bridge the two zinc ions together [139]. Monocrotophos degradation was reported for dicrotophos, which is first demethylated to monocrotophos converted to methyl amine [140]. At the same time with other OP compounds, the first degradation step of monocrotophos should involve hydrolysis, which could produce dimethyl phosphate and N-methyl acetoacetamide [141]. Promote additional degradation of N-methyl acetoacetamide formed acetic acid in *Bacillus megaterium* and valeric acid in *Arthrobacter atrocyaneus* [142]. Acetic acid is the key intermediate of the glycolytic pathway in microorganisms. The path of monocrotophos degradation is shown in Figure 11.1.

OP compounds bind to the binuclear metal center within the active site via coordination of the phosphoryl oxygen to the b-metal ion. This interaction weakens the binding of the linking hydroxide to the b-metal. The metal–oxygen interaction polarizes the phosphoryl oxygen bond and creates a more electrophilic phosphorus center. Subsequent nucleophilic attack by the bound hydroxide is aided by proton abstraction from Asp 301. The hydroxide attacks the phosphorus center, resulting in weakening of the bond to the leaving group as mentioned by Raushel [143]. The role of one metal ion in the active site of OPH is to increase the electrophilicity of the phosphorus center through coordination with the nonester oxygen atom of the substrate, whereas the second metal ion acts as a promoter of the attacking nucleophile [144]. The pKa value of the bridging solvent is believed to be determined by variation of the kinetic parameters with pH. In addition, it has still to be resolved whether the two Zn ions within the active site have distinct functions or whether they act in tandem [143]. It hydrolyzes P—O, P—F, and P—S bonds to have a wide range of substrate specificity. The lowest specificity is for the P—S bond. However, the enzyme does not catalyze the cleavage of carbonyl groups such as those found in p-nitrophenyl acetate. Similarly, organophosphate diesters are very poor substrates [145]. The effects of metal substitution on the catalytic activity of OPH were studied by removing the native metal (Zn) from purified OPH and reconstitution with a series of divalent cations, which include Co, Cd, Cu, Fe, Mn, and Ni [139, 146]. Divalent cations increased the activity of enzymes by assisting folding of expressed enzyme in the medium [147]. The recent finding of OP- encoding genes in a field soil, which has never been exposed to this group of pesticides, supports this hypothesis [148]. Another possibility is that this enzyme has evolved new substrate specificity from preexisting enzymes as it has been shown that an OPH (phosphotriesterase) could acquire phosphodiesterase activity by alteration of only one amino acid [149]. Urease has been found to have carbamylated lysine as a bridging ligand with binuclear Ni at the active site [150].

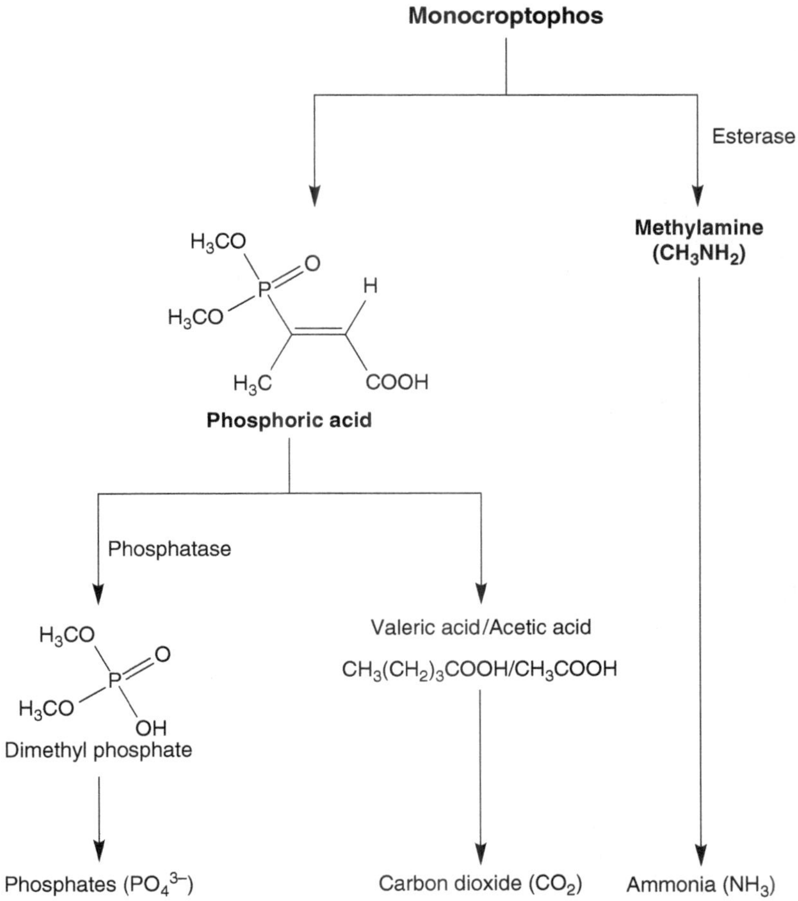

Figure 11.1 Monocrotophos biodegradative pathways [142].

The binuclear center of urease and OPH was found to be remarkably similar. However, the chemical nature of the active sites of these enzymes is quite different [143]. A larger group of enzymes with similar active-site architecture has been identified [151]. Interestingly this superfamily also includes atrazine chlorohydrolase. A similar enzyme, OPDA, has been isolated from *Agrobacterium radiobacter* and was found to have 90% homology to OPH at the amino acid level and similar overall secondary structure [152, 153]. Regardless of these similarities, the two enzymes have dissimilar substrate specificities. There is about a 30-sequence difference between OPH and OPDA. Yang et al. [153] stated that alteration in the active site of OPH gave substrate specificity and represented the progressive natural evolution of the enzyme from OPH to OPDA. A highly active OPAA (organophosphorus acid anhydrolase) from *Alteromonas undina* was isolated and purified and is comprised of a single polypeptide with molecular weight 53 kDa [81]. OPAA isolated from *Alteromonas* sp. JD6.5 is composed of 517 amino acids with a molecular weight of 60 kDa. However, one from *Alteromonas haloplanktis* contains a similar enzyme with 440 amino acids with molecular weight 50 kDa [154–157]. The 10 kDa difference between OPAAs of these two *Alteromonas* spp. was found to be due to the presence of an extended C-terminal region in the JD 6.5 enzyme [158]. The three-dimensional structure of this enzyme is not yet known. It possesses low catalytic activity against P—O but high activity against P—F bonds. OPAA also displays stereoselectivity toward the chiral phosphorus center by displaying preference for the Rp-enantiomers. It also exhibits an additional preference for the stereochemical configuration at the chiral carbon center of the soman analogue [159]. OPAAs from different species of *Alteromonas* have shown wide variation in catalytic activity, with the highest activity observed with the enzyme obtained from *Alteromonas* sp. J.D.6.5 [158]. OPAA from *Alteromonas* sp. JD6.5 has a high degree of homology at the amino acid level with *Escherichia coli* X-Pro dipeptidase (48%) and *E. coli* aminopeptidase P (31%). Further molecular and biochemical analyses of OPAA have established that this enzyme is a prolidase; a type of dipeptidase cleaving dipeptide bond with a prolyl residue at the carboxyl terminus [155]. Molecular modeling studies with Soman and Leu-Pro [158] revealed that the three-dimensional structure and electrostatic density maps of the two are nearly identical. This explained why several dipeptidase enzymes have catalytic activity against OP compounds. Determination of the crystal structure of OPAA confirmed this hypothesis. Both prokaryotes and several eukaryotes have been found to possess this enzyme [159–161], which suggests that OPAA is not a newly evolved enzyme [155]. Three unique parathion hydrolases were isolated, purified, and characterized from gram-negative bacterial isolates. One cytosolic hydrolase described as an ADPase (aryl di-alkyl phosphatase) from *Nocardia* sp. strain B-1 was composed of a single sub-unit of approximately 43 kDa [162]. Another hydrolase from strain SC was membrane bound and was composed of four identical subunits of 67 kDa. While having some common features such as, constitutive production and similar temperature optima around 40 °C, it was found that the substrate specificity and structure of these enzymes differed one from another, and also from the other known OPHs [163]. Laccase was observed to be capable of complete rapid degradation of VX and Russian VX [164]. Several white-rot fungi are capable of OP degradation. A novel OP degrading fungal enzyme (A-OPH) was isolated from *Aspergillus niger* ZHY256 that could hydrolyze a range of P—S bonds containing OP compounds [9]. A-OPH does not require divalent cations for activation; however, Cu21 was found to activate its activity. Another novel OP-hydrolyzing enzyme was purified from

Penicillium lilacinum BP303. Interestingly, this penicillium OPH (P-OPH) was found to degrade various OP compounds by cleaving both P—O and P—S linkages [9].

11.7 Pesticide–Antibiotic Cross Resistance

Antibiotics uses have been widespread since the middle of the twentieth century. A major challenge in addressing antibiotic resistance is that both human and microbial factors contribute to its persistence in an evolutionary context, considering use of antibiotics in both human and veterinary medicine [15]. Many bacteria have a built-in resistance to a number of antibiotics and the genes that provide this resistance can be passed among the bacterial community by vertical or horizontal gene transfer [31]. Contaminants resulting from the direct or indirect introduction of drugs and chemicals can cause changes at all levels of biological organizations [10]. A common relationship between antibiotic input and resistance expansion is still in debate, either supporting or declining the contribution of antibiotics. The antibiotics that are thought to confer drug resistance are divided into different classes such as ß-lactams, quinolones, tetracycline, macrolides, and sulfonamides [30, 165]. When a bacterial strain develops resistance against a pesticide through gene mutations and if the same strain resists a range of antibiotics, the newly developed resistance to antibiotics is called a pesticide–antibiotic cross resistance. The evolution of pesticide degradation pathways along with genetic mutations led to the increased level of multidrug resistance among soil bacteria due to the continuous supply of pesticides [31, 34, 36]. They may be transferred into human or animal systems through food or other routes and resides in the gut of the host (Figure 11.2).

Consequently, the pesticide-resistant bacteria develop as multidrug resistant bacteria in the host through cross resistance. Therefore, the excess exposure to pesticides adversely affects active site enzymes often exploited in antibiotic resistant pesticide. Antibiotic cross resistance is acquired by adapting some common strategies against pesticides and antibiotics: (i) bacteria-forming biofilms that evade the action of xenobiotics (pesticide and antibiotics); (ii) by having analogous active sites for pesticides and antibiotics; (iii) by up-regulated expression of genes of stress-tolerant enzymes; (iv) increase in copy numbers of genes essential to tolerate the stress; (v) acquiring new antibiotic resistant genes along the gene cluster (for pesticide resistance) that transfer into the soil bacteria, these antibiotic resistant genes and pesticide resistant genes are entirely different, yet they were acquired together; and (vi) acquiring antibiotic resistant genes along with pesticide resistant genes through transposon-mediated horizontal transfer. These are some of the demonstrated strategies that confer cross resistance to soil

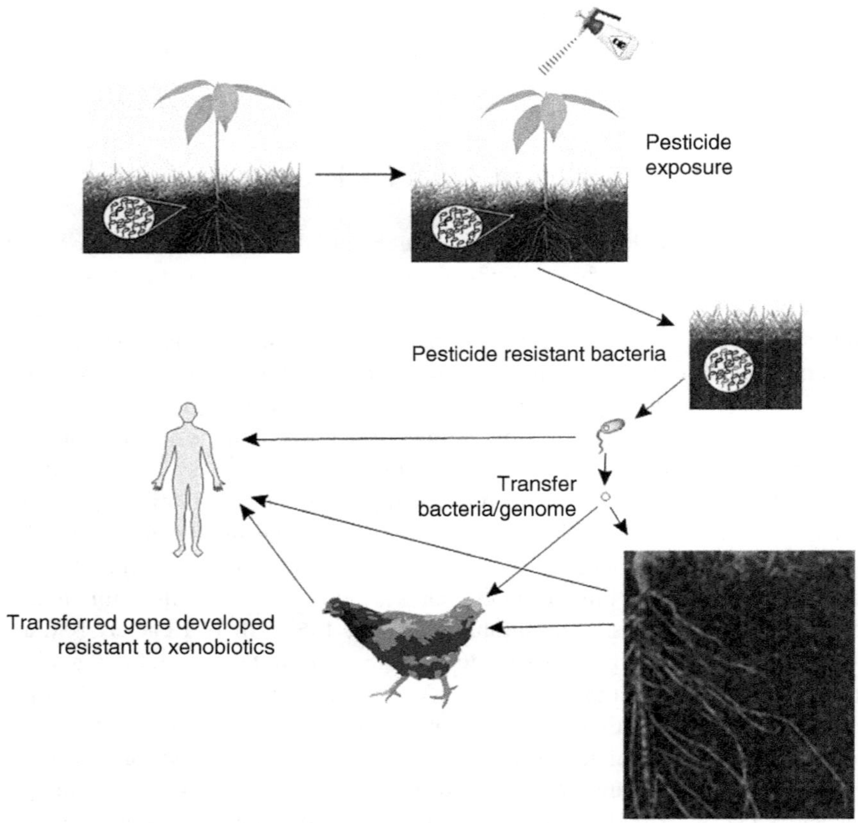

Figure 11.2 Bioaccumulation of pesticide.

microbial flora against antibiotics [37]. Wild type bacterial species (*E. coli*, *Salmonella typhimurium*) were exposed to sublethal levels of herbicide (Kamba, Roundup, and 2,4-D) continuously, subsequently they developed resistance against antibiotics like ampicillin, tetracyclin, ciplofloxacin, kanamycin, and chloramphenicol [34]. Despite these reports, there is no solid research that supports the claim to an elaborated mechanism of cross resistance. This specific area needs a focus to develop therapeutic strategies in future to overcome pesticide–antibiotic cross resistance. In the evolutionary relationship, Darwin's theory defines evolution as descent with modifications [166]. Development of antibiotic and pesticide resistance is often presented as a modern example of evolution by mutations, as a clear evidence for Darwinism [31]. The risks associated with use of certain pesticides and antibiotics have direct and indirect effects on human health [32, 33]. The multiple antibiotic resistances in disease-causing bacteria were developed by herbicides, 2,4-dichlorophenoxy acetic acids and dicamba are good examples of pesticides that can induce multidrug resistance [34]. The observed ecological impacts are also in accordance with Darwin's theory. The parallel break down of antibiotics/pesticides or parallel resistance can also be due to behavioral change as a result of humans changing the environment. The multidrug resistance among soil bacteria is a natural selection process where bacteria adapt to a new environment [36]. Through the evolution of innovative pesticide degradation pathways, continuous application and persistence of pesticides may lead to increased multidrug resistance among soil bacteria, and genetic mutation has shown the existence of pesticides to select populations that are immune to both pesticides and antibiotics [8]. Continuous exposure to pesticides would contribute constant pressure among the genes thereby developing drug resistance. The prolonged application of disinfectants at suboptimal concentrations provokes the adaption of microorganisms to those subconcentrations and possibly the development of true resistance [32, 33]. The loss of porin proteins, which may facilitate the permeation of molecules through the cell membrane, seems to contribute to the resistance as mentioned. The expulsion of toxic materials through active secretion is another way that living cells protect themselves from poisoning. The enzyme is nonspecific for low-molecular weight aldehydes and this may explain the cross resistance to other aldehydes and may also explain the cross resistance to other aldehydes or formaldehyde releasing compounds [37, 165]. Bacteria also pose multidrug resistance by producing higher copies of target molecules so that the previous antibiotics concentration is no longer sufficient for metabolic processes. Mere transfer of transposable elements or plasmid simply stretched out cytoplasm and has contributed multidrug resistance to many other bacteria habituating the soil. It has been proved that strains treated with SDS were able to revert to antibiotic sensitivity. Studies have quoted a direct comparison between plasmid-containing and plasmid-cured cells proving a loss of drug resistance among plasmid-cured bacterial cells [37]. Multidrug resistant populations were quite common among pesticide degrading soil flora due to a gene carried on self-transmissible genes that can jump between plasmid and chromosomes [108]. Resistance gene cassettes have been found for most classes of pesticides and the gene products are involved in various resistance mechanisms. Integron movements allow transfer of the gene cluster associated resistance gene from one DNA replicon to another. Horizontal transfer of resistance genes can be achieved when an integron is incorporated into a broad host range of plasmids. The resistance gene encoding gene establishes a resistance gene clusters, which represents a potential source for gene horizontal transfer between bacteria. The horizontal gene transfer between bacterial species such as pesticide resistance genes or transposable elements has been identified.

Continuous application/accumulation of pesticide in the agricultural field, even at a low level could exert selective pressure toward selection of bacteria. Bacteria enzymes slowly develop cross resistance to antibiotics, this has been proved by removing plasmid from the selected isolates that subsequently lost the ability to grow/use the pesticides [92, 109].

There are previous reports of OP hydrolase from *Geobacillus* sp. docked with monocrotophos and antibiotics such as streptomycin, chloramphenicol, ampicillin, and cephatoxime [37, 167]. There are neither identical sequences available nor the nearest neighbor in the blast-and-docking analysis. Hence, an alternative method for finding the homologous protein is required. The identification of antibiotics that fit to hydrolase was identified, based on the docking score and the number of hydrogen bonds that show a high range of interaction in chloramphenicol among antibiotics. The OPH isolated from *Pseudomonas* sp. strains catalyze the degradation of the organophosphate pesticide MP (methyl parathion) with a high efficiency [168, 169]. The structure of OPH (PDB ID: 1P9E) has been resolved, revealing a typical αβ/βα sandwich fold structure of the metallo-β-lactamase superfamily [170]. In parallel with MPH (Methyl parathion hydrolase), a newly identified OPH, OPHC2, isolated from *Pseudomonas pseudoalcaligenes*, is a thermo stable OPH [171] in β-lactamase superfamily, which shares 46% sequence identity with MPH. PoOPH, a newly discovered lactonase from *Pseudomonas oleovorans*, exhibited high lactonase and esterase activities, but low OPH activity [172]. PoOPH shares 98.5% sequence identity with OPHC2. By simple double mutations His250Ile/Ile263Trp, PoOPH was switched into an efficient OPH, displaying 6962-fold improvement in catalytic efficiency toward MP [172]. The study proves the emergence of efficient and robust enzymes for OP detoxification by OPH activity evolution in the β-lactamase superfamily. There are reports that a distinctive putative OPH (AbOPH) identified from *Acinetobacter* sp. exhibited high lactonase activity with latent OPH activity. Compared with OPHC2 and MPH in the β-lactamase superfamily, AbOPH acts as a separate branch in the evolutionary tree and there are four additional amino acids (Gly206, Thr207, Val208, and Glu209) in its sequence analysis. Locating the key residues that affect its OPH activity was investigated by molecular docking studies [171–174].

11.8 Conclusions

An excessive and indiscriminate usage of chemical pesticide serves as environmental stress and disposes the microbial community to become resistant to such xenobiotics. Various factors are involved in cross resistance through pesticide-resistant genes or enzymes and phenotype being transferred to wild microbial flora, which may develop different pesticide-resistant properties. It is understood the indiscriminate use of pesticide should be regulated in order to avoid the emergence of multidrug resistance bacteria. Antibiotics used in the agriculture and poultry industry contribute to multidrug resistance to a greater extent than which was expected. A future approach to combat multidrug resistance would rely on the examining of metagenomics DNA and modification of DNA with respect to exposure against various xenobiotics. Parallel to analyzing changes in the DNA, functional evaluation of the different enzymes and evaluating catalytic property of the enzymes would reveal the possible mechanism of antibiotic degradation by the nonspecific cleavage.

Acknowledgments

The authors acknowledge the Periyar University, Salem, Tamil Nadu, India, for a University Research Fellowship and DST-FIST Grant no. SR/FST/LSI-640/2015(c).

References

1 Jiries, A.G., Nasir, F.M., and Beese, F. (2002). Pesticide and heavy metals residue in wastewater, soil and plants in waste disposal site near Al-Lajoun Valley, Karak/Jordan. *Water, Air, and Soil Pollution* 133: 97–107.

2 Chen, H., He, X., Rong, X. et al. (2009). Adsorption and biodegradation of carbaryl on montmorillonite, kaolinite and goethite. *Applied Clay Science* 46: 102–108.

3 Seo, J.S., Keum, Y.S., Harada, R.M. et al. (2007). Isolation and cauterization of bacteria capable of degrading polycyclic aromatic hydrocarbons (PAHs) and organophosphorus pesticides from PAH-contaminated soil in Hilo, Hawali. *Journal of Agricultural and Food Chemistry* 55: 5383–5389. https://doi.org/10.1021/jf0637630.

4 Damalas, C.A. (2009). Understanding benefits and risks of pesticides use. *Scientific Research and Essay* 4: 945–949.

5 Zhang, W., Jiang, F., and Ou, J. (2011). Global pesticide consumption and pollution: with China as a focus. *International Academy of Ecology Environment Sciences* 2: 125–144.

6 Quin, L.D. (2000). *A Guide to Organophosphorus Chemistry*. Hoboken, NJ: Wiley.

7 Yadav, S., Kumar, S., Hotam, V. et al. (2015). Isolation and characterization of organophosphate pesticides degrading bacteria from contaminated agricultural soil. *Online Journal of Biological Sciences* 15: 113–125. https://doi.org/10.3844/ojbsci.2015.113.125.

8 Kirubakaran, R., Murugan, A., Chinnathambi, P. et al. (2017). Influence of residual pesticide on plant growth promoting bacteria isolated from agriculture field. *Journal of Basic Applied Plant Sciences* 1 (2): 110.

9 Liu, C.J., Men, W.J., and Liu, Y.J. (2002). The pollution of pesticides in soils and its bio-remediation. *System Sciences and Comprehensive Studies in Agriculture* 18: 295–297.

10 Livingstone, D.R. (1998). The fate of organic xenobiotics in aquatic ecosystems: quantitative and qualitative differences in biotransformation by invertebrates and fish. *Comparative Biochemistry and Physiology* 120: 43–49. https://doi.org/10.1016/S1095-6433.

11 Bataillon, S., Tattevin, P., Maillard, J. et al. (2016). Efflux pump induction by quaternary ammonium compounds and fluoroquinolone resistance in bacteria. *Future Microbiology* 11 (1): 81–92. https://doi.org/10.2217/fmb.15.131.

12 Lambert, R.J.W. (2004). Comparative analysis of antibiotic and antimicrobial biocide susceptibility data in clinical isolates of methicillin-sensitive *Staphylococcus aureus*, methicillin-resistant. *Applied Microbiology* 97: 699–711. https://doi.org/10.1111/j.1365-2672.2004.02345.x.

13 Palumbi, S.R. (2001). Evolution human as the world's greatest evolutionary force. *Science* 293: 1786–1790. https://doi.org/10.1126/science.293.5536.1786.

14 Pan-Hou, H.S.K., Nishimoto, M., and Imura, N. (1981). Possible role of membrane proteins in mercury resistance of *Enterobacter aerogenes*. *Archives of Microbiology* 130: 93–95.

15 Orzech, K.M. and Nichter, M. (2008). From resilience to resistance: political ecological lessons from antibiotic and pesticide resistance. *Annual Review of Anthropology* 37: 267–282. https://doi.org/10.1146/annurev.anthro.37.081407.085205.

16 Zhang, R., Cui, Z., Jiang, J. et al. (2005). Diversity of organophosphorus pesticides degrading bacteria in a polluted soil and conversation of their organophosphorus hydrolase genes. *Canadian Journal of Microbiology* 5: 337–343.

17 Ortiz-Hernandez, M.L. and Sanchez-Salinas, E. (2010). Biodegradation of the organophosphate pesticide tetrachlorvinphos by bacteria isolated from agricultural soils in Mexico. *Revista Internacional de Contaminacion Ambiental* 26: 27–38.

18 Shah, B.P. and Devkota, B. (2009). Obsolete pesticides: their environmental and human health hazards. *The Journal of Agriculture and Environment* 10: 51–56.

19 Soto, A., Gonzalez, H., Mora, M. et al. (2017). Effects of oxytetracycline on the performance and activity of biomixtures: removal of herbicides and mineralization of chlorpyrifos. *Journal of Hazardous Materials* 321 (5): 1–8. https://doi.org/10.1016/j.jhazmat.2016.08.078.

20 Pandit, V., Seshadri, S., Rao, S.N. et al. (2011). A case of organophosphate poisoning presenting with seizure and unavailable history of parenteral suicide

attempt. *Journal of Emergencies, Trauma, and Shock* 4 (1): 132–134. https://doi.org/10.4103/0974-2700.76825.

21 Tallur, P.N., Megadi, V.B., and Ninnekar, H.Z. (2008). Biodegradation of cypermethrin by *Micrococcus* sp. strain CPN. *Biodegradation* 19: 77–82.

22 Pandit, N.P., Ahmad, N., and Maheshwari, S.K. (2012). Vermicomposting biotechnology: an eco-loving approach for recycling of solid organic wastes into valuable biofertilizers. *Journal of Agricultural Science and Food Research* 2: 113.

23 Pandit, S.V. and Mundhe, A.Y. (2013). Monocrotophos induced behavioral stress, biochemical and histological alterations in *Lamellidens marginalis* (Lamarck). *The International Quality Journal of Life Science* 8: 1053–1056.

24 Berasategui, A., Shukla, S., Salem, H. et al. (2015). Potential applications of insect symbionts in biotechnology. *Applied Microbiology and Biotechnology* 100: 1567–1577. https://doi.org/10.1007/s00253-015-7186-9.

25 Heinzel, M. (1998). Phenomena of biocide resistance in microorganisms. *International Biodeterioration and Biodegradation* 41: 225–234.

26 Hussaini, S.Z., Shaker, M., and Iqbal, M.A. (2013). Isolation of bacterial for degradation of selected pesticides. *Advance Bioresource* 4: 82–85. http://dx.doi.soeagra.com/abr/abr.htm.

27 Hussaini, S.Z., Shaker, M., and Iqbal, M.A. (2013). Isolation of fungal isolates for degradation of selected pesticides. *Bulletin of Environment, Pharmacology and Life Sciences* 2 (4): 50–53.

28 Nieradko-Iwanicka, B. and Borzęcki, A. (2008). Effect of bifenthrin on memory processes, movement activity, and coordination in mice exposed to transient cerebral oligaemia. *Bulletin of the Veterinary Institute in Pulawy* 52: 291–294.

29 Khalid, S., Hashmi, I., and Khan, S.J. (2016). Bacterial assisted degradation of chlorpyrifos: the key role of environmental conditions, trace metals and organic solvents. *Journal of Environmental Management* 168: 1–9. https://doi.org/10.1016/j.jenvman.2015.11.030. PMID: 26692411.

30 Ding, C. and He, J. (2010). Effect of antibiotics in the environment on microbial populations. *Applied Microbiology and Biotechnology* 87: 925–941.

31 Bergman, J. (2003). Does the acquisition of antibiotic and pesticide resistance provide evidence for evolution? *Journal of Creation* 17: 26–32.

32 Thi-Tu, H., Frederic-Silvestre, Y.Z., Nguyen-Thanh, P.Z. et al. (2010). Effects of pesticides and antibiotics on penaeid shrimp with special emphases on behavioral and biomarker responses. *Environmental Toxicology and Chemistry* 29: 929–938.

33 Anjum, R. and Krakat, N. (2015). Improper antibiotic utilization evokes the dissemination of resistance in biotic environments-a high risk of health hazards. *Pharmaceutica Analytica Acta* 6: 12.

34 Kurenbach, B., Marjoshi, D., Amabile-Cuevas, C.F. et al. (2015). Sublethal exposure to commercial formulations of the herbicides dicamba,

2,4-dichlorophenoxyacetic acid, and glyphosate cause changes in antibiotic susceptibility in *Escherichia coli and Salmonella entericaserovar*, *Typhimurium*. *MBio* 6 (2): e00009–e00015.

35 Riya, P. and Jagapati, T. (2012). Biodegradation and bioremediation of pesticides in soil: its objectives, classification of pesticides, factors and recent developments. *World Journal of Science and Technology* 2: 36–41.

36 Anjum, R. and Krakat, N. (2016). Detection of multiple resistances, biofilm formation and conjugative transfer of *Bacillus cereus* from contaminated soils. *Current Microbiology* 72 (3): 321–328. https://doi.org/10.1007/s00284-015-0952-1.

37 Kirubakaran, R., Murugan, A., Natarajan, D. et al. (2017). Emergence of multi drug resistance among soil bacteria exposing to insecticides. *Microbial Pathogenesis* 105: 153–165. https://doi.org/10.1016/j.micpath.2017.02.011.

38 Mayanglambam, T., Vig, K., and Singh, D.K. (2005). Quinalphos persistence and leaching under field conditions and effects of residues on dehydrogenase and alkaline phosphomonoesterases activities in soil. *Bulletin of Environmental Contamination and Toxicology* 75 (6): 1067–1076.

39 Ramakrishnan, B., Megharaj, M., Venkateswarlu, K. et al. (2011). Mixtures of environmental pollutants: effects on microorganisms and their activities in soils. *Reviews of Environment Contamination and Toxicology* 211: 63–120.

40 Tabatabai, M.A. (1994). Soil enzymes. In: *Method of Soil Analysis. Part 2. Microbiological and Biochemical Properties* (eds. R.M. Weaver, S. Angle, P. Bottomley, et al.), 775–833. Madison, WI: Soil Science Society of America.

41 Tabatabai, M.A. (1982). Soil enzymes in method of soil analysis. Part 2. Chemical and microbiological properties. In: *Agronomy Monograph*, vol. 9 (eds. A.L. Page, R.H. Miller and D.R. Keeney), 903–947.

42 Antonious, G.F. (2003). Impact of soil management and two botanical insecticides on urease and invertase activity. *Journal of Environmental Science and Health. Part. B* 38: 479–488.

43 Nannipieri, P. (1994). The potential use of soil enzymes as indicators of productivity, sustainability and pollution. In: *Soil Biota: Management in Sustainable Farming Systems* (eds. C.E. Pankhurst, B.M. Doube, V.V.S.R. Gupta, et al.), 238–244. East Melbourne: CSIRO.

44 Schuster, E. and Schroeder, D. (1990). Side effects of sequentially and simultaneously applied pesticides on non-target soil organisms: laboratory experiments. *Soil Biology and Biochemistry* 22: 375.

45 Bucket, J.Z. and Dick, R.P. (1998). Microbial and soil parameters in relation to N mineralization in soils of diverse genesis under differing management systems. *Biology and Fertility of Soils* 27: 430–438.

46 Dick, R.P., Doran, J.W., Coleman, D.C. et al. (1994). Soil enzyme activities as indicators of soil quality. *Soil Enzymology* 7: 119–148. https://doi.org/10.1007/978-3-642-14225-3_7.

47 Kucharski, J., Jastrzebska, E., Wyszkowska, J. et al. (2000). Effect of pollution with diesel oil and leaded petrol on enzymatic activity of the soil. *Zeszyty Problemowe Postępów Nauk Rolniczych* 472: 457–464.

48 Trasar-Cepeda, C., Leiros, M.C., Seoane, S. et al. (2000). Limitation of soil enzymes as indicators of soil pollution. *Soil Biology and Biochemistry* 32: 1867–1875.

49 Fernandez-Perez, M., Garrido-Herrera, F.J., and Gonzalez-Pradas, E. (2011). Alginate and lignin-based formulations to control pesticides leaching in a calcareous soil. *Journal of Hazardous Materials* 190 (1): 794–801.

50 Burns, R.G. (1982). Enzyme activity in soil: location and a possible role in microbial ecology. *Soil Biology and Biochemistry* 14: 423–427.

51 Ismail, B.S., Yapp, K.F., and Omar, O. (1998). Effects of metsulfuron-methyl on amylase, urease, and protease activities in two soils. *Australian Journal of Soil Research* 36: 449–456.

52 Malkomes, H.P. (1989). Identification of herbicide effects on microbial activities in soil samples after storage. *Microbiogical Research* 144: 389–398.

53 Menon, P., Gopal, M., and Parsad, R. (2005). Effects of chlorpyrifos and quinalphos on dehydrogenase activities and reduction of Fe 3þ in the soils of two semi-arid fields of tropical India. *Agriculture, Ecosystems and Environment* 108: 73–83.

54 Monkiedje, A. and Spiteller, M. (2002). Effects of the phenylamide fungicides, mefenoxam and metalaxyl, on the biological properties of sandy loam and sandy clay soils. *Biology and Fertility of Soils* 35: 393–398.

55 Perucci, P. and Scarponi, L. (1994). Effects of the herbicide imazetapir on soil microbial biomass and various soil enzyme activities. *Biology and Fertility of Soils* 17: 237–240.

56 Tu, C.M. (1992). Effect of some herbicides on activities of microorganisms and enzymes in soil. *Journal Environmental Science and Health, Part B* 27: 695–709.

57 Megharaj, M., Singleton, I., Kookana, R. et al. (1999). Persistence and effects of fenamiphos on native algal populations and enzymatic activities in soil. *Soil Biology and Biochemistry* 31 (11): 1549–1553.

58 Shukla, A.K. (1997). Effect of herbicides butachlor, fluchloralin, 2,4-d and oxyfluorfen on microbial population and enzyme activities of rice field soil. *Indian Journal of Ecology* 24: 189–192.

59 Kanazawa, S. and Filip, Z. (1986). Distribution of microorganisms, total biomass, and enzyme activities in different particles of brown soil. *Microbial Ecology* 12: 205–215.

60 Malkomes, H.P. (1997). Applications of ecotoxicity tests to assess side effects of pesticides in soils. In: *Soil Ecotoxicology* (eds. J. Tarradellas, G. Bitton and D. Rossel), 386. Boca Raton, FL: Lewis Publishers.

61 Kruger, M., Schledorn, P., Schrodl, W. et al. (2014). Detection of glyphosate residues in animals and humans. *Journal of Environmental and Analytical Toxicology* 4: 1–5.

62 Nannipieri, P., Ceccanti, B., and Gregos, S. (1990). Ecological significance of the biological activity in soil. In: *Soil Biochemistry* (eds. J.M. Bollag and G. Stotzky), 293–355. New York;: Marcel Dekker.

63 Skujins, J. (1976). Extracellular enzymes in soil. *Critical Reviews in Microbiology* 4: 383–421.

64 Kandeler, E., Luxhi, J., Tscherko, D. et al. (1999). Xylanase, invertase and protease at the soil–litter interface of a loamy sand. *Soil Biology and Biochemistry* 31: 1171–1179.

65 Lalande, B., Gagon, B., and Simard, R.R. (1998). Microbial biomass C and alkaline phosphatase activity in two compost amended soils. *Canadian Journal of Soil Science* 78: 581–587.

66 Bergstrom, D.W., Montreal, C.M., Millette, J.A. et al. (1998). Spatial dependence of soil enzyme activities along a slope. *Soil Science Society of America Journal* 62: 1302–1308.

67 Ross, D.J. and Speir, T.W. (1984). Temporal fluctuations in biochemical properties of soil under pasture. II. Nitrogen mineralization and enzyme activities. *Australian Journal of Soil Research* 22: 319–330.

68 Tscherko, D., Kandeler, E., and Jones, T.H. (2001). Effect of temperature on below-ground N-dynamics in a weedy model ecosystem at ambient and elevated atmospheric CO_2 levels. *Soil Biology and Biochemistry* 33: 491–501.

69 Wu, X., Ernst, F., and Conkle, J.L. (2013). Comparative uptake and translocation of pharmaceutical and personal care products (PPCPs) by common vegetables. *Environment International* 60: 15–22. https://doi.org/10.1016/j.envint.2013.07.015.

70 Pancholy, S.K. and Rice, E.L. (1973). Soil enzymes in relation to old field succession: amylase, cellulase, invertase dehydrogenase and urease. *Soil Science Society of America Journal* 37: 47–50.

71 Chen, S.K., Edwards, C.A., and Subler, S. (2001). A microcosm approach for evaluating the effects of the fungicides benomyl and captan on soil ecological processes and plant growth. *Applied Soil Ecology* 18: 69–82.

72 Sukul, P. (2006). Enzymes activities and microbial biomass in soil as influenced by metalaxyl residues. *Soil Biology and Biochemistry* 38: 320–326.

73 Greaves, M.P., Davies, H.A., Marsh, J.A.P. et al. (1976). Herbicides and soil microorganisms. *CRC Critical Reviews in Microbiology* 5 (1): 1–38. https://doi.org/10.3109/10408417609102308.

74 Anderson, R.A., Armstrong, S., and Smith, N. (1981). Methods to evaluate pesticide damage to the biomass of the soil microflora. *Soil Biology and Biochemistry* 13 (2): 149–153. https://doi.org/10.1016/0038-0717 (81)90011-0.

75 Sarathchandra, S.U., Perrott, K.W., and Upsdell, M.P. (1984). Microbiological and biochemical characteristics of a range of New Zealand soils under established pasture. *Soil Biology and Biochemistry* 16: 177–183.

76 Ingram, C.W., Coyne, M.S., and Williams, D.W. (2005). Effects of commercial diazinon and imidacloprid on microbial urease activity in soil. *Journal of Environmental Quality* 34: 1573–1580.

77 Nowak, A., Nowak, J., Klodka, D. et al. (2004). Changes in the microflora and biological activity of the soil during the degradation of isoproturon. *Journal of Plant Disease Protection* 19: 1003–1016.

78 Singh, B.K., Walker, A., Morgan, J.A.W. et al. (2003). Role of soil pH in the development of enhanced biodegradation of fenamiphos. *Applied and Environmental Microbiology* 69: 7035–7043.

79 Klose, S.S., Acosta-Martinez, V., and Ajwa, H.A. (2006). Microbial community composition and enzyme activities in a sandy loam soil after fumigation with methyl bromide or alternative biocides. *Soil Biology and Biochemistry* 38 (6): 1243–1254.

80 Sogorb, M.A., Vilanova, E., and Carrer, V. (2004). Future applicans of phosphotriesterases in the prophylaxis and treatment of organophosphorus insecticide and nerve agent poisonings. *Toxicology Letters* 151: 219–233. https://doi.org/10.1016/j.toxlet.2004.01.022.

81 Cheng, D., Guo, Z., Riegler, M. et al. (2017). Gut symbiont enhances insecticide resistance in a significant pest, the oriental fruit fly *Bactrocera dorsalis* (Hendel). *Microbiome* 5: 13. https://doi.org/10.1186/s40168-017-0236-z.

82 Webber, M.A., Ricci, V., and Whitehead, R. (2013). Clinically relevant mutant DNA gyrase alters supercoiling, changes the transcriptome, and confers multidrug resistance. *MBio* 23: 00273–00213.

83 Webber, M.A., Whitehead, R.N., Mount, M. et al. (2015). Parallel evolutionary pathways to antibiotic resistance selected by biocide exposure. *Journal of Antimicrobial Chemotherapy* 70: 2241–2248. https://doi.org/10.1093/jac/dkv109.

84 Kazama, H., Hamashima, H., and Sasatsu, M. (1998). Distribution of the antiseptic-resistance gene qacE Delta l in Gram-positive bacteria. *FEMS Microbiology Letters* 165: 295–299.

85 Davison, J. (1999). Genetic exchange between bacteria in the environment. *Journal of Plasmid* 42: 73–91.

86 Bernier, S.P. and Surette, M.G. (2013). Concentration-dependent activity in natural environments. *Frontiers in Microbiology* 4 (2) https://www.frontiersin.org/articles/10.3389/fmicb.2013.00020/full (

87 Torres-Duarte, C., Roman, R., Tinoso, R. et al. (2009). Halogenated pesticide transformation by a laccase mediator system. *Chemosphere* 77 (5): 687–692.

88 Makia, R.S., Ismail, M.C.H., and Fadhil, A.M.A. (2013). Detection of plasmid DNA profile in bacterial and fungal isolates from catheterized patients and its

relation with antibiotic susceptibility. *Journal of Biotechnology Research Center* 7: 51–61.

89 Cho, M.H., Mulchandani, A., and Chen, W. (2002). Bacterial cell surface display of organophosphorus hydrolase for selective screening of improved hydrolysis of organophosphate nerve agents. *Applied and Environment Microbiology* 68: 2026–2030.

90 Richins, R., Mulchandani, A., and Chen, W. (2000). Expression, immobilization and enzymatic characterization of cellulose-binding domain-organophosphorus hydrolase fusion enzymes. *Biotechnology and Bioengineering* 69: 591–596.

91 Wyndham, R.C., Cashore, A.E., Nakatsu, C.H. et al. (1994). Catabolic transposons. *Biodegradation* 5: 323–342.

92 Lorenz, M.G. and Wackernagel, W. (1994). Bacterial gene transfer by natural genetic transformation in the environment. *Microbiological Reviews* 58: 563–602.

93 Popowska, M. and Krawczyk-Balska, A. (2013). Broad-host-range IncP-1 plasmids and their resistance potential. *Frontiers in Microbiology* 4: 44. https://doi.org/10.3389/fmicb.2013.00044.

94 Chen, C.C., Huang, C., Wu, M.T. et al. (2014). Multidrug resistance gene variants, pesticide exposure, and increased risk of DNA damage. *BioMed Research International* 2014: 965729. https://doi.org/10.1155/2014/965729.

95 Shi, H., Pei, L., Gu, S. et al. (2012). Glutathione S-transferase (GST) genes in the red flour beetle, *Tribolium castaneum* and comparative analysis with five additional insects. *Genomics* 100: 327–335.

96 Zhongli, C., Shunpeng, L., and Guoping, F. (2001). Isolation of methyl parathion-degrading strain m6 and cloning of the methyl parathion hydrolase gene. *Applied and Environmental Microbiology* 67: 4922–4925.

97 Shimizu, H. (2002). Metabolic engineering-integrating methodologies of molecular breeding and bioprocess systems engineering. *Journal of Bioscience and Bioengineering* 94: 563–573.

98 Ehiaghe, F.A., Ehiaghe, I.J., Agbonlahor, D.E. et al. (2013). Plasmid profiling and curing analysis of fluoroquinolone multidrug resistant *Pseudomonas aeruginosa* in Benin City, Nigeria. *Open Journal of Medical Microbiology* 3: 201–205. https://doi.org/10.4236/ojmm.33030.

99 Wang, Y. (2017). Spatial distribution of high copy number plasmids in bacteria. *Plasmid* 91: 2–8.

100 Roy, P.H. (1995). Integrons: novel mobile genetic elements mediating antibiotic resistance on enterobacteria and *Pseudomonas*. *Alliance Prudent Use Antibiotics Newsletter* 13: 1–6.

101 Roy, B., Taneja, R., and Chambon, P. (1995). Synergistic activation of retinoic acid (RA)-responsive genes and induction of embryonal carcinoma cell differentiation by an RA receptor a(RARa), RARb, or RARg – selective ligand in combination with a retinoid X receptor-specific ligand. *Molecular and Cellular Biology* 15 (12): 6481–6487.

102 Salisbury, V., Hedges, R.W., and Datta, N. (1972). Two modes of curing transmissible bacterial plasmids. *General Microbiology* 70: 443–445. https://doi.org/10.1099/00221287-70-3-443.

103 Patyka, V., Buletsa, N., Pasichnyk, L. et al. (2016). Specifics of pesticides effects on the phytopathogenic bacteria. *Ecological Chemistry and Engineering* 23 https://doi.org/10.1515/eces-2016-0022.

104 Dzidic, S., Suskovic, J., and Kos, B. (2008). Antibiotic resistance mechanisms in bacteria biochemical and genetic aspects. *Food Technology and Biotechnology* 46: 11–21.

105 Heinemann, J. (2000). Do antibiotics maintain antibiotic resistance. *Drug Discovery Today*: 195–204.

106 Gonzalez-Lopez, J., Martinez-Toledo, M.V., Rodelas, B. et al. (1993). Studies on the effects of the insecticides phorate and malathion on soil microorganisms. *Environmental Toxicology and Chemistry* 12: 1209–1214. https://doi.org/10.1002/etc.5620120709.

107 Carattoli, A., Zankari, E., Garcia-Fernández, A. et al. (2014). In silico detection and typing of plasmids using Plasmid Finder and plasmid multilocus sequence typing. *Antimicrobial Agents and Chemotheraphy* 58 (7): 3895–3903. https://doi.org/10.1128/AAC.02412-14.

108 Shafiani, S. and Malik, A. (2003). Tolerance of pesticides and antibiotic resistance in bacteria isolated from wastewater-irrigated soil. *World Journal of Microbiology and Biotechnology* 19: 897–901.

109 Stevenson, P.C., Isman, M.B., and Belmain, S.R. (2017). Belmain Pesticidal plants in Africa: a global vision of new biological control products from local uses. *Industrial Crops and Products* 110 (30)): 2–9.

110 Qiu, X., Zhong, Q., Li, M. et al. (2007). Biodegradation of p-nitrophenol by methyl parathion degrading *Ochrobactrum* sp. B2. *International Journal of Biodeteriotion and Biodegradation* 59: 297–301.

111 Dua, M., Singh, A., Sethunathan, N. et al. (2002). Biotechnology and bioremediation: successes and limitations. *Applied Microbiology and Biotechnology* 59: 143–152.

112 Araya, M. and Lakhi, A. (2004). Response to consecutive nematicide applications using the same product in Mussa AAAcv. Grande naine originated from in vitro; propagative material and cultivated in virgin soil. *Nematologia Brasileira* 28: 55–61.

113 Aislabie, J. and Jones, G.L. (1995). A review of bacterial-degradation of pesticides. *Soil Research* 6: 925–942.

114 Iranzo, M., Sainz-Pardo, I., Boluda, R. et al. (2001). The use of microorganisms in environmental remediation. *Annals of Microbiology* 51 (2): 135–144.

115 Vischetti, C., Cristiano, C., and Piero, P. (2002). Relationship between changes of soil microbial biomass content and imazamox and benfluralin degradation. *Biology and Fertility of Soils* 35 (1): 13–17.

116 Jeon, C.O. and Madsen, E.L. (2012). In situ microbial metabolism of aromatic-hydrocarbon environmental pollutants. *Current Opinion in Biotechnology* 24: 474–481.

117 Sun, X. and Wang, X. (2010). Acetylcholinesterase biosensor based on prussian blue-modified electrode for detecting organophosphorous pesticides. *Biosensors and Bioelectronics* 25 (12): 2611–2614. https://doi.org/10.1016/j.bios.2010.04.028.

118 Sun, W., Chen, Y., Liu, L. et al. (2010). Conidia immobilization of T-DNA inserted Trichoderma atroviride mutant AMT-28 with dichlorvos degradation ability and exploration of biodegradation mechanism. *Bioresource Technology* 101: 9197–9203.

119 Ramakrishnan, R., Siuter, D.R., Nakatsu, C.H. et al. (2000). Feeding inhibition and mortality in *Reticulitermes flavipes* (Isoptera: Rhinotermitidae) after exposure to imidacloprid-treated soils. *Journal of Economic Entomology* 93 (2): 422–428.

120 Briceno, G., Palma, G., and Duran, N. (2007). Influence of organic amendment on the biodegradation and movement of pesticides. *Critical Reviews in Environmental Science and Technology* 37: 233–271.

121 Amer, A. and Aly, E. (2012). Characterization of a strain of *Pseudomonas putida* isolated from agricultural soil that degrades cadusafos (an organophosphorus pesticide). *World Journal of Microbiology and Biotechnology* 3: 805–814.

122 Vasquez, M. and Reyes, W. (2002). *Degradacion de Aroclor 1242 por Pseudomas sp*. Peru: Biblioteca Nacionaldel Peru.

123 Senthilkumar, S., Anthinisamy, A., Arunkumar, S. et al. (2011). Biodegradation of methyl parathion and endosulfan using Pseudomonas aeruginosa and *Trichoderma viridae*. *Journal of Environmental Science & Engineering* 53 (1): 115–122.

124 Fournier, P.E., Fujita, H., Takada, N. et al. (2002). Genetic identification of rickettsiae isolated from ticks in Japan. *Journal of Clinical Microbiology* 40: 2176–2181.

125 Pal, R., Chakrabarti, K., Chakraborty, A. et al. (2006). Degradation and effects of pesticides on soil microbiological parameters – a review. *International Journal of Agricultural Research* 1: 240–258.

126 Bass, C. and Field, L.M. (2011). Gene amplification and insecticide resistance. *Pest Management Science* 67: 886–890.

127 Trigo, A., Valencia, A., and Cases, I. (2009). Systemic approaches to biodegradation. *FEMS Microbiology Reviews* 33: 98–108.

128 Scott, C., Pandey, G., Hartley, C.J. et al. (2008). The enzymatic basis for pesticide bioremediation. *Indian Journal of Microbiology* 48: 65–79.

129 Eerd, L.L.V., Hoagland, R.E., Zablotowicz, R.M. et al. (2003). Pesticide metabolism in plants and microorganisms. *Weed Science* 51 (4): 472–495. https://doi.org/10.1614/0043-1745(2003)051[0472:PMIPAM]2.0.CO;2.

130 Komori, K., Toda, K., and Ohtake, H. (1990). Effects of oxygen stress on chromate reduction in Enterobacter cloacae strain HO1. *Journal of Fermentation and Bioengineering* 69 (1): 67–69.

131 Li, F.B., Li, X.M., Zhou, S.G. et al. (2010). Enhanced reductive dechlorination of DDT in an anaerobic system of dissimilatory iron-reducing bacteria and iron oxide. *Environmental Pollution* 158 (5): 1733–1740.

132 Li, J., Liu, J., Shen, W. et al. (2010). Isolation and characterization of 3,5,6-trichloro-2-pyridinol-degrading Ralstonia sp. strain T6. *Bioresource Technology* 101: 7479–7483. https://doi.org/10.1016/j.biortech.2010.04.030.

133 Amy, P.S., Schulke, J.W., Frazier, L.M. et al. (1985). Characterization of aquatic bacteria and cloning of genes specifying partial degradation of 2,4-dichlorophenoxyacetic acid. *Applied and Environmental Microbiology* 49 (5): 1237–1245.

134 Sander, P., Wittich, R.M., Fortnagel, P. et al. (1991). Degradation of 1,2,4-trichloro- and 1,2,4,5-tetrachlorobenzene by *Pseudomonas* strains. *Applied and Environmental Microbiology* 57: 1430–1440.

135 Susanna, A., Stromeyer, S.B., Hermann, R. et al. (1993). Dichloromethane as the sole carbon source for an acetogenic mixed culture and isolation of a fermentative degrading bacterium. *Applied and Environmental Microbiology* 58: 3790–3797.

136 Bordeleau, L.M., Rosen, J.D., and Bartha, R. (1972). Herbicide-derived chloroazobenzene residues: pathway of formation. *Journal of Agricultural and Food Chemistry* 20: 573.

137 Hammel, K.E. and Tardone, P.J. (1988). The oxidative 4-dechlorination of polychlorinated phenols is catalyzed by extracellular fungal lignin peroxidase. *Biochemistry* 27: 6563–6568.

138 Dumas, D.P., Caldwell, S.R., Wild, J.R. et al. (1989). Purification and properties of the phosphotriesterase from *Pseudomonas diminuta*. *The Journal of Biological Chemistry* 264: 19659–19665.

139 Benning, M.M., Sims, H., Raushel, F.M. et al. (2001). High resolution X-ray structures of different metal-substituted forms of phosphotriesterase from *Pseudomonas diminuta*. *Biochemistry* 40: 2712–2722.

140 Eto, M. (1974). *Organophosphorous Pesticides: Organic and Biological Chemistry*. Ohio: CRC Press, Inc.

141 Beynon, K.I., Hutson, D.H., and Wright, A.N. (1973). The metabolism and degradation of vinyl phosphate insecticides. *Residue Reviews* 47: 55–142.

142 Bhadbhade, B.J., Sarnaik, S.S., and Kanekar, P.P. (2002). Bioremediation of an industrial effluent containing monochrotophos. *Current Microbiology* 45: 346–349. https://doi.org/10.1007/s00284-002-3681-1.

143 Raushel, F.M. (2002). Bacterial detoxification of organophosphate nerve agents. *Current Opinion in Microbiology* 5: 288–295.

144 Efrmenko, E.N. and Sergeeva, V.S. (2001). Organophosphate hydrolase – an enzyme catalyzing degradation of phosphorus-containing toxins and pesticides. *Russian Chemical Bulletin* 50: 1826–1832.

145 Omburo, G.A., Kuo, J.M., Mullins, L.S. et al. (1992). Characterization of zinc binding site of bacterial phosphotriestearase. *Computational Biology and Chemistry* 267: 13278–13283.

146 DiSioudi, B.D., Miller, C.E., Lai, K.H. et al. (1999). Rational design of organophosphorus hydrolase for altered substrate specificities. *Chemical Biology Interaction* 120: 211–223. https://doi.org/10.1016/S0009-2797(99)00030-7.

147 Manavathi, B., Pakala, S.B., Gorla, P. et al. (2005). Influence of zinc and cobalt on expression and activity of parathion hydrolase from *Flavobacterium* sp. ATCC27551. *Pesticide Biochemistry and Physiology* 83: 37–45.

148 Singh, S. and Singh, D.K. (2003). Utilization of monocrotophos as phosphorus source by *Pseudomonas aeruginosa* F10B and *Clavibacter michiganense* sub sp. *inisidiosum* SBL. *Canadian Journal of Microbiology* 49: 101–109.

149 Shim, H., Hong, S.B., and Raushel, F.M. (1998). Hydrolysis of phosphodiesters through transformation of the bacterial phosphotriesterase. *The Journal of Biological Chemistry* 272: 17445–17450.

150 Hausinger, R.P. and Fukumori, F. (1995). Characterization of the first enzyme in 2, 4-dichlorophenoxyacetic acid metabolism. *Environmental Health Perspectives* 103 (5): 37–39.

151 Holm, L. and Sander, C. (1997). An evolutionary treasure: unification of a broad set of amidohydrolase related to urease. *Proteins* 28: 72–82.

152 Horne, I., Sutherland, T.D., Harcourt, R.L. et al. (2002). Identification of an opd (organophosphate degradation) gene in an *Agrobacterium* isolate. *Applied Environmental Microbiology* 68: 3371–3376.

153 Yang, H., Carr, P.D., McLoughlin, S.Y. et al. (2003). Evolution of an organophosphate-degrading enzyme: a comparison of natural and directed evolution. *Protein Engineering* 16: 135–145.

154 Cheng, T., Liu, L., Wang, B. et al. (1997). Nucleotide sequence of a gene encoding an organophosphorus nerve agent degrading enzyme from *Alteromonas haloplanktis*. *Journal of Industrial Microbiology and Biotechnology* 18 (1): 49–55.

155 Cheng, T.C., DeFrank, J.J., and Rastogi, V.K. (1999). *Alteromonas* prolidase for organophosphorus G-agent decontamination. *Chemico-Biological Interactions* 119: 455–462.

156 Cheng, T.C., Harvey, S.P., and Chen, G.L. (1996). Cloning and expression of a gene encoding a bacterial enzyme for decontamination of organophosphorus nerve agents and nucleotide sequence of the enzyme. *Applied and Environmental Microbiology* 62 (5): 1636–1641.

157 Cheng, T.C., Harvey, S.P., and Stroup, A.N. (1993). Purification and properties of a highly active organophosphorus acid anhydrolase from *Alteromonas undina. Applied and Environmental Microbiology* 59 (9): 3138–3140.

158 DeFrank, J.J. and White, W.E. (2002). Phosphofluoridates: biological activity and biodegradation. In: *Organofluorines*, The Handbook of Environmental Chemistry, vol. 10 (ed. A.H. Neilson), 295–343. Berlin: Springer.

159 Hill, C.M., Li, W.S., Cheng, T.C. et al. (2001). Stereochemical specificity of organophosphorus acid anhydrolase toward p-nitrophenyl analogs of soman and sarin. *Bioorganic Chemistry* 29: 27–35.

160 Mazur, A. (1946). An enzyme in animal tissues capable of hydrolysing the phosphorus-fluorine bond of alkyl fluorophosphates. *The Journal of Biological Chemistry* 164: 271.

161 Hoskin, F.C.G., Walker, J.E., and Mello, C.M. (1999). Organophosphorus acid anhydrolase in slime mold duckweed and mug bean: a continuing search for a physiological role and a natural substrate. *Chemico-Biological Interactions* 199 (120): 399–404.

162 Mulbry, W.W. (1992). The aryldialkylphosphatase-encoding gene adpB from Nocardia *sp.* strain B-1: cloning, sequencing and expression in *Escherichia coli. Gene* 121 (1): 149–153.

163 Mulbry, W.W. and Karns, J.S. (1989). Purification and characterization of three parathion hydrolase from Gram negative bacterial strains. *Applied Environmental Microbiology* 55: 289–293.

164 Amitai, G., Adani, R., Sod-Moriah, G. et al. (1998). Oxidative biodegradation of phosphorothiolates by fungal laccase. *FEBS Letters* 438 (3): 195–200.

165 Henry, L.M., Maiden, M.C., Ferrari, J. et al. (2015). Insect life history and the evolution of bacterial mutualism. *Ecology Letters* 18: 516–525. https://doi.org/10.1111/ele.12425.

166 Crews, F.C. (2001). Saving us from Darwin. *The New York; Review of Books* (October 4).

167 Thomsen, R. and Christensen, M.H. (2006). MolDock: a new technique for high accuracy molecular docking. *Journal of Medical Chemistry* 49 (11): 3315–3321. https://doi.org/10.1021/jm051197e.

168 Neti, N. and Zakkula, V. (2013). In silico structural analysis and binding of organophosphorus hydrolase of *kocuria sp* with chloropyrifos. *International Journal of Computer Applications* 66: 0975–8887. https://doi.org/10.5120/11078-6015.

169 Jorgensen, M., Givney, R., Pegler, M. et al. (1996). Typing multidrug-resistant *Staphylococcus aureus*: conflicting epidemiological data produced by genotypic and phenotypic methods clarified by phylogenetic analysis. *Journal of Clinical Microbiology* 34 (2): 398–403.

170 Dong, Y.J., Bartlam, M., Sun, L. et al. (2005). Crystal structure of methyl parathion hydrolase from *Pseudomonas* sp. WBC-3. *Journal of Molecular Biology* 353: 655–663.

171 Gotthard, G., Hiblot, J., Gonzalez, D. et al. (2013). Structural and enzymatic characterization of the phosphotriesterase OPHC2 from *Pseudomonas pseudoalcaligenes*. *PLoS One* 8 (11): e77995. https://doi.org/10.1371/journal.pone.0077995.

172 Luo, X.J., Kong, X.D., Zhao, J. et al. (2014). Switching a newly discovered lactonase into an efficient and thermostable phosphotriesterase by simple double mutations His250Ile/Ile263Trp. *Biotechnology and Bioengineering* 111: 1920–1930.

173 Kirubakaran, R., Murugan, A., Parray, J.A. et al. (2019). Pesticide residues in the soil cause cross-resistance among soil bacteria. In: *Plant Growth Promoting Rhizobacteria for Sustainable Stress Management*, vol. II, 205–218. Singapore: Springer Nature Singapore Pvt Ltd https://doi.org/10.1007/978-981-13-6536-2_11.

174 Kirubakaran, R., Murugan, A., Natarajan, D. et al. (2018). Pesticide-degrading, naturally multidrug-resistant bacteria flora. *Microbial Pathogenesis* 114: 304–310.

12

Bioremediation of Heavy Metals and Other Toxic Substances by Microorganisms

Dhaneshwar Padhan[1, 4]*, Pragyan Paramita Rout*[2]*, Ritesh Kundu*[1]*, Samrat Adhikary*[1]*, and Purbasha Priyadarshini Padhi*[3]

[1] *Department of Agricultural Chemistry and Soil Science, Mohanpur, West Bengal, India*
[2] *Institute of Agricultural Sciences, Siksha O Anusandhan University, Odisha, India*
[3] *Department of Soil Science and Agricultural Chemistry, Indira Gandhi Krishi Vishwavidyalaya, Raipur, Chhattisgarh, India*
[4] *Scientist-B, Central Silk Board, Bangalore, Karnataka, India*

12.1 Introduction

In the current global scenario, increased industrialization, modernization with technological advancements, and improper use of natural resources pose severe threats to the environment as well as to humans in terms of pollution to soil, air, and water. Anthropogenic as well as natural sources of contaminants (heavy metals and other organic as well as inorganic pollutants) are potential threats causing deadly diseases and other problems in humans and animals. Plants exposed to these contaminants for longer periods may result in a reduction in yield and these contaminants could be transferred to food grains through biomagnifications, ultimately hampering the food chain.

Soil is considered as a source, as well sink, for potential contaminants. Lead (Pb), chromium (Cr), arsenic (As), zinc (Zn), cadmium (Cd), copper (Cu), mercury (Hg), and nickel (Ni) although essential for the growth and development of plants are considered as potential heavy metal contaminants when exceeding certain critical levels in the environment. These heavy metals and metalloids emanating from: mine tailings; burning leaded fuels; a range of paints; chemical fertilizers, pesticides, compost, and sewage water applied to agricultural fields; coal burning; and spillage of petrochemicals are potential sources of heavy metal contamination in soils.

Soil Bioremediation: An Approach Towards Sustainable Technology, First Edition.
Edited by Javid A. Parray, Abeer Hashem Abd Elkhalek Mahmoud, and Riyaz Sayyed.

Soil is the natural medium for plant growth. It sustains productivity and could act as a buffer system for a number of chemical reactions. The presence of humus as an intrinsic component of the soil organic matter complex could bind heavy metal cations because of its high affinity and could extract them from the water that passes through the soil. Roots of crops and other plants, when grown in such heavy metal contaminated soils, pick up these elements along with water and pass them on to plants and then to animals. Heavy metals, being positively charged, are also retained in the soil by adsorption on mineral particles present in the soil and also undergo precipitation reactions [1]. Once these contaminants enter the food chain, being generally nonbiodegradable, they accumulate in the food materials and pose a potential threat of multiple health hazards such as, damage to lungs, kidneys, liver, pancreas, and the nervous system [2, 3]. For instance, reports have found damage to DNA due to the accumulation of Cr(VI) and its reduced counterpart Cr(III) in the cells. Similar adverse health effects have also been reported for other heavy metals and contaminants. Hence, the United States Environmental Protection Agency (EPA) and World Health Organization (WHO) have set a maximum permissible limit for every heavy metal in various systems in order to prevent the negative effects in human and animals [3].

So far several methodologies have been developed and tested for removing these contaminants from soil, water, and air. Chemical precipitation, coagulation, flocculation, filtration, reverse osmosis, ion exchange mechanism, membrane process, biosorption, aerobic, and anaerobic microbial degradation have been extensively explored for their effectiveness in the removal of heavy metals and other potential contaminants from the environment [4]. However, among these physical and chemical mechanisms, biosorption is considered as an innovative technology and a remedial strategy because of its low cost and ecofriendly nature [2, 5]. But, there are certain reports on possibilities of the generation of secondary contaminants if physical and chemical methods of remediation are being followed. Noteworthy findings on the use of soil organic matter for adsorption of heavy metals or reduction of the metal bioavailability using microorganisms have been reported [6, 7]. This promoted the process of bioremediation, which is considered as a green cleanup technology for the removal of heavy metal contaminants and other organic and inorganic pollutants from soil and water bodies. In this chapter the possible mechanisms of remediation are discussed although the major focus lies on the process of bioremediation and its mechanism, possibilities, and field applicability.

12.2 Sources of Heavy Metals

Lead (Pb), chromium (Cr), arsenic (As), zinc (Zn), cadmium (Cd), copper (Cu), mercury (Hg), and nickel (Ni) are commonly considered as heavy metals. These metal(loid)s are deposited in soils and water bodies as a result of natural processes and human activities to high even poisonous levels. Weathering of minerals, erosion, and volcanic activity are the most significant natural sources while anthropogenic sources include mining, smelting, electroplating, use of pesticides and (phosphate) fertilizers as well as biosolids in agriculture, sludge dumping, industrial discharge, and atmospheric deposition [8, 9]. Heavy metals in the soil from anthropogenic sources tend to be more mobile, hence more bioavailable, than pedogenic or lithogenic ones.

A summary of the different sources of heavy metal(loid) contamination in soils and water bodies is discussed.

12.2.1 Natural Sources

Weathering of different types of rocks in the earth surface, volcanic eruptions, mineral deposits erosion, oceanic evaporation, and general pedogenic processes are natural phenomena that contribute heavily to the accumulation of heavy metals in soils [10]. Although it has been reported that in soils with high concentrations of heavy metals the actual concentrations vary with soil type; the predominant factor for such occurrence could be the chemical composition of the parent rocks and the speciation of metals affected by the prevailing climate and soil organisms of the particular area [11]. It may happen that these heavy metals occur as sulfides of Pb, Co, Fe, As, Pb—Zn, Ag, and Ni, and oxides of Se, Al, Mn, and Sb.

12.2.2 Anthropogenic Sources

Anthropogenic sources of heavy metals and other toxicants contamination in the environment might be the result of industrialization, urbanization, and intensive agricultural practices [12]. Industries like tanneries, textiles, metallurgy, galvanizing factories, distilleries, and factories producing pesticides, fertilizers, paints, varnishes, and pharmaceuticals are the most common sources of pollutants in the environment [13, 14]. Extraction, processing, and the use of metals in metallurgical industries produce direct contamination however, most industries produce pollution indirectly for example, through the release of metals due to the burning of fossil fuels [9]. During the process of metal ore mining, large amounts of waste rocks containing low concentrations of heavy metals are being produced for every gram of metal extracted and these metals could reach the ground and surrounding water bodies by chemical or biological leaching ultimately causing pollution in the food chain [15]. Textile industries and tanneries generate highly polluted effluents that reach water courses and affect even very remote areas.

Urbanization generates large amounts of domestic waste, wastewater, and sewage sludge, which are potential sources of heavy metals in soils irrigated with these sewage-sludge fed water. Many reports confirmed that metal concentrations in this waste and sewage-sludge fed water are low; but long-term irrigation with these waters could lead to the accumulation of heavy metals in soils [16] creating possibilities of contamination to the crops grown in a particular area. Inadequate disposal of municipal solid waste produces contaminated leachate containing high concentrations of Fe(III), Mn(II), and Cr(VI) that could pollute the groundwater and surface water, which may be used for drinking water [17]. Agricultural practices such as, indiscriminate use of chemical fertilizers, pesticides, and different soil amendments without appropriate soil testing could be one of the major sources of heavy metal contamination to soil, water bodies, and food grains. Use of crude or natural rock phosphate as a source of phosphate fertilizer for crop production is a source for heavy metals (Ni(II), Cd(II), Zn(II), Pb(II), As(V/III), and Cr(VI)) as it contains the former as impurities [18]. In the same way, pesticides contain heavy metals like As (V/III), Hg(II), Pb(II), and Cu(II) as active ingredients and could be a potential threat as metal contaminants in the ecosystem. The level of toxicity generated by these metals depends upon their persistence in soils and oxidation states.

The use of biosolids (animal manures, metropolitan sewage sludge) and other manures for agricultural crop production has been practiced since the beginning of cultivation. These products contain traces of heavy metals (Pb, Ni, Cd, Cr, Cu, and Zn) in sizeable quantities that then enter food materials through plant systems.

Another potential source of contamination of heavy metals and other organic and inorganic pollutants are those generated through airborne processes. Airborne pollutants are particulates, biological molecules, or other harmful materials that exist in the atmosphere. The entry into the atmosphere may occur as a volatile gas (ozone, benzene), as liquid droplets (sulfuric acid, nitrogen dioxide), or as particulate matter (diesel exhaust, aromatic hydrocarbons). These airborne toxicants are potential threats for many deadly diseases creating direct problems in immune systems.

The anthropogenic sources of metal(loid)s contamination into five groups [19]:

i) Agriculture (Zn, As, Pb, Cd, Cu, Se, U).
ii) Metalliferous mining and smelting (Cd, Pb, As, and Hg).
iii) Industrial developments (Cd, Hg, As, Cr, Cu, Co, Ni, and Zn).
iv) Waste disposal and sewage sludge fed water (As, Pb, Cu, Cd, Cr, Zn, and Hg).
v) Atmospheric deposition (As, Pb, Cr, Hg, Cu, Cd, and U).

12.2.3 Electronic Waste

Electronic waste or e-waste describes discarded electrical or electronic devices. It is created when an electronic product is discarded at the end of its useful life. The rapid expansion of technology is generating a very large amount of e-waste. Electronic scrap components, such as CPUs, contain potentially harmful materials such as lead, cadmium, beryllium, or brominated flame retardants. When these materials are disposed of into soil, the contaminants are retained in soils and create toxic environments for living organisms. Many reports have confirmed the presence of these contaminants in top soils when they are disposed of without proper treatments [20].

A simple mass balance of the heavy metals in the soil as suggested by Alloway [21] and Lombi and Gerzabek [22] can be expressed as follows:

$$M_{total} = (M_p + M_a + M_f + M_{ag} + M_{ow} + M_{ip}) - (M_{cr} + M_l)$$

Where

"M" is the heavy metal
"p" is the parent material
"a" is the atmospheric deposition
"f" is the fertilizer sources
"ag" are the agrochemical sources
"ow" are the organic waste sources
"ip" are other inorganic pollutants
"cr" is crop removal; and
"l" is the losses by leaching and volatilization.

12.3 Effects of Heavy Metals on Plant, Microorganisms, and Human Health

It is quite well established that plants require certain heavy metals for their growth and development, but excessive quantities of these metals become toxic to plants. Plants can accumulate the essential as well nonessential metals simultaneously because of nutrient uptake mechanism [23]. Once the concentration of these metals within plants exceeds the optimal limit, they adversely affect plant growth and development. The effects can be direct, resulting in inhibition of cytoplasmic enzymes and damage to cell structures due to oxidative stress [24]. However, these metals replace the essential nutrients at cation exchange sites of plants [25], showing an indirect effect. Further, the negative influence heavy metals have on the growth and activities of soil microorganisms may also indirectly affect the growth of plants. For instance, a reduction in the number of beneficial soil microorganisms due to high metal concentration may lead to a decrease in organic matter decomposition leading to a decline in soil nutrients. Enzyme activities useful for plant metabolism may also be hampered due to heavy metal interference with activities of soil microorganisms. These toxic effects (both direct and indirect) lead to a decline in plant growth, which sometimes results in the death of the plant [26]

The effect of heavy metal toxicity on the growth, biochemistry, and physiology of plants varies according to the particular heavy metal involved in the process. For metals such as Pb, Cd, Hg, and As, which do not play any beneficial role in plant growth, adverse effects have been recorded at very low concentrations of these metals in the growth medium. Kibra [27] recorded significant reduction in the height of rice plants growing on a soil contaminated with 1 mgHg/kg. Reduced tiller and panicle formation also occurred at this concentration of Hg in the soil. For Cd, reduction in shoot and root growth in wheat plants occurred when Cd in the soil solution was as low as 5 mg/l [28]. Most of the reduction in growth parameters of plants growing on polluted soils can be attributed to reduced photosynthesizing activities, plant mineral nutrition, and reduced activity of some enzymes [29]. Microorganisms in soils require certain quantities of metals for their functioning and growth. But reports on harmful effects of long-term heavy metal contamination on soil microbial activity, especially microbial respiration, are not uncommon [30]. Aside from long-term metal-mediated changes in soil enzyme activity, many reports have shown large reductions in microbial activity due to short-term exposure to toxic metals [31]. High concentrations of heavy metals and other toxic contaminants could reduce the activity of certain microflora and fauna. Moreover, habitats that have high levels of metal contamination show lower numbers of microbes than uncontaminated habitats [32]. Although, toxicity of heavy metals to microbes is a well-established phenomenon, the effects of those metals upon specific enzymatic systems at lower ("subacute") concentrations are not well known.

Although some heavy metals are required for life's physiological processes (e.g., components of metallo-enzymes), their excessive accumulation in living organisms is always detrimental. Generally, toxic metals cause enzyme inactivation, damage cells by acting as antimetabolites, or form precipitates or chelates with essential metabolites. The impact of some toxic metals on human health has been reported by Forstner [33]. Human diseases have resulted from consumption of cadmium-contaminated foods [34]. Likewise, Pb exposure can cause seizures, mental retardation, and behavioral disorders. The threat that heavy metals pose to human and animal health is aggravated by their low environmental mobility, even under high precipitations, and their long-term persistence in the environment. Also, the average biological half-life of Cd, another accumulation poison similar to Pb, has been estimated to be about 18 years [33]. As a result of low environmental mobility of those metals, a single contamination episode could set the stage for a long-term exposure of microbial community to metal, necessitating a long-term monitoring effort to assess metal effects.

Ideally a pesticide must be lethal to the target pests, but not to nontarget species, including humans. Unfortunately, this is not the case and inadequate management of pesticides constitutes potential occupational hazards and environmental risks for ecosystems [35]. Besides, pesticide residues retained in crops can directly influence public health via food consumption.

12.4 Current Approaches for Remediation of Heavy Metals

Remediation of toxic heavy metals is mandatory so as to prevent the hazardous effects on the environment and preserve the environment for future generations [36]. A variety of methodologies have been employed for the elimination of toxic heavy metals. Eliminating heavy metals from the environment is considered as a challenging job with respect to technical complexity and cost [37]. A number of strategies to clean-up soil have been used over the years. These strategies are categorized as biological, chemical, and physical approaches (Figure 12.1). Metal contaminated soils are traditionally cleaned-up via on-site management or excavation and disposal to landfill sites. However, such approaches only shift the problems elsewhere. Additionally, problems may also arise during the transportation of contaminants to adjacent compartments of the environment [39]. Another strategy to eliminate soil metal contaminants involves soil washing, which produces pollutant rich residues that require further treatment. However, such soils become unsuitable for plant growth due to the removal of biological activities [40, 41]. Chemical and physical methods have limitations, e.g., changes in native soil flora, irreversible alterations in soil properties, intensive labor, and high cost. Likewise, chemical methods are expensive, generate secondary pollution, and produce large quantities of sludge [41, 42].

Bioremediation is an innovative and promising technology available for the removal of heavy metals and recovery of heavy metals in polluted water and land. Since microorganisms have developed various strategies for survival in heavy metal polluted habitats, these organisms are known to develop and adopt different detoxifying mechanisms such as biosorption, bioaccumulation, biotransformation, and biomineralization. Generally, bioremediation involves the adsorption of the soil contaminants on organic matter or reduction of the metal bioavailability by microorganisms [6, 7, 43]. However, most researchers suggest the preferential use of microorganisms over organic matter to remove/reduce heavy metal concentration from contaminated sites. Such an inherent property of microbial communities has been effectively utilized to reduce, degrade, or immobilize

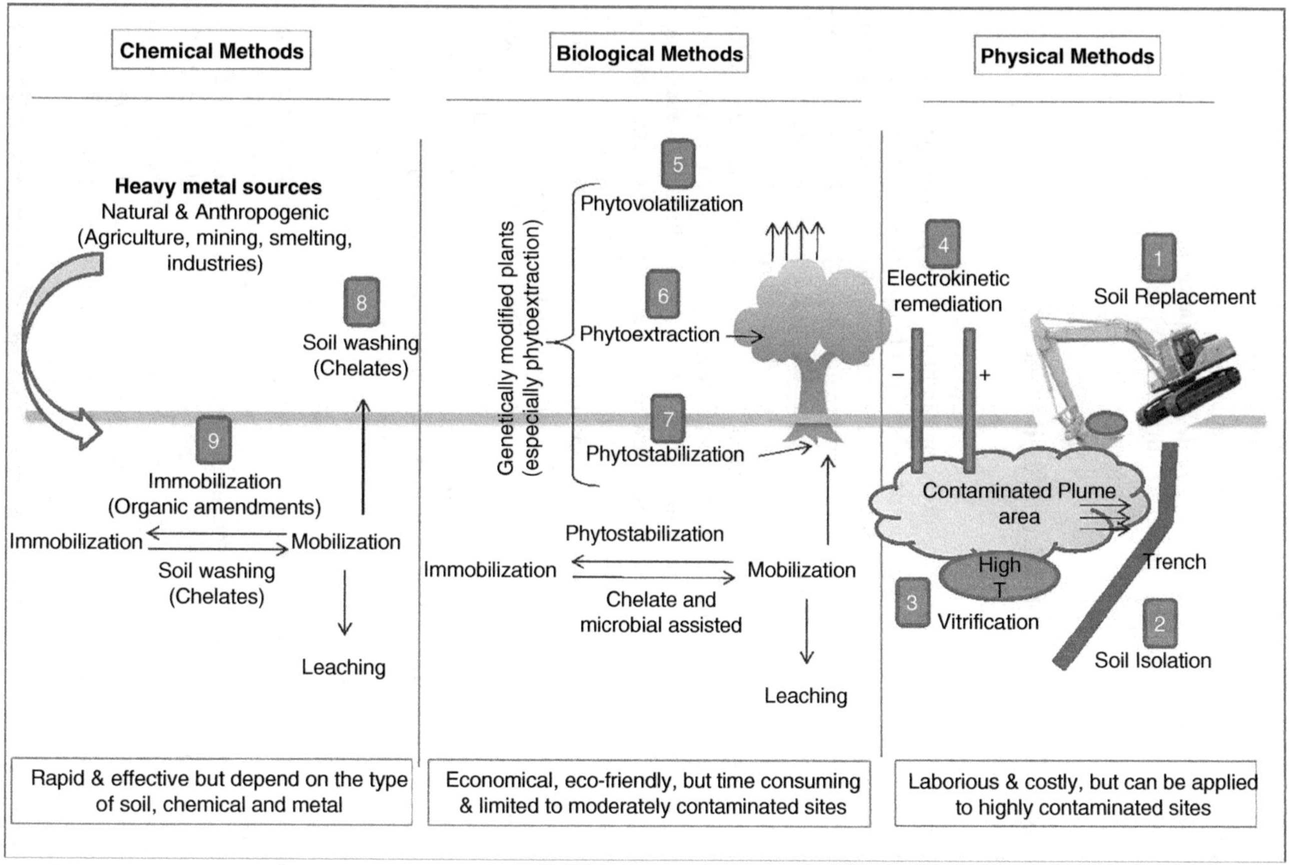

Figure 12.1 Physical, chemical, and biological methods of remediation of heavy metals and other toxic compounds in contaminated sites. (*Source:* Adopted from Khalid et al. [38].)

hazardous contaminants and clean-up polluted environments. The following sections elucidate the various mechanistic aspects of bioremediation processes in detail.

12.4.1 Physico-chemical Approaches for Remediation of Heavy Metals

Heavy metal contaminated or polluted soils can be remediated by physico-chemical and biological techniques. Each of these techniques can be carried out using an *ex-situ* and or an *in-situ* method. *Ex-situ* remediation methods involve excavation of the contaminated top soil to somewhere else for treatment while *in-situ* methods involve remediating the contaminated soil on site [44, 45]. Some remediation methods can be carried out by *in-situ* and *ex-situ* methods. The choice of remediation method depends on the site characteristics, contaminant concentration, type of contaminants or pollutants to be remediated, and the final use of the contaminated soil [46]. Generally, *in-situ* remediation technique is often preferred over *ex-situ* because the former is always cheaper and less destructive to the ecosystem [47]. However, all these methods of remediation either operating *in situ* or *ex situ* have some advantages as well as disadvantages; one can select the methods according to the availability of resources and intensity of contamination (Table 12.1).

12.4.1.1 Soil Replacement

The process involves partial or complete replacement of contaminated soils with noncontaminated soil. Previously, excavation, off-site disposal, and soil replacement were the most commonly employed method for cleaning up contaminated sites. The basic idea operating in this process is dilution of the concentration of heavy metal(loid)s in the soil, and in turn increasing soil functionality. Replaced soils can be further treated to remove heavy metal(loid)s or in some cases dumped in other places. Soil replacement is carried out by (i) soil spading and (ii) new soil importing. In soil spading, the contaminated site is dug deeply and the heavy metal(loid)s are spread into the deep sites, thus achieving the aim of metal diluting. New soil importing refers to adding clean soil to the heavy metal(loid) polluted soil. The added soil is either covered on the surface or mixed to dilute the metal concentration. The soil replacement method can effectively isolate the contaminated soil and ecosystem, thus minimizing its effect on the environment. However, this technique is costly due to high involvement of labor, and is appropriate for heavily contaminated soils of a small area. Costs for long-distance transport of excavated soil may be substantially higher than following other clean-up technology. Moreover, this technique may not be applicable to agricultural sites because there is a risk of loss of soil fertility.

Table 12.1 Physical, chemical, and biological remediation techniques for the removal of contaminants and pollutants from heavy metal contaminated soils; mechanism involved, advantages, and disadvantages with their applicability.

Methods	Mechanism involved	Advantages	Disadvantages	Applicability
Physical remediation				
Soil replacement	Excavation of contaminated soil and refill with non-contaminated soil	Effective isolation of heavy metals from highly contaminated sites	Labor intensive and costly, chances of generation of secondary wastes	Small scale and long term
Soil isolation	Isolation of contaminated soil from the uncontaminated soil using subsurface barriers	Off-site transport of heavy metals checked and suitable for highly contaminated soils	Costly, soil clean-up depends on type of subsurface barrier	Small scale, and short to long term
Vitrification	Reduction in metal bioavailability by forming vitreous material using high temperatures	Easy application, applicable to a variety of contaminants	High cost due to energy requirement	Small scale but long term
Electrokinetic remediation	Removal of heavy metals from soil via electrophoresis, or electro-migration	Cost effective and easy to operate; nature of soils restored	Applicable in soils with low permeability and pH to be controlled	Small scale but long term
Chemical remediation				
Immobilization techniques	Reduction in metal mobility and bioavailability forming stable and immobile complexes with addition of immobilizing agents	Fast and easy applicability, relatively low costs, covers a broad spectrum of inorganic pollutants	Temporary solution and permanent monitoring is necessary	Small to medium scale and short term
Soil Washing	Removal of heavy metals from soil by using organic or inorganic extractants forming stable and mobile complexes	Cost effective and completely removes metals; reduces long-term liability	Chances of contamination, effectiveness varies with soil, metal and extractant used	Small scale but can be long term

Biological remediation				
Phytovolatilization	Heavy metal uptake by plants from soil and release in vapor form to atmosphere	Economical and less disruptive	Applicable for volatile metals, may cause other environmental issues and no control after metal is released into the atmosphere	Small to medium scale and long term
Phytostabilization	Use of plants to decrease metal bioavailability and mobility in soils via sequestration in plant roots	Economical and less disruptive	Temporary solution, effectiveness varies with soil, plant and metal type	Small to medium scale and shortterm
Phytoextraction	Use of hyperaccumulator plants to uptake, translocate, and concentrate heavy metals from soil to the above-ground harvestable plant parts	Highly economical, ecofriendly, less disruptive	Effectiveness depends on growing conditions, tolerance of the plant, and bioavailability of metals in soil	Large scale and long term
Chelate assisted phytoextraction	Use of organic and inorganic ligands to enhance phytoextraction capacity of plants	Low time of remediation, enhance metal uptake and translocation	Costly, can be disruptive, effective for low to moderately contaminated soils, groundwater contamination risk	Small to medium scale and long term, low to moderate levels of metal
Microbial assisted phytoextraction	Use of microorganisms to enhance phytoextraction capacity of plants	Economical, low time of remediation, enhance plant growth and metal uptake and translocation	Depends on microorganism, soil, plant, and metal type	Large scale and long term

Source: Adopted from Khalid et al. [38].

12.4.1.2 Soil Isolation and Containment

Soil isolation is an *in-situ* remediation method where contaminated soils are capped and contained by installing different types of barriers to prevent the metal contaminant from spreading to nearby water bodies and to prevent leakage to the environment, which could potentially contaminate the uncontaminated areas. Barrier walls, usually made up of impermeable materials such as steel, cement, and bentonite, could be used for capping either for vertical and/or horizontal containment. Such practices, when followed, can reduce water filtration; the vertical walls minimize movement of contaminated groundwater from the contaminated site or limit the flow of uncontaminated water onto the contaminated site, while horizontal barriers are potentially useful in restricting the downward movement of metal contaminants [48]. Although the method doesn't operate directly it can significantly reduce the migration of metals into the groundwater [46, 49].

12.4.1.3 Solidification and Stabilization

Solidification and stabilization are twin technologies that use binders and additives to reduce metal contaminants' mobility. The process of solidification involves the mixing of contaminated soils with physical binding agents such as cement, bitumen, asphalt, thermoplastic binders in order to form a crystalline, glassy, or polymeric mass that has low leaching rates. In the process of stabilization, chemical reactions are induced between the stabilizing agent and metal contaminants, to bind the contaminants to substrate or to yield less-mobile compounds, which contain the metal contaminants. Stabilization does not necessarily yield a solid, but does yield a more chemically stable compound. The resultant compound can thereafter be removed from the contaminated environment if desired [46].

12.4.1.4 Vitrification

Vitrification is a solidification/stabilization process that operates through heating and subsequently melting the contaminated soils at a sufficiently high temperature using thermal energy or electricity forming a vitreous material [50]. These melted products form inert gases when cooled. The process involves the insertion of graphite electrodes into the contaminated soil, at sufficiently close spacing, and energizing with a high electrical resistance heating (>1700 °C) that causes the soil to melt to a molten pool. On cooling, a glassy solid is formed, which immobilizes the metals and this can be removed from the site [48]. Vitrification is comparatively easy to apply compared to other physical remediation methods. Vitrification can be applied to the majority of soils contaminated with inorganic (heavy metal(loid)s) and organic contaminants. Vitrification can be performed both *in situ* and *ex situ*, although the *in-situ* method is preferred due to comparatively low cost and energy requirements. *Ex-situ* vitrification processes contain various stages such as, excavation, mixing, pretreatment, melting feeding, and casting of the melted product [51]. *Ex-situ* vitrification requires high energy input for melting, and is therefore costly. In some case the vitrified material is mixed with additives such as clay, native soil, or sand to create a product with certain characteristics, which may improve the effectiveness of this technology. Solidification/stabilization and vitrification techniques are suitable for pollution of shallow depts and large volume [49].

12.4.1.5 Soil Washing

Soil washing is an *in-situ* remediation method for removing contaminants from the soil by the use of water or appropriate washing solution. Natural acids, bases, chelating agents like EDTA (Ethylenediaminetetraacetic acid), EDDS (Ethylenediamine-N,N'-disuccinic acid), humic substances, and surfactants are added to water and used as extraction or flushing solutions to recover metals, organic, and oil contaminants like phenol [46]. These extractants are applied to the soils through surface flooding, sprinklers, surface trenches, horizontal drains, or vertical drains. Contaminants are mobilized by solubilization, formation of emulsions, or a chemical reaction with the extraction solution [48]. The proficiency of soil washing relies upon the capacity of the extractant to break up the metal(loid)s in the soil. In this manner, extractants, which can break up abnormal states of metals, would be appropriate for soil cleaning. Among the extractants, engineered chelates such as EDTA and EDDS are considered the best and appropriate for soil washing. Makino et al. [52] stated that $FeCl_3$ was highly effective for soil washing of Cd-contaminated paddy soils because of cost effectiveness, its Cd extraction efficiency, and its lower environmental impact. It is a rapid method that can meet specific criteria without any long-term liability [53] and hence, is considered to be one of the best, economic, soil remediation technologies.

12.4.1.6 Electrokinetic Remediation

Soil electrokinetic remediation is another economic physical strategy for the remediation of metal(loid)s. Electrokinetic remediation (electro-reclamation) involves passing a low-intensity electric current between ceramic electrodes that are divided into a cathode and an anode array imbedded in the contaminated soil. Ions and small charged particles, in addition to water, are transported between the two electrodes. Anions move toward the positive electrodes while cations move toward the negative electrodes. An electric gradient initiates movement by electro-migration (charged chemicals movement), electro-osmosis (movement of fluid), electrophoresis (charged particles movement), and electrolysis (chemical reactions due to an electric field) [49]. Buffer solutions are used to maintain the pH at the electrodes. Contaminant metals arriving at the electrodes can be removed by electroplating or electrodeposition, precipitation or co-precipitation, adsorption, complexing with ion exchange resins, or pumping of water or any other fluid near the electrodes by using ion exchange resistance [46]. Metals that occur as soluble ions and bound to soil such as, oxides, hydroxides, and carbonates are removed by this remediation process. This method can be effective for removing contaminants in clay soils of a low permeability. It is suitable for saturated soils with low groundwater flow rates.

However, electrokinetic remediation method operates well for soils having low permeability. This method is economically effective because it is easy to install and operate, moreover, electrokinetic remediation does not abolish the original nature of the soil. However, the main limiting factor for direct electrokinetic remediation is fluctuation in soil pH because it cannot maintain soil pH value. In some case, adding buffer solutions in cathode and anode, using complexant or ion exchange membrane, controls soil pH.

12.4.1.7 Immobilization Techniques

Immobilization techniques involved reducing metal versatility, bioavailability, and bioaccessibility of substantial metal(loid)s by adding immobilizing agents to contaminated soils. Substantial metal(loid)s can be immobilized in soil by complexation, precipitation, and adsorption responses. These procedures cause redistribution of substantial metal(loid)s in soils. Metal(loid)s immobilization in soil is, by and large, done by utilizing natural and inorganic correction to soils. Incorporation of corrections like the addition of concrete, earth, zeolites, phosphates, minerals, and microorganisms could substantially achieve the goal. Chemical fertilizers and organic manures with low concentrations of heavy metal are now being utilized for metal immobilization in soils. For instance, application of diammonium phosphate (DAP) could immobilize Cu, Cd, and Zn while farmyard manure application is helpful in immobilizing Fe, Cr, Ni, Mn, and Pb in soils. Similarly, use of biochar in soil changes physical, chemical, and biological properties resulting particularly in an increment of pH that could help in precipitation of metal components and thereby heavy metal immobilization in soils.

12.4.2 Biological Approaches/Bioremediation for Remediation of Heavy Metals

The concept of bioremediation emerged initially with the idea of using microorganisms to clean soil contaminated with petroleum derivatives as developed by Tausz and Donath in 1930. Today, bioremediation is a commonly used method for the restoration of the natural and useful values of contaminated sites by microorganisms that are able to degrade, transform, or chelate various toxic compounds [54]. Microorganisms can break down organic pollutants by using them as a source of carbon and energy, or by cometabolism. Heavy metals cannot be degraded or destroyed biologically but can undergo transformation from one oxidative state or organic complex to another. It changes their water solubility and decreases their toxicity [55]. Bioremediation is ecofriendly, noninvasive, and is cheaper than conventional remediation methods that involve degradation or transformation of environmental contaminants into harmless or less toxic forms [56, 57]. Soil bioremediation can be carried out at the place of contamination (*in situ*), or in a specially prepared place (*ex situ*).

12.5 Mechanisms Involved in Bioremediation

The bioremediation process employs an immobilization mechanism, which limits the mobility of microbial cells or their enzymes with a simultaneous preservation of their viability and catalytic functions [58]. This mechanism may use the natural ability of microorganisms to form biofilms on the surface of various materials, which is commonly observed in the environment. Immobilization significantly reduces costs of bioremediation processes and improves their efficiency. There are four main mechanisms of immobilization: (i) biosorption, (ii) binding on a surface (electrostatic or covalent), (iii) entrapment, and (iv) encapsulation.

12.5.1 Biosorption

Microbes can accumulate heavy metals by adsorption, which is the main mechanism to increase metal ions in soil [59]. The biosorption mechanism involves the overall sorption phenomenon, including precipitation, chemical adsorption, ion exchange, surface precipitation, formation of stable complexes with organic ligands, and redox reaction.

Adsorption involves complexation of heavy metals on the cell surface, from which they can be absorbed into the cell [9]. Because of the cell surface's structure, principally cell wall and mucus layer, heavy metals can be adsorbed and absorbed relatively easily. Many ions in the cell surface functional groups, such as nitrogen, oxygen, sulfur, and phosphorus can be complexed with metal ions as coordination atoms. In addition, phosphoric acid anions and carboxyl anionic groups on the surface of the microbial cell wall are negatively charged, and most heavy metal surfaces carry a cationic group that interacts with the cell wall and allows the metal ions to bind or pass through the cell membrane [60]. It has been found that at pH 7.2, Bacillus can absorb 60% of its Cu^{2+} capacity within the first minute and reach adsorption equilibrium within 10 minutes [61]. Heavy metal ions bind to the surface of the cell not only by electrostatic interaction and complexation but also by ion exchange to the cell surface; for example, the nonliving brown algae (*Ascophyllum nodosum*) exchanges the original cell wall adsorption of K^+, Ca^{2+} and Mg^{2+} to adsorb Co^{2+} [62]. Brady and Duncan [60] showed that yeast releases approximately 70% of K^{+} rapidly and 60% of Mg^{2+} slowly in the process of adsorbing Cu^{2+}. Furthermore, studies have shown that ion exchange can occur with complexation.

12.5.2 Binding on a Surface

Electrostatic binding on a surface is very similar to physical adsorption, but the probability of microorganisms leaking is lower. This method requires washing the surface of the carrier with a buffer solution to obtain a hydrophilic surface that can attract the negatively charged cells or enzymes [63]. The procedure for immobilization is different in the case of covalent binding, because it requires the presence of a binding agent. Immobilization can be performed only on chemically activated carriers enriched with amide, ether, and carbamate bonds. This method is mainly used for the immobilization of enzymes, because binding agents are often toxic to cells, and for that reason microbial viability and activity are lowered. The advantage of the covalent bonds is that they are strong enough to prevent the leaking of molecules into the environment [64].

12.5.3 Entrapment

Entrapment of microorganisms is well known and widely used in bioremediation. After the entrapment, microbial cells are able to move only within a carrier. This prevents the cells from leaking into the environment but may limit the exchange of nutrients and metabolites. Microorganisms entrapped in the heterogeneous carrier are physiologically diverse. The cells located near the surface have high metabolic activity in contrast to starved cells in the interior of the carrier [65]. Entrapment is a rapid, nontoxic, inexpensive, and versatile method. Entrapped microorganisms are protected against environmental factors. The most important parameter in entrapment of microorganisms is the ratio of the size of the pores of the carrier to the size of the cells. In a situation where the pores are larger than the immobilized cells, they leak into the environment [66].

12.5.4 Encapsulation

Encapsulation is very similar to entrapment, but in this case immobilized particles are separated from the external environment with a semi-permeable membrane. The biggest advantage of this method is the significant protection of biological material against the adverse conditions of the external environment. However, due to the limited permeability of the used membrane and the probability of its damage by growing cells, encapsulation is rarely used in *ex-situ* bioremediation [58].

12.6 Types of Bioremediation

On the basis of removal and transport of wastes from contaminated sites, there are basically two methods of bioremediation (i) *in-situ* bioremediation and (ii) *ex-situ* bioremediation.

12.6.1 *In-situ* Bioremediation

In-situ bioremediation involves the decontamination of sites in the place itself and is followed when there is no possibility to transfer polluted soil, or when contamination affects an extensive area [67].Such a method is most desirable due to lower cost and less disturbance since they provide the remediation treatment in place avoiding excavation and transport of contaminants. The depth of the soil that can be effectively treated limits *in-situ* treatment. In many soils effective oxygen diffusion for desirable rates of bioremediation extend to a range of only a few centimeters to about 30 cm into the soil, although depths of 60 cm and greater have been effectively treated in some cases [57]. *In-situ* biodegradation involves the method of bioventing and biosparging for supplying oxygen and nutrients by circulating aqueous solutions through contaminated sites to stimulate naturally occurring microbes to degrade organic contaminants. It can be used for soil and groundwater. Generally, this technique includes conditions such as the infiltration of water-containing nutrients and oxygen or other electron acceptors for groundwater treatment.

There are three basic methods of *in-situ* bioremediation with microorganisms: natural attenuation, biostimulation, and bioaugmentation. Natural attenuation is connected with the degradation activities of indigenous microorganisms. This method avoids damaging the habitat, allows the ecosystem to revert to its original condition, and enables detoxification of toxic compounds [68]. Removal of contaminations by natural attenuation takes a long time because degrading microorganisms in soil represent only about 10% of total population.

The increase of bioremediation efficiency *in situ* may be realized in the bioaugmentation process, in which specific degraders are introduced into the soil [69]. This method is applied when the indigenous microflora are unable to break down pollutants, or when the population of microorganisms capable of degrading contaminants is not sufficiently large. To make the process of bioaugmentation successful, microorganisms introduced into the polluted environment, as a free or immobilized inoculum, should be able to degrade specific contamination and survive in a foreign and unfriendly habitat, be genetically stable and viable, and move through the pores in the soil. Nonindigenous strains or genetically modified microorganisms can also be incorporated into the remediated soil. To accelerate *in-situ* bioremediation processes, biostimulation is used in order to modify the physical and chemical parameters of the soil [70]. For this purpose, compounds such as nutrients (e.g., biogas slurry, manure, spent mushroom compost, rice straw, and corncob) or electron acceptors (phosphorus, nitrogen, oxygen, carbon) are introduced into the soil [71].

12.6.2 *Ex-situ* Bioremediation

Ex-situ methods of bioremediation involve the excavation or removal of contaminated soil from the ground [72]. During the *ex-situ* processes, the contaminated medium is excavated or extracted and moved to the process location. Liquids can be cleaned in constructed wetlands while semi-solid or solid wastes in slurry bioreactors. Solid contaminations are biodegraded through land farming, composting, and biopiles [73]. These techniques allow more efficient removal of pollutants, by controlling the physico-chemical parameters, resulting in a shortening of the total time of reclamation [74]. The disadvantages are additional costs and risks connected with the possibility of dispersion of the contamination during transport.

Land farming is one of the most widely used soil bioremediation technologies. In this technology, excavated contaminated soils are spread out in a thin layer on the ground surface. Aerobic microbial activity within the soil is stimulated through aeration and addition of minerals, nutrients, and moisture [75]. Land farming is a relatively simple technology however, it is inexpensive and effective for easily biodegradable contaminants only at low concentration [76]. Composting is a controlled biological process that treats agricultural and municipal solid wastes and sewage sludge using microorganisms under thermophilic and aerobic conditions [77]. Through composting, it is possible to reduce the volume of residues in landfills. Composting has been applied to bioremediation of soils contaminated with petroleum hydrocarbons, solvents, chlorophenols, pesticides, herbicides, polycyclic aromatic hydrocarbons (PAHs), and nitro-aromatic derivatives [78]. More advanced systems of composting are biopiles that are hybrids of land farming and composting [79]. Typically used for treatment of surface contamination with petroleum hydrocarbons they are a refined version of land farming that tends to control physical losses of the contaminants by leaching and volatilization [73]. Biopiles provide a favorable environment for indigenous aerobic and anaerobic microorganisms.

12.7 Strategies of Bioremediation

Bioremediation involves a number of species of microbes or higher plants having different mechanisms depending upon the environmental factors and the nature of chemicals [43] either to degrade or to remove the toxic contaminants from the environment. The use of microbes or higher plants to change the concentration of heavy metals and other organic contaminants in soil and improve the ability of growing plants to deal with elevated metals concentrations has significant economic and ecological benefits. Based on the living biota used for decontamination strategies of bioremediation are classified into two groups: (a) microbe-based remediation and (b) plant-based (phyto)remediation.

12.7.1 Microbe-Based Bioremediation

Microbial bioremediation is a natural way to mitigate environmental pollution, which includes the activities of aerobic and anaerobic microorganisms (Table 12.2). The fundamental principle lies with the fact that despite having different nutritional requirements, both life forms have a basic need of an organic carbon and energy source. Higher microbial counts and diversity is directly associated with the higher content of organic matter in soil [43] however, the presence of organic substances of high molecular weight such as PAHs may limit microbial activity [94]. On the contrary, Hesham et al. [95] reported the degradation of chrysene and benzo(a)pyrene (PAHs of high molecular weight) by yeast fungi up to 98.5 and 95.2% respectively in soil. Generally, degrading microbes rely upon associations with their neighboring organisms for nutrition, energy, as well as shelter [96]. For example, arbuscular mycorrhizal fungi and plant-growth promoting rhizobia live in association, protecting the host organisms from several environmental toxic chemicals [97, 98]. Biostimulation (enhancing the activity of native microbes), bioaugmentation (increasing the viable microbial counts), bioaccumulation (storage of toxic or nontoxic elements by microbes) [99, 100], biosorption (removal of elements from the environment through adsorption) [101], and the use of biofilms [102] are some of the recent advancements in microbial bioremediation techniques. The various strategies followed in microbe-based remediation follow.

12.7.1.1 Use of Indigenous Microbes

Natural attenuation involves the use of indigenous microorganisms for bioremediation. The native soil microbes such as bacteria and fungi are capable of either degrading or transforming contaminants to nontoxic forms. However, the metabolic pathways and mechanism involved in degrading inorganic and organic contaminants differs.

Microbial Remediation of Heavy Metals: It is imperative to note that microbes in any case cannot degrade the heavy metal ions and are only able to transform metallic ions from higher to lower oxidation states to stabilize them [67]. Microorganisms, which are mostly prokaryotic, participate in redox reactions and change the valence of heavy metals, thereby changing their activity, which can

Table 12.2 Microorganisms involved in sorption and accumulation of heavy metals.

Metal sorbed	Bacteria	Fungi	Algae	Yeast	References
Chromium(IV)	*Bacillus megaterium*	*Termitomyces clypeatus*	*Spirogyra*	*Phanerochaete crysosporium*	[80–83]
Copper	*Pseudomonas cepacia*	*Penicillin ochrochloron*	*Sargassam* sp.	*Candida tropicalis*	[84–87]
Lead	*Pseudomonas putida*	*Aspergillus parasiticus*	–	*Saccharomyces cerevisiae*	[88–90]
Mercury	–	Agaricus macrosporus	–	*Saccharomyces cerevisiae subsp. uvarum*	[91, 92]
Cadmium	*Pseudomonas putida*	*Agaricus macrosporus*	–	*Candida tropicalis CBL-1*	[83, 91, 93]

affect their mobility or toxicity [103]. For example, Hg^{2+} once reduced by aerobic bacteria to Hg^0, can evaporate from the contaminated sites [104]. Toxic and water-soluble Cr^{6+} can be reduced by *Coryne* bacterium into nontoxic and poorly water-soluble Cr^{3+} and *Bacillus licheniformis* can reduce Pb^{2+} to Pb^0 [105].

Further microbes are able to metabolize, detoxify, and accumulate metals mostly in cell walls. Microbes releasing chelating agents and acids altering physico-chemical properties such as redox potential (Eh) of surroundings [106] bring substantial changes by increasing the bioavailability of metal ions [107]. When the content of heavy metals in the soil reaches toxic levels, the mucus secreted by the fungal cell wall can combine with the polyphosphate and organic acid ions in the fungal tissue to combine with the heavy metal ions and reduce the mobility. Arbuscular mycorrhizal fungus produces polysaccharide ligands that chelate with heavy metals to form stable complexes [108]. In addition, physico-chemical reaction such as biosorption is the first step in the interaction between microbial cell and metal ions [109] followed by the physical adsorption, ion exchange, and complexation. *Rhizopus* can adsorb Zn, Cu, Cd, Pb, and other heavy metal ions [110], and *Thiobacillus* can absorb heavy metal ions as well as inorganic ions such as S, which combines with the metal ions to form precipitate that can be separated from the soil [111].

Biochemical reactions during microbial transformation of metal ions involve the action of specific enzymes for oxidation, reduction, methylation, dealkylation, and precipitation [109]. Microbes have diversified conditions for their growth with ability to adapt to the changing environmental conditions, e.g., Naraian et al. [112] reported that the multiple-drug-resistant strain of *Pseudomonas aeruginosa* isolated from tannery effluent had plasmid-encoded chromate ion resistance, exhibiting the adaptability feature of such microbes.

Microbial Remediation of Organic Contaminants: The native soil microbes have proven ability to degrade most of the (organic contaminants) OCs such as PCBs (polychlorinated biphenyls), PAHs, and polybrominated biphenyl ethers (PBEs) in the environment [113, 114]. Microbial remediation involves the catabolic activity of microbes transforming them into less or completely nontoxic residues [115]. The microbes use the organic substrate as the source of organic carbon and energy. In general, single microbial species hardly degrade any organic substrate in isolation [116] and they work well in community. Interactions between microbes in the community ensure the exchange of genetic information between microbial species conferring resistance, tolerance, and chemical-degrading ability. In general, microbes can degrade a variety of organic substances depending upon the plasmid-encoded catabolic processes and chromosomal genes (in case of bacteria) or extracellular enzymatic activity (fungal degradation process) [117]. Bacterial chemotaxis is one of the mechanisms whereby the bacterial cells move with the change of chemical gradients [118], e.g., *Pseudomonas putida* is attracted by naphthalene [119].

However, fungi have several advantages over bacterial degrading processes as the former can grow at higher concentrations of toxic organics [120], e.g., *Phanerochaete chrysosporium* metabolize lignocellulosic compounds by the activity of degrading enzymes such as laccase, lignin peroxidase, and manganese peroxidase [121]. Similarly, yeast fungi to cleave the benzene ring [95]; Gadd [122] noted that they produce catechol 2,3-dioxygenase.

12.7.1.2 Bioaugmentation

The increase in efficiency of *in-situ* bioremediation is realized in the bioaugmentation process, in which the specific degraders are introduced into the soil [69]. This method is applied when the indigenous microflora are unable to break down pollutants, or when the population of microorganisms capable of degrading contaminants is not sufficiently large. To make the process of bioaugmentation successful, microorganisms introduced into the polluted environment, as a free or immobilized inoculum, should be able to degrade specific contamination and survive in a foreign and unfriendly habitat, be genetically stable and viable, and move through the pores in the soil [70]. However, the result of bioaugmentation depends on the interaction between exogenous and indigenous populations of microorganisms because of the competition, mainly for nutrients.

12.7.1.3 Biostimulation

To accelerate *in-situ* bioremediation processes, biostimulation is used in order to modify the physical and chemical parameters of the soil. For this purpose, compounds such as nutrients (e.g., biogas slurry, manure, spent mushroom compost, rice straw, and corn cob) or electron acceptors are introduced into the soil [71]. *Bioventing* is the most common method in biostimulation that involves a supply of air and nutrients through wells to contaminated soil to stimulate the indigenous microbes [68]. Bioventing employs low air-flow rates and provides only the amount of oxygen necessary for the biodegradation while minimizing volatilization and release of contaminants to the atmosphere. It works for simple hydrocarbons and can be used where the contamination is deep under the surface. Biosparging involves the injection of air under pressure below the water table to increase groundwater oxygen concentrations and enhance the rate of biological degradation of contaminants by naturally occurring microbes. It increases mixing in the saturated zone and thereby increases the contact between soil and groundwater.

12.7.1.4 Biotechnological Approaches Involving Genetically Modified Microorganisms

In the scientific perspective, genetic engineering is one of the elite biotechnological branches offering infinite combinations of genetic codes imparting the knowledge of the design of potentially degrading and substrate-specific microbes [123–125].

For a successful bioengineered microbe, it is mandatory to identify the relevant species and their substrate-specific strains [126]. In addition, genetic modification with DNA reshuffling has provided enormous combinations enabling microbial cells to degrade even the most recalcitrant organic compounds [127].

Advancements in microbial remediation techniques and studies on plasmid-encoded biochemical information and genetic engineering helped in designing new strains of known bacterial species for targeted bioremediation, e.g., recombinant *Escherichia coli* expressing metallothionein gene (*Neurospora crasa*) for Cd uptake, which was more rapid than the gene donor microbial species [128]. *Staphylococcus xylosus* and *Staphylococcus carnosus* were designed by the introduction of polyhistidyl peptides encoding genes conferring Ni-binding capacity to these microbes [129].

Technological advancements in microbial bioremediation include the conversion of slow biodegradation processes into a rapidone by means of several modern methods including microbial fuel cells (MFCs) [130], bioreactors [131], biofilms [102], and the use of microbial consortia for the degradation of recalcitrant organic compounds.

12.7.2 Plant Based Bioremediation/Phytoremediation

The plant-based bioremediation otherwise known as "phytoremediation" makes use of green plants to remove pollutants from the environment or to render them harmless [132]. This concept was first proposed by Chaney [133] and then developed through the study of plant species' ability to remove pollutants from environment. The basic principle lies with plant roots that either break down the contaminants in the soil or suck up the contaminants in its stems and leaves. Plant responses to heavy metals can be as: (i) metal excluders (prevent metals from entering into plant parts); (ii) metal indicators (actively accumulate metals in their aerial tissues and reflect metal level in soil); and (iii) metal accumulators (concentrate metals in aerial parts to levels far exceeding than soil) [134]. The phytoavailability of contaminants to plants is a crucial factor regulating the efficiency of phytoremediation process, which depends strongly on their specific chemical forms and soil properties. For heavy metal contaminated soil, four processes of phytoremediation are recognized: phytoextraction, phytostabilization, phytovolatilization, and rhizofiltration.

Plant species differ widely in their ability to accumulate heavy metals [135] that vary between species and even between genotypes of a species [136]. Such diversity in metal uptake among plant species is mainly attributed to several factors, (i) root architecture, (ii) water use efficiency, (iii) rhizosphere chemistry, (iv) expression and affinity of root surface transporter proteins for metals, and (v) xylem loading of metals and translocation within the plant. Also, the age and the growth stage of the plant can affect the concentration of metals in plants. Plant species for phytoremediation are selected based on their root depth, the nature of the contaminants, soil, and regional climate. It has been reported that for phytoremediation, grasses are the most commonly evaluated plants attributed to characteristic rapid growth, large amount of biomass, strong resistance, effective stabilization to soils, and an ability to remediate different types of soils [137]. The use of indigenous plant species is generally favored because they show tolerance to imposed stress conditions, require less maintenance, and present fewer environmental and human risks than nonnative or genetically altered species. Karimi [138] has investigated the potential of alfalfa and sorghum for phytoremediation of soil contaminated with chromium and reported that the concentration of chromium in soil after phytoremediation with alfalfa was decreased by 60–74%, and with sorghum by 51.2–69.5%. Other plant species such as *Eicchornia* (water hyacinth), willow plant, *Acacia*, and *Atriplex* are the commonly used plant species for phytoremediation purpose. Hence, phytoremediation is a promising green technology that can be used to remediate heavy metal contaminated soils. The complexity of factors that control the efficiency of this technique such as soil properties, plant species, and climatic conditions need to be identified. More species that have remediative abilities need to be identified and research should focus on developing agricultural techniques to enhance phytoremediation efficiency and reduce time and cost of heavy metal removal from soils.

12.8 Factors Affecting Bioremediation

Various biotic and abiotic factors affect the behavior and growth of microbial cells, biological processes occur in a microbial community and thereby influence the process of bioremediation. The bioremediation process is subjected to multiphasic heterogeneous surroundings influencing the rate of biochemical reactions involved [43]. Hence, a detailed understanding of microbial ecology is required for improvement in microbiological process and to predict a successful bioremediation process [117].

12.8.1 Environmental Factors

There is no direct evidence of any likely impacts of climate change on bioremediation processes however, research studies on soil microbes and global climate [139–141] suggested changes in the physico-chemical properties of the microbial niche, which may alter microbial metabolic processes and thereby bioremediation. Environmental factors such as atmospheric temperature and CO_2

have considerable effects on microbial extracellular enzyme production as well as the ability of *in-situ* microbes to degrade toxic contaminants [141].

12.8.1.1 Atmospheric Temperature

It has been predicted that the rise in soil temperature due to global climate change, could cause substantial increases in microbial enzymatic activity [142]. Frey et al. [143] reported increased utilization efficiency of recalcitrant substrate in soils by microbes at higher temperature as a positive feedback to climate. Although feedback from ecosystems to the climate change have reportedly been regulated by microbial communities [144].

The solubility of PAHs increases with a rise in atmospheric temperature, which ultimately, enhances its bioavailability to microbes [145]. The microorganisms that degrade PAHs in crude oil are most active in specified temperature ranges that govern the production of enzymes; however, most studies have reported higher biodegrading efficiency at mesophilic temperatures of 20–35 °C rather than very low or high temperatures [146]. At lower temperatures, oil viscosity increases and the volatility of lower chain hydrocarbons decreases thereby enhancing its solubility and making it less prone to biodegradation. On the contrary, naphthalene and phenanthrene degradation from crude oil in seawater at temperatures as low as 0 °C has also been cited [147]. The laccase and manganese peroxidase enzymes of ligninolytic fungi were reported to have a temperature optimum of between 50 and 75 °C respectively. In addition, oxygen solubility decreases with increasing temperature, which reduces metabolic activity of aerobic microbes [148].

12.8.1.2 Atmospheric Carbon Dioxide

Soil microbial community plays an important role in the cycling of carbon as microbes exhibit increased ability to decompose soil organic matter at elevated CO_2 in ecosystems [140]. Elevated ambient CO_2 level has been found associated with the increase in bacterial abundance [139] and decreased fungal abundance [149]. Decrease in fungal biomass at warm and dry conditions favored due to elevated atmospheric CO_2 level decreases the fungal: bacterial biomass ratio in soil [141], indicating decreased magnitude of carbon cycling in an ecosystem that further disturbs the natural degradation process cometabolically facilitated by naturally available carbon.

12.8.2 Soil Factors

Soil factors greatly influence the bioremediation process as the soil serves as the medium for the activity of microbes and the occurrence of biodegrading pathways. Its factors include texture, permeability, water holding capacity, pH, redox potential, temperature, nutrient, and oxygen content. To enable the bioremediation process, soil is mixed with amendments or bulking materials (straw, sawdust, etc.), as the bioremediation processes rely on microbial activity, and microorganisms require oxygen inorganic nutrients, water, and optimal temperature and pH to support cell growth and sustain biodegradation.

12.8.2.1 Physico-chemical Parameters

Physico-chemical parameters include pH, redox potential (Eh), ionic strength, the presence or absence of electron acceptors and donors, and the age of organometallic ions. Biosorption, a major mechanism in toxic metal removal by microbes, is a pH-dependent phenomenon [127]. The pH influences the isoelectric point in a solution, affecting the net negative charge on the microbial cell surface. Additionally, the change in pH also brings changes in ionic state of ligands, e.g., carboxyl, phosphoryl, amino acid, and groups [150]. In addition, lower pH values are required for metal ions to attached to the microbial cell surface [151] whereas alkaline medium favors precipitation of metal ions [109].

Solubility plays an important role in the degradation of OCs as those are hydrophobic or sparingly soluble compounds that persist in the environment for a long time and are biounavailable [116]. Bioremediation techniques involve the reduction of metal ions to an insoluble form by microorganisms from higher to lower oxidation state (pH dependent) as soluble metal ions can only undergo enzymatic reduction. Higher oxidation states of metals are generally soluble [67]. The pH affects the solubility of metal ions, which increases with a decrease in the pH of the medium, thereby affecting adsorption by microbial cells [152]. Organic acid-producing microorganisms *Pseudomonas* sp. and *Burkholderia cepacia* produce gluconic acid and *Rhizobium* sp. and *Bacillus* produce ketogluconic acid [131], which lower the pH of systems and increase the solubility of metal ions. Organometalic compounds increase mobilization of metal ions [153].

Bioremediation is strongly inhibited by toxic metal ions, such as Zn^{2+}, which acts as a respiratory inhibitor in microbes [154]. The presence of electron acceptors, e.g., oxygen in aerobic microbes and NO_3, SO_4, and Fe(III) oxides in case of anaerobic microbes, also affects biodegradation processes [155].

12.8.2.2 Moisture

Moisture is required for all biological processes to help transport nutrients, foods, and waste products in and out of microorganisms. For oceans, lakes, and other surface water environments, this poses no immediate problems. For soil environments, some moisture must be present for degradation to occur. Too much water can impede the aeration of soil, and the process may turn anaerobic. The optimum ratio of moisture depends on the climate and soil type. The aerobic biodegradation of hydrocarbons in soils is greatest in ranges of 50–70% water holding capacity [156] whereas waves and tidal actions are useful in supplying aerated sea water to beaches and marshes, rainfall is useful in the biodegradation of inland soils by supplying moisture and dissolved oxygen to the microbes.

12.8.2.3 Aeration

It is now well established that bioremediation of organic contaminants such as PAHs can proceed under both aerobic and anaerobic conditions. During aerobic PAH metabolism, oxygen is integral to the action of mono- and dioxygenase enzymes in the initial oxidation of the aromatic ring [157]. For surface contamination, soil tilling and/or mixing can aerate the sites well enough to allow PAH transformation to proceed. There is still debate as to whether the benefits of anaerobic bioremediation are outweighed by the negatives, with the aeration of contaminated anaerobic aquifers successfully used to stimulate aerobic microbial communities resulting in significant reductions in PAH concentrations in groundwater. This has been accomplished using hydrogen peroxide, sodium nitrate [158], and perchlorate [159]. In addition, the aerobic biodegradation of hydrocarbons has been reported to be up to an order of magnitude higher relative to anaerobic biodegradation. However, it has also been reported that rates of anaerobic PAH degradation under denitrifying conditions were comparable to those under aerobic conditions [160]. Though it appears that the future of anaerobic bioremediation is promising, there are several drawbacks to the promotion of anaerobic bioremediation.

12.8.2.4 Organic Matter

The presence of naturally occurring organic matter in soil affects the bioremediation process by indirectly influencing water retention and temperature. It also plays a major role in the biosorption process, a major mechanism in toxic metal removal by microbes which in turn affects the bioavailability of contaminants to microbes. Contaminants targeted for biodegradation impart resistance to alter upon enzymatic attack due to sorption onto organic matter. Addition of organic matter slows the natural rate of biodegradation due to increased competition, but long term increases infiltration, permeability, and porosity of soil. The addition of organic matter is also helpful in biodegradation under low moisture retaining environments.

12.8.2.5 Nutrient Availability

In addition to a readily degradable carbon source, biodegrading microorganisms also require mineral nutrients such as N, P, and K for cellular metabolism and therefore successful growth. The capacity of the microorganism to grow in a given system depends on the organism's ability to utilize any available nutrient. The types and quantities of nutrients present in the soil system play an important role in limiting the rate of hydrocarbon degradation. The organic carbon content in a hydrocarbon contaminated site is found to be very high attributed to a constant input of hydrocarbons. Because of the high carbon content of oil and the low level of other nutrients essential for microbial growth, the rate, and extent of degradation are, in general, limited by the low availability of nitrogen and phosphorus. Consequently, growth of hydrocarbon-degrading bacteria and hydrocarbon degradation can be strongly enhanced by fertilization with inorganic N and P. Aerobic microorganisms utilize various types of nutrients including nitrogen, phosphorus, and trace amounts of potassium, calcium, sulfur, magnesium, iron, and manganese. Nitrogen and phosphorus are vital nutrients especially since oil contains very little of either. Many studies have shown that an inadequate supply of nutrients may result in a slower rate of degradation for hydrocarbons [161].

Therefore, it is common practice to supplement contaminated land with nutrients, generally nitrogen, and phosphates to stimulate the *in-situ* microbial community and therefore enhance bioremediation. The amounts of N and P required for optimal microbial growth and hence bioremediation is estimated from the ratio of C : N : P in microbial biomass. However, a recent study has shown that optimal microbial growth and creosote biodegradation occurred in soil with a much higher C : N ratio of 25 : 1 [162]. It was investigated that the external addition of KNO_3 as a source of inorganic nitrogen resulted in 10-fold increase in degradation of phenanthrene by natural micro flora present in seawater samples [163]. It is worth noting that fungi are able to effectively recycle nutrients (specifically nitrogen), and that excessively high nutrient loadings may in fact inhibit microbial metabolism. In addition, the high molecular weight PAH-oxidizing ligninolytic enzymes of fungi are produced under nutrient deficient (often low N) conditions [164].

It is therefore imperative that the nutrient status of contaminated sites should be quantified prior to the supplementation of additional nutrients. Even though microbial metabolism may be temporarily increased, the long-term inhibition of functionally important organisms may result in the failure of bioremediation of high molecular weight PAHs. This in turn necessitates external supplementation of nutrients to support biological activity, and hence bioremediation.

12.8.3 Bioavailability of Contaminants

The term "bioavailability" refers to the fraction of contaminants in soil that can be utilized or transformed by living microorganisms [165]. This is attributed to the effect of physico-chemical and microbiological factors on the rate and extent of biodegradation [166] and is believed to be one of the most important factors affecting bioremediation.

The bioavailability of contaminants is dependent on physical state, hydrophobicity, sorption onto soil particles, volatilization, and solubility, etc. Mainly, after the entry of contaminants in soil, biodegradation, leaching or volatilization may reduce an organic contaminant, or it may accumulate within soil biota or be sequestered and complexed within the mineral and organic matter fractions. PAH compounds with low water solubility have low bioavailability, and are classed as hydrophobic organic contaminants [167]. The larger molecular weight of PAHs lowers their solubility, which in turn reduces their accessibility for metabolism by microbial cells. In addition, PAHs can undergo rapid sorption to surfaces of mineral and organic matter (i.e., humic and fulvic acids) in soil matrix reducing their bioavailability.

The rate at which microorganisms degrade the contaminants depends on rate of uptake, the rate of transfer to microbial cells, and metabolism. It is controlled by a number of physical–chemical processes such as sorption/desorption, diffusion, and dissolution. The mass transfer of a contaminant determines microbial bioavailability. Put another way, the bioavailability of a pollutant is defined as the ratio of its mass transfer and biota intrinsic activities in soil. The longer a contaminant is in contact with soil, more irreversible the sorption and lower the degradation of contaminants [168]. This phenomenon is known as "ageing" of the contaminant. Therefore, the bioavailability of pollutants is linked to their persistence in a given environment [169].

12.8.4 Characteristics of Contaminants

The rate and kinetic of degradation of hydrocarbons and other toxic contaminants by microorganisms depends upon their chemical structure and concentration [170]. Initiation of microbial degradation of xenobiotics depends on the toxic pressure exerted by contaminants on microbes that induce enzymatic modifications [118]. Petroleum hydrocarbons contain a complex mixture of compounds, which degrade at different rate. The n-alkanes of intermediate length (C_{10}-C_{25}) are the preferred substrates for microorganisms and tend to be the most readily degradable. The shorter chain compounds (carbon backbone less than C_{10}) and longer chain alkanes (C_{25}-C_{40}) are more toxic due to greater hydrophobicity and poor water solubility as well as bioavailability [171]. The highly condensed branched chain alkanes and cycloalkanes are also degraded more slowly than the corresponding normal alkanes.

The rate of microbial uptake and biodegradation of hydrocarbons is dependent on solubility of hydrocarbons in the surrounding solvent. The degradation rates of hydrocarbons highly soluble in water are almost proportional to their concentration but it differs for less aqueous hydrocarbons [172]. It has also been reported that high solubility may also impart detrimental effects on degrading microbes.

Usually an oil loading greater than 5% leads to a decrease in the microbial activity [173]. In addition to this high concentration may disturb the C : N : P ratio and create oxygen limitations.

12.8.5 Biological Factors

Since the bioremediation process involves the biodegradation of toxic compounds mediated by microorganisms, the influence of biological factors is often realized while implementing bioremediation techniques. Among biological factors, consortium of microbes play a major role as toxic hydrocarbons vary in size from monomeric to polymeric carbon molecules. No single species of microorganism is capable of degrading a range of hydrocarbons, rather microbial communities of complex multispecies interactive networks with different enzymes and metabolic pathways are required for such degradation [174].

The main factor that affects the rate of microbial decomposition of hydrocarbons is the bioavailability of contaminants to microorganisms that have the catabolic ability to degrade them. The soil microorganism's number is usually in the range 10^4–10^7 CFU. For successful biodegradation this number should not be lower than 10^3 per gram of soil [175]. Microorganism numbers lower than 10^3 CFU (colony-forming unit) per gram of soil indicate the presence of toxic concentrations of organic or inorganic contaminants.

Maphosa et al. [176] defined bioremediation as a function of metagenomics or community genomics wherein the metagenomics sequencing helps in providing data of member species of consortia that support substrate degradation. Certain microbes utilize other sources of organic carbon, electron acceptor, and energy as cometabolic substrate and degrade the recalcitrant toxic chemicals, e.g., *Mycobacterium gilvum* reportedly degrade pyrene aerobically in the rhizospheric zone of *Phragmites australis* simultaneously with the degradation of benzopyrene [177].

There are several inherent characters of microbes, e.g., plasmid-encoded genes that aid specificity for substrates and encode the specific enzymes (proteins) for microbial degradation. However, it has been observed in nature that bacterial cells have diversified specificity for different substrates than other microbial communities [178]. Bacterial chemotaxis is an advantageous behavior of bacteria for the degradation of recalcitrant organic compounds [118]. The factors affecting metabolism, composition, and abundance of microbial communities may also alter the biodegradation pathways [179]. The allelopathic effect of terrestrial plants and other inhibiting microbes negatively influences the degradation potential of microbial consortium [180].

12.9 Pros and Cons of Applicability of Bioremediation Approaches Under Field Conditions

Bioremediation is a natural process and is therefore perceived by the public as an acceptable waste treatment process for contaminated soil. The residues generated as a byproduct through bioremediation are usually harmless products and include carbon dioxide, water, and cell biomass. Theoretically, bioremediation is useful for the complete destruction of a wide variety of contaminants. Many compounds that are legally considered to be hazardous can be transformed to harmless products. Instead of transferring contaminants from one environmental medium to another, it enables *in-situ* destruction of target pollutants. This also eliminates the need to transport quantities of waste off site and the potential threats to human health and the environment that can arise during transportation. Bioremediation can prove less expensive than other technologies that are used for the clean-up of hazardous waste.

Bioremediation, although considered a boon in the midst of present-day environmental situations, can also be considered problematic because, while additives are added to enhance the functioning of microbes, it may be disruptive to other organisms inhabiting that same environment when done *in situ* [181]. Bioremediation is generally very costly, is labor intensive, and can take several months to achieve acceptable levels. Bioremediation is limited to those compounds that are biodegradable. Being biological processes, it is highly specific and requires the presence of metabolically capable microbial populations, suitable environmental growth conditions, and appropriate levels of nutrients and contaminants [182].

12.10 Conclusion and Future Prospects

Heavy metals and toxic organic pollutants are ever increasing in the environment due to rapid industrialization, urbanization, and mechanized agricultural practices. Rapid environmental clean-up and stabilization of contaminants is highly desirable to support the sustainability of natural ecosystems. The presence of plants and associated microbes in contaminated areas is due to the xenobiotic-resistant genes they naturally contain; these genes help chelate, traffic, and sequestrate them. The biological approaches involving the use of efficient plants and rhizobacteria/genetically engineered microbes are always safer and preferable over costly physico-chemical methods of environmental remediation. The bottleneck in using such biological techniques is the slow growth kinetics of microbial metabolism, which may be caused by environmental factors or even inhibited by pollutants. To overcome these challenges, future research must be directed toward developing efficiently growing rhizomicrobial and plant species via rhizosphere engineering using natural gene transfer or recombinant DNA/RNA technology. Synergistic use of suitable plants and efficient microbes can greatly enhance remediation efficiency. Furthermore, inoculating seeds/plants with specific endophytic and rhizospheric microbes can also increase this partnership. Microbial strains having both features plant-growth-promoting as well as contaminant-degrading abilities offer an efficient, cost-effective, and environmentally safer option for bioremediation and phytoremediation.

References

1 Manafi, Z., Abdollahi, H., Majd, M.Z. et al. (2012). Isolation and identification of microorganisms causing microbial degradation of organic phase of the solvent extraction unit in the copper industries. *Internatinal Journal of Mineral Processing* 112: 43–48.

2 Arivalagan, P., Singaraj, D., Haridass, V. et al. (2014). Removal of cadmium from aqueous solution by batch studies using Bacillus cereus. *Ecological Engineering* 71: 728–735.

3 Kumar, P., Kim, K.-H., Bansal, V. et al. (2017). Progress in the sensing techniques for heavy metal ions using nanomaterials. *Journal of Industrial and Engineering Chemistry* 54: 30–43.

4 Carolin, C.F., Kumar, P.S., Saravanan, A. et al. (2017). Efficient techniques for the removal of toxic heavy metals from aquatic environment: a review. *Journal of Environmental Chemical Engineering* 5: 2782–2799.

5 Congeevaram, S., Dhanarani, S., Park, J. et al. (2007). Biosorption of chromium and nickel by heavy metal resistant fungal and bacterial isolates. *Journal of Hazardous Materials* 146: 270–277.

6 Kuppusamy, S., Thavamani, P., Venkateswarlu, K. et al. (2017). Remediation approaches for polycyclic aromatic hydrocarbons (PAHs) contaminated soils: technological constraints, emerging trends and future directions. *Chemosphere* 168: 944–968.

7 Mukherjee, T., Chakraborty, S., Biswas, A.A. et al. (2017). Bioremediation potential of arsenic by non-enzymatically biofabricated silver nanoparticles adhered to the mesoporous carbonized fungal cell surface of Aspergillus foetidus MTCC8876. *Journal of Environmental Management* 201: 435–446.

8 Fulekar, M., Singh, A., and Bhaduri, A.M. (2009). Genetic engineering strategies for enhancing phytoremediation of heavy metals. *African Journal of Biotechnology* 8: 529–535.

9 Wuana, R.A. and Okieimen, F.E. (2011). Heavy metals in contaminated soils: a review of sources, chemistry, risks and best available strategies for remediation. *ISRN Ecology* 2090-4614: 1–20.

10 Zeng, L., Li, N., Shao, D. et al. (2014). Concentrations, sources, and risk assessment of polychlorinated biphenyls in vegetables near a waste-incinerator site, South China. *Archives of Environmental Contamination and Toxicology* 67: 78–86.

11 Alvarez, A., Saez, J.M., Costa, J.S.D. et al. (2017). Actinobacteria: current research and perspectives for bioremediation of pesticides and heavy metals. *Chemosphere* 166: 41–62.

12 Shahid, M., Khalid, S., Abbas, G. et al. (2015). Heavy metal stress and crop productivity. In: *Crop Production and Global Environmental Issues* (ed. K.R. Hakeem), 1–25. Cham: Springer International Publishing.

13 Chabukdhara, M. and Nema, A.K. (2012). Assessment of heavy metal contamination in Hindon River sediments: A chemometric and geochemical approach. *Chemosphere* 87: 945–953.

14 Srinivasa Gowd, S., Ramakrishna Reddy, M., and Govil, P. (2010). Assessment of heavy metal contamination in soils at Jajmau (Kanpur) and Unnao industrial areas of the Ganga Plain, Uttar Pradesh, India. *Journal of Hazardous Materials* 174: 113–121.

15 Li, Q., Liu, H., Alattar, M. et al. (2015). The preferential accumulation of heavy metals in different tissues following frequent respiratory exposure to PM2.5 in rats. *Scientific Reports* 5: 16936. https://doi.org/10.1038/srep16936.

16 Kothe, E., Dimkpa, C., Haferburg, G. et al. (2010). Streptomycete heavy metal resistance: extracellular and intracellular mechanisms. In: *Soil Biology* (eds. I. Sherameti and A. Varma), 225–235. Berlin/Heidelberg: Springer.

17 Fernández, R., Fernández-Fuego, D., Bertrand, A. et al. (2014). Strategies for Cd accumulation in Dittrichia viscosa (L.) Greuter: roleofthe cell wall, non-protein thiols and organic acids. *Plant Physiology and Biochemistry* 78: 63–70. https://doi.org/10.1016/j.plaphy.2014.02.021.

18 Nziguheba, G. and Smolders, E. (2008). Inputs of trace elements in agricultural soils via phosphate fertilizers in European countries. *Science of the Total Environment* 90 (1): 53–57.

19 Ross, S.M. (ed.). (1994). Sources and forms of potentially toxic metals in soil-plant systems. In: Toxic Metals in Soil Plant Systems, 3–25.

20 Wu, Q., Leung, J.Y.S., Geng, X. et al. (2015). Heavy metal contamination of soil and water in the vicinity of an abandoned e-waste recycling site: implications for dissemination of heavy metals. *Science of the Total Environment* 506–507: 217–225. https://doi.org/10.1016/j.scitotenv.2014.10.121.

21 Alloway, B.B.J. (1990). *Heavy Metals in Soils*. New York;, NY: Blackie and Son.

22 Lombi, E. and Gerzabek, M.H. (1998). Determination of mobile heavy metal fraction in soil: results of a pot experiment with sewage sludge. *Communications in Soil Science and Plant Analysis* 29 (17–18): 2545–2556.

23 Djingova, R. and Kuleff, I. (2000). Instrumental techniques for trace analysis. In: *Trace Elements: Their Distribution and Effects in the Environment* (ed. J.P. Vernet), 137–185. London: Elsevier.

24 Jadia, C.D. and Fulekar, M.H. (2009). Phytoremediation of heavy metals: recent techniques. *African Journal of Biotechnology* 8 (6): 921–928.

25 Taiz, L. and Zeiger, E. (2002). *Plant Physiology*. Sunderland, MA: Sinauer Associates.

26 Schaller, A. and Diez, T. (1991). Plant specific aspects of heavy metal uptake and comparison with quality standards for food and forage crops. In: *Der Einfluß von festen Abfallen auf Boden, Pflanzen* (eds. D. Sauerbeck and S. Lubben), 92–125. Julich: KFA.

27 Kibra, M.G. (2008). Effects of mercury on some growth parameters of rice (Oryza sativa L.). *Soil and Environment* 27 (1): 23–28.

28 Ahmad, I., Akhtar, M.J., Zahir, Z.A. et al. (2012). Effect of cadmium on seed germination and seedling growth of four wheat (Triticum aestivum L.) cultivars. *Pakistan Journal of Botany* 44 (5): 1569–1574.

29 Kabata-Pendias, A. (2001). *Trace Elements in Soils and Plants*, 3e. Boca Raton, FL: CRC Press.

30 Doelman, P. and Haanstra, L. (1984). Short-term and long-term effects of Cd, Cr, Cu, Ni, Pb, and Zn on microbial respiration in relation to abiotic soil factors. *Plant and Soil* 79: 317–321.

31 Hemida, S.K., Omar, S.A., and Abdel-Mallek, A.Y. (1997). Microbial populations and enzyme activity in soil treated with heavy metals. *Water Air Soil Pollution* 95: 13–22.

32 Kandeler, E., Tscherko, D., Bruce, K.D. et al. (2000). Structure and function of the soil microbial community in microhabitats of a heavy metal polluted soil. *Biology and Fertility of Soils* 32: 390–400.

33 Forstner, U. (1995). Land contamination by metals: global scope and magnitude of problem. In: *Metal Peciation and Contamination of Soil* (eds. H.E. Allen, C.P. Huang, G.W. Bailey, et al.), 1–33. Boca Raton, FL: CRC Press.

34 Nogawa, K., Honda, R., Kido, T. et al. (1987). Limits to protect people eating cadmium in rice, based on epidemiological studies. *Trace Substances in Environmental Health* 21: 431–439.

35 Lake, I.R., Hooper, L., Abdelhamid, A. et al. (2012). Climate change and food security: health impacts in developed countries. *Environmental Health Perspectives* 120: 1520–1526.

36 Taj, Z.Z. and Rajkumar, M. (2016). Perspectives of plant growth-promoting actinomycetes in heavy metal phytoremediation. In: *Plant Growth Promoting*

Actinobacteria (eds. A. Sathya, R. Vijayabharathi and S. Gopalakrishnan), 213–231. Singapore: Springer.

37 Mahar, A., Wang, P., Ali, A. et al. (2016). Challenges and opportunities in the phytoremediation of heavy metals contaminated soils: A review. *Ecotoxicology and Environmental Safety* 126: 111–121.

38 Khalid, S., Shahid, M., Khan, N. et al. (2017). A comparison of technologies for remediation of heavy metal contaminated soils. *Journal of Geochemical Exploration* 182: 247–268.

39 Thakur, S., Singh, L., Wahid, Z.A. et al. (2016). Plant-driven removal of heavy metals from soil: uptake, translocation, tolerance mechanism, challenges, and future perspectives. *Environmental Monitoring and Assessment* 188: 206. https://doi.org/10.1007/s10661-016-5211-9.

40 Gaur, A. and Adholeya, A. (2004). Prospects of arbuscular mycorrhizal fungi in phytoremediation of heavy metal contaminated soils. *Current Science* 86 (4): 528–534.

41 Tangahu, B.V., Sheikh Abdullah, S.R., Basri, H. et al. (2011). A review on heavy metals (As, Pb, and Hg) uptake by plants through phytoremediation. *International Journal of Chemical Engineering* 31: 1687-806x. https://doi.org/10.1155/2011/939161.

42 Zubair, M., Shakir, M., Ali, Q. et al. (2016). Rhizobacteria and phytoremediation of heavy metals. *Environmental Technology Reviews* 5: 112–119.

43 Boopathy, R. (2000). Factors limiting bioremediation technologies. *Bioresource Technology* 74: 63–67.

44 Maenpaa, K.A., Kukkonen, J.V.K., and Lydy, M.J. (2002). Remediation of heavy metal- contaminated soils using phosphorus: evaluation of bioavailability using an earthworm bioassay. *Archives of Environmental Contamination and Taxicology* 43: 389–398.

45 Rahimi, M., Farhardi, R., and Yusef, P.H. (2012). Phytoremediation of arsenic. *International Journal of Plant Production* 3 (7): 230–233.

46 Jankaite, A. and Vasarevicius, S. (2005). Remediation technologies for soils contaminated with heavy metals. *Journal of Environmental Engineering and Landscape Management* 12 (2): 109a–113va.

47 Khan, A.G., Kuek, C., Chaudhry, T.M. et al. (2000). Role of plants, mycorrhizae and phytochelators in heavy metal contaminated land remediation. *Chemosphere* 21: 197–207.

48 Dada, E.O., Njoko, K.L., Osuntoki, A.A. et al. (2015). A review of current techniques of in situ physico-chemical and biological remediation of heavy metal polluted soil. *Ethiopian Journal of Environmental Studies and Management* 8 (5): 606–615.

49 Mulligan, C.N., Yong, R.N., and Gibbs, B.F. (2001). Remediation technologies for metal-contaminated soils and groundwater: an evaluation. *Engineering Geology* 60 (1–4): 193–207.

50 Mallampati, S.R., Mitoma, Y., Okuda, T. et al. (2015). Dynamic immobilization of simulated radionuclide 133Cs in soil by thermal treatment/vitrification with nanometallic Ca/CaO composites. *Journal of Environmental Radioactivity* 139: 118–124.

51 Dellisanti, F. (2016). In-field remediation of tons of heavy metal-rich waste by Joule heating vitrification. *International Journal of Mineral Processing* 93: 239–245.

52 Makino, T., Takano, H., Kamiya, T. et al. (2008). Restoration of cadmium-contaminated paddy soils by washing with ferric chloride: Cd extraction mechanism and bench-scale verification. *Chemosphere* 70: 1035–1043.

53 Park, B. and Son, Y. (2016). Ultrasonic and mechanical soil washing processes for the removal of heavy metals from soils. *Ultrasonics Sonochemistry* 35 (B): 640–645.

54 Mosa, K.A., Saadoun, I., Kumar, K. et al. (2016). Potential biotechnological strategies for the cleanup of heavy metals and metalloids. *Frontiers of Plant Science* 7: 303. https://doi.org/10.3389/fpls.2016.00303.

55 Garbisu, C. and Alkorta, I. (2001). Phytoextraction: A cost-effective plantbased technology for the removal of metals from the environment. *Bioresource Technology* 77 (3): 229–236.

56 Kulik, N., Goi, A., Trapido, M. et al. (2006). Degradation of polycyclic aromatic hydrocarbons by combined chemical pre-oxidation and bioremediation in creosote contaminated soil. *Journal of Environmental Management* 78: 382–391.

57 Perelo, L.W. (2010). In situ and bioremediation of organic pollutants in aquatic sediments. *Journal of Hazardous Materials* 177: 81–89. https://doi.org/10.1016/j.jhazmat.2009.12.090.

58 Kourkoutas, Y., Bekatorou, I., Banat, M. et al. (2004). Immobilization technologies and support materials suitable in alcohol beverages production: a review. *Food Microbiology* 21 (4): 377–397.

59 Alloway, B.B.J. (1995). *Heavy Metals in Soils*. Glasgow, UK: Blackie Academic and Professional.

60 Brady, D. and Duncan, J.R. (1994). Binding of heavy metals by the cell walls of Saccharomyces cerevisiae. *Enzyme and Microbial Technology* 16 (7): 633–638.

61 He, M. and Tebo, B.M. (1998). Surface charge properties of and Cu(II) adsorption by spores of the marine Bacillus sp. strain SG-1 LEE. *Applied and Environmental Microbiology*: 1123–1129.

62 Freitas, P.S., Clarke, L.J., Kennedy, H. et al. (2006). Environmental and biological controls on elemental (Mg/Ca, Sr/Ca and Mn/Ca) rations in shells of the king scallop. *Geochimica et Cosmochimica Acta* 70: 5119–5133.

63 Hudson, S., Magner, E., Cooney, J. et al. (2005). Methodology for the immobilization of enzymes onto mesoporous materials. *Journal of Physical Chemistry B* 109: 19496–19506.

64 Yu, C.M., Yen, M.J., and Chen, L.C. (2010). A bioanode based on MWCNT/protein-assisted coimmobilization of glucose oxidase and 2,5-dihydroxybenzaldehyde for glucose fuel cells. *Biosensors and Bioelectronics* 25: 2515–2521.

65 Bayat, Z., Hassanshahian, M., and Cappello, S. (2015). Immobilization of microbes for bioremediation of crude oil polluted environments: A mini review. *Open Microbiology Journal* 9: 48–54.

66 Guzik, U., Hupert-Kocurek, K., Krysiak, M. et al. (2014). Degradation potential of protocatechuate 3,4-dioxygenase from crude extract of Stenotrophomonas maltophilia strain KB2 immobilized in calcium alginate hydrogels and on glyoxyl agarose. *BioMed. Research International* 2014 (3): 1–8.

67 Garbisu, C. and Alkorta, I. (2003). Basic concepts on heavy metal soil bioremediation. *European Journal of Mineral Processing abnd Environmental Protection* 3: 58–66.

68 Simarro, R., González, N., Bautista, L.F. et al. (2013). Assessment of the efficiency of in situ bioremediation techniques in a creosote polluted soil: change in bacterial community. *Journal of Hazardous Materials* 262: 158–167. https://doi.org/10.1016/j.jhazmat.2013.08.025.

69 Pimmata, P., Reungsang, A., and Plangklang, P. (2013). Comparative bioremediation of carbofuran contaminated soil by natural attenuation, bioaugmentation and biostimulation. *International Biodeterioration and Biodegradation* 85: 196–204. https://doi.org/10.1016/j.ibiod.2013.07.009.

70 Hamdi, H., Benzarti, S., Manusadžianas, L. et al. (2007). Bioaugmentation and biostimulation effects on PAH dissipation and soil ecotoxicity under controlled conditions. *Soil Biology and Biochemistry* 39: 1926–1935.

71 Kauppi, S., Sinkkonen, A., and Romantschuk, M. (2011). Enhancing bioremediation of dieselfuel-contaminated soil in a boreal climate: comparison of biostimulation and bioaugmentation. *International Biodeterioration and Biodegredation* 65: 359–368. https://doi.org/10.1016/j.ibiod.2010.10.011.

72 Xu, Y. and Lu, M. (2010). Bioremediation of crude oil-contaminated soil: comparison of different biostimulation and bioaugmentation treatments. *Journal of Hazardous Materials* 183: 395–401.

73 Tomei, M.C. and Daugulis, A.J. (2013). Ex situ bioremediation of contaminated soils: an overview of conventional an innovative technologies. *Critical Reviews in Environmental Science and Technology* 43: 2107–2139.

74 Mohan, S.V., Reddy, B.P., and Sarma, P.N. (2009). Ex situ slurry phase bioremediation of chrysene contaminated soil with the function of metabolic function: process evaluation by data enveloping analysis (DEA) and Taguchi design of experimental methodology (DOE). *Bioresource Technology* 100: 164–172.

75 Paudyn, K., Rutter, A., Rowe, R.K. et al. (2008). Remediation of hydrocarbon contaminated soils in the Canadian Arctic by landfarming. *Cold Regions Science and Technology* 3: 102–114.

76 Maila, M.P. and Cloete, T.E. (2004). Bioremediation of petroleum hydrocarbons through landfarming: are simplicity and cost-effectiveness the only advantages? *Reviews in Environmental Science and Bio/Technology* 3: 349–360.

77 Rawe, J., EPA, Hodge, V. et al. (2006). In situ and ex situ biodegradation technologies for remediation of contaminated sites. *Engineering* 625/R-06/015.

78 Golodyaev, G.P., Kostenkov, N.M., and Oznobikhin, V.I. (2009). Bioremediation of oilcontaminated soils by composting. *Eurasian Soil Science* 42: 926–935.

79 von Fahnestock, F.M., Wickramanayake, G.B., Kratzke, K.J. et al. (1998). *Biopile Design, Operation, and Maintenance Handbook for Treating Hydrocarbon Contaminated Soil.* Columbus, OH: Battelle Press.

80 Srinath, T., Verma, T., Ramteke, P. et al. (2002). Chromium (VI) biosorption and bioaccumulation by chromate resistant bacteria. *Chemosphere* 48: 427–435.

81 Das, S.K. and Guha, A.K. (2007). Biosorption of chromium by Termitomyces clypeatus. *Colloids and Surfaces B* 60: 46–54.

82 Kiran, B., Kaushik, A., and Kaushik, C.P. (2007). Response surface methodological approach for optimizing removal of Cr (VI) from aqueous solution using immobilized cyanobacterium. *Chemical Engineering Journal* 126: 147–153.

83 Marandi, R. (2011). Biosorption of hexavalent chromium from aqueous solution by dead fungal biomass of Phanerochaete crysosporium: batch and fixed bed studies. *Canadian Journal of Chemical Engineering* 2: 8–22.

84 Savvaidis, I., Hughes, M.N., and Poole, R.K. (2003). Copper biosorption by *Pseudomonas cepacia* and other strains. *World Journal of Microbiology and Biotechnology* 19: 117–121.

85 Stokes, P. and Lindsay, J.E. (1979). Copper tolerance and accumulation in Penicillium ochrochloron isolated from copper plating solution. *Mycologia* 11: 796–806.

86 Antunes, W.M., Luna, A.S., Henriques, C.A. et al. (2003). An evaluation of copper biosorption by a brown seaweed under optimized conditions. *Electronic Journal of Biotechnology* 6: 174–184.

87 Junghans, K. and Straube, G. (1991). Biosorption of copper by yeasts. *Biology of Metals* 4: 233–237.

88 Uslu, G. and Tanyol, M. (2006). Equilibrium and thermodynamic parameters of single and binary mixture biosorption of lead (II) and copper (II) ions onto eudomonas putida: effect of temperature. *Journal of Hazardous Materials* 135: 87–93.

89 Akar, T., Tunali, S., and Cabuk, A. (2007). Study on the characterization of lead (II) biosorption by fungus Aspergillus parasiticus. *Applied Biochemistry and Biotechnology* 136: 389–405.

90 Al-Saraj, M., Abdel-Latif, M.S., El-Nahal, I. et al. (1999). Bioaccumulation of some hazardous metals by sol–gel entrapped microorganisms. *Journal of Non-Crystaline Solids* 248: 137–140.

91 Garcia, M., Alonso, J., and Melgar, M. (2005). Agaricus macrosporus as a potential bioremediation agent for substrates contaminated with heavy metals. *Journal of Chemical Technology and Biochemistry* 80: 325–330.

92 Yavuz, H., Denizli, A., Güngüneş, H. et al. (2006). Biosorption of mercury on magnetically modified yeast cells. *Separation and Purification technology* 52: 253–260.

93 Pardo, R., Herguedas, M., Barrado, E. et al. (2003). Biosorption of cadmium, copper, lead and zinc by inactive biomass of Pseudomonas putida. *Analytical and Bioanalytical Chemistry* 376: 26–32.

94 Nam, K., Rodriguez, W., and Kukor, J.J. (2001). Enhanced degradation of polycyclic aromatic hydrocarbons by biodegradation combined with a modified Fenton reaction. *Chemosphere* 45: 11–20.

95 Hesham, A.L., Khan, S., Tao, Y. et al. (2012). Biodegradation of high molecular weight PAHs using isolated yeast mixtures: application of metagenomic methods for community structure analyses. *Environmental Science and Polution Research (International)* 19 (8): 3568–3578.

96 Badri, D.V., Weir, T.L., Van-der-Lelie, D. et al. (2009). Rhizosphere chemical dialogues: plant-microbe interactions. *Current Opinion in Biotechnology* 20: 642–650.

97 Bais, H.P., Weir, T.L., Perry, L.G. et al. (2006). The role of root exudates in rhizosphere interactions with plants and other organisms. *Annual Review of Plant Biology* 57: 233–266.

98 Harrison, M.J. (2005). Signaling in arbuscular mycorrhizal symbiosis. *Annual Review of Microbiology* 59: 19–42.

99 Tyagi, M., da Fonseca, M.M.R., and de Carvalho, C.C.C.R. (2011). Bioaugmentation and biostimulation strategies to improve the effectiveness of bioremediation processes. *Biodegradation* 22: 231–241.

100 Vasudevan, P., Padmavathy, V., Tewari, N. et al. (2001). Biosorption of heavy metal ions. *Journal of Scientific and Industrial Research* 60: 112–120.

101 Gupta, R., Ahuja, P., Khan, S. et al. (2000). Microbial biosorbents: meeting challenges of heavy metal pollution in aqueous solution. *Current Science – India* 78: 967–973.

102 Singh, R., Paul, D., and Jain, R.K. (2006). Biofilms: implications in bioremediation. *Trends in Microbiology* 14 (9): 389–397.

103 Gavrilescu, M. (2004). Removal of heavy metals from the environment by biosorption. *Engineering Life Sciences* 4: 219–232.

104 Danis, U., Nuhoglu, A., and Demirbas, A. (2008). Ferrous ion-oxidizing in Thiobacillus ferrooxidans batch cultures: Influence of pH, temperature and initial concentration of Fe2+. *Fresenius Environmental Bulletin* 17 (3): 371–377.

105 Goyal, N., Jain, S.C., and Banerjee, U.C. (2003). Comparative studies on the microbial adsorption of heavy metals. *Advances in Environmental Research* 7: 311–319.

106 Abou-Shanab, R.A., Angle, J.S., Delorme, T.A. et al. (2003). Rhizobacterial effects on nickel extraction from soil and uptake by Alyssum murale. *New Phytologist* 158: 219–224.

107 Abou-Shanab, R.A.I., Angle, J.S., and Chaney, R.L. (2006). Bacterial inoculants affecting nickel uptake by Alyssum murale from low, moderate and high Ni soils. *Soil Biology and Biochemistry* 38: 2882–2889.

108 Wang, F. (2006). Hotspots in arbuscular mycorrhiza-assisted phytoremediation of heavy metal-contaminated soils. *Ecological Environment* 15: 1086–1090.

109 Han, X. and Gu, J. (2010). Sorption and transformation of toxic metals by microorganisms. In: *Environmental Microbiology* (eds. R. Mitchell and J. Gu), 153–176. Hoboken, NJ: Wiley Blackwell.

110 Abd-Alla, M.H., Morsy, F.M., El-Enany, A.W.E. et al. (2012). Isolation and characterization of a heavy-metal-resistant isolate of *Rhizobium leguminosarum* bv. *viciae* potentially applicable for biosorption of Cd^{2+} and Co^{2+}. *International Biodeterioration and Biodegredation* 67: 48–55.

111 Leusch, A., Holan, Z.R., and Volesky, B. (1995). Biosorption of heavy metals (Cd, Cu, Ni, Pb, Zn) by chemically-reinforced biomass of marine algae. *Journal of Chemical Technology and Biotechnology* 62: 279–288.

112 Naraian, R., Ram, S., Kaistha, S.D. et al. (2012). Occurrence of plasmid linked multiple drug resistance in bacterial isolates of tannery effluent. *Cellular and Molecular Biology* 58 (1): 134–141.

113 Fennell, D.E., Du, S., Liu, H. et al. (2011). Dehalogenation of polychlorinated dibenzo-p-dioxins and dibenzofurans, polychlorinated biphenyls and brominated flame retardants and potential as a bioremediation strategy. In: *Comprehensive Biotechnology*, 2e (eds. M. Moo-Young and M. Butler), 136–149. Oxford: Pergamon Press.

114 Hiraishi, A. (2008). Biodiversity of dehalorespiring bacteria with special emphasis on polychlorinated biphenyl/dioxin dechlorinators. *Envirnomental Microbiology* 23: 1–12.

115 Margesin, R., Zimmerbauer, A., and Schinner, F. (2000). Monitoring of bioremediation by soil biological activities. *Chemosphere* 40: 339–346.

116 Pieper, D.H. and Reineke, W. (2000). Engineering bacteria for bioremediation. *Current Opinion in Biotechnology* 11: 262–270.

117 Watanabe, K. (2001). Microorganisms relevant to bioremediation. *Current Opinion in Biotechnology* 12: 237–241.

118 Pandey, G. and Jain, R.K. (2002). Bacterial chemotaxis towards environmental pollutants: role in bioremediation. *Applied and Environmental Microbiology* 68 (12): 5789–5795.

119 Grimm, A.C. and Harwood, C.S. (1997). Chemotaxis of Pseudomonas sp. to the polycyclic aromatic hydrocarbon, naphthalene. *Applied and Environmental Microbiology* 63: 4111–4115.

120 Fragoeiro, S. (2005). *Use of Fungi in Bioremediation of Pesticides*. Bedford, UK.: Cranfield University.

121 Schmidt, K.R., Chand, S., Gostomski, P.A. et al. (2005). Fungal inoculum properties and its effect on growth and enzyme activity of Trametes versicolor in soil. *Biotechnology Progress* 21: 377–385.

122 Gadd, G.M. (2001). *Fungi in Bioremediation*. Cambridge: Cambridge University Press.

123 Ang, E.L., Zhao, H., and Obbard, J.P. (2005). Recent advances in the bioremediation of persistent organic pollutants via biomolecular engineering. *Enzyme and Microbial Technology* 37: 487–496.

124 Chaloupkova, R., Sykorova, J., Prokop, Z. et al. (2003). Modification of activity and specificity of haloalkane dehalogenase from Sphingomonas paucimobilis UT26 by engineering of its entrance tunnel. *Journal of Biological Chemistry* 278: 52622–52628.

125 Li, W., Shi, J., Wang, X. et al. (2004). Complete nucleotide sequence and organization of the naphthalene catabolic plasmid pND6-I from Pseudomonas sp. strain ND6. *Gene* 336: 231–240.

126 Wood, T.K. (2008). fmediation. *Current Opinion in Biotechnology* 19: 572–578.

127 Srivastava, J., Naraian, R., Kalra, S.J.S. et al. (2014). Advances in microbial bioremediation and the factors influencing the process. *International Journal of Environmental Science and Technology* 11: 1787–1800.

128 Pazirandeh, M., Chrisey, L.A., Mauro, J.M. et al. (1995). Expression of the Neurospora crassa metallothionein gene in Escherichia coli and its effects on heavy-metal uptake. *Applied Microbiology and Biotechnology* 43: 1112–1117.

129 Samuelson, P., Wernerus, H., Svedberg, M. et al. (2000). Staphylococcal surface display of metal binding polyhistidyl peptides. *Applied and Environmental Microbiology* 66: 1243–1248.

130 Moris, J.M. and Jin, S. (2012). Enhanced biodegradation of hydrocarbon contaminated sediments using microbial fuel cells. *Journal of Hazardous Materials* 30 (213–214): 474–477.

131 Robles-Gonzalez, I.V., Fava, F., and Poggi-Varaldo, H.M. (2008). A review on slurry bioreactors for bioremediation of soils and sediments. *Microbial Cell Factories* 7 (5): 1–16.

132 Laghlimi, M., Baghdad, B., Hadi, H. et al. (2015). Phytoremediation mechanisms of heavy metal contaminated soils: a review. *Open Journal of Ecology* 5 (8): 375–388.

133 Chaney, R.L. (1983). Plant uptake of inorganic waste constituents. In: *Land Treatment of Hazardous Wastes* (eds. J.F. Parr, P.B. Marsh and J.M. Kla), 50–76. Park Ridge, NJ: Noyes Data Corp.

134 Ghosh, M. and Singh, S.P. (2005). A review on phytoremediation of heavy metals and utilization of its byproducts. *Applied Ecology and Environmental Research* 3 (1): 1–18.

135 Nouri, J., Khorasani, N., Loretami, B. et al. (2009). Accumulation of heavy metals in soil and uptake by plant species with phytoremediation potentials. *Environmental Earth Sciences* 59: 315–323.

136 Hamon, R., Wundke, J., McLaughlin, M. et al. (1997). Availability of zinc and cadmium to different plant species. *Australian Journal of Soil Research* 35: 1267–1277.

137 Pivetz, B.E. (2001). *Phytoremediation of Contaminated Soil and Ground Water at Hazardous Sites. Ground Water Issue*, EPA/540/S-01/500.

138 Karimi, N. (2013). Comparative phytoremediation of chromium-contaminated soils by alfalfa (Medicago sativa) and Sorghum bicolor (L) Moench. *International Journal of Scientific Research in Environmental Sciences* 1 (3): 44–49.

139 Castro, H.F., Classen, A.T., Austin, E.E. et al. (2010). Soil microbial community responses to multiple experimental climate change drivers. *Applied and Environmental Microbiology* 76 (4): 999–1007.

140 Nie, M., Pendall, E., Bell, C. et al. (2013). Positive climate feedbacks of soil microbial communities in a semi-arid grassland. *Ecology Letters* 16 (2): 234–241.

141 Sowerby, A., Emmett, B., Beier, C. et al. (2005). Microbial community changes in heath land soil communities along a geographical gradient: interaction with climate change manipulations. *Soil Biology and Biochemistry* 37 (10): 1805–1813.

142 Baldrian, P., Snajdr, J., Merhautova, V. et al. (2013). Responses of the extracellular enzyme activities in hardwood forest to soil temperature and seasonality and the potential effects of climate change. *Soil Biology and Biochemistry* 56: 60–68.

143 Frey, S.D., Lee, J., Melillo, J.M. et al. (2013). The temperature response of soil microbial efficiency and its feedback to climate. *Nature Climate Change* 3: 395–398.

144 Bardgett, R.D., Freeman, C., and Ostle, N.J. (2008). Microbial contributions to climate change through carbon cycle feed backs. *ISME Journal* 2: 805–814.

145 Genther, B.R.S., Townsend, G.T., Lantz, S. et al. (1997). Persistence of polycyclic aromatic hydrocarbon components of creosote under anaerobic enrichment conditions. *Archives of Environmental Contamination and Toxicology* 32 (1): 99–105.

146 Zekri, A.Y. and Chaalal, O. (2005). Effect of Temperature on biodegradation of crude oil. *Energy Sources* 27 (1–2): 233–244.

147 Zaidi, T.S., Lyczak, J., Preston, M. et al. (1999). Cystic fibrosis transmembrane conductance regulator-mediated corneal epithelial cell ingestion of Pseudomonas aeruginosa is a key component in the pathogenesis of experimental murine keratitis. *Infection and Immunity* 67 (3): 1481–1492.

148 Zavala, M.A.L., Funamizu, N., and Takakuwa, T. (2004). Temperature effect on aerobic biodegradation of feces using sawdust as a matrix. *Water Research* 38: 2406–2416.

149 Frey, S.D., Drijber, R., Smith, H. et al. (2008). Microbial biomass, functional capacity, and community structure after 12 years of soil warming. *Soil Biology and Biochemistry* 40: 2904–2907.

150 Sag, Y., Ozer, D., and Kustal, T. (1995). A comparative study of the biosorption of lead (II) ions to Zoogloea ramigera and Rhizopus arrhizus. *Process Biochemistry* 30: 169–174.

151 Rajendran, P., Muthukrishnan, J., and Gunasekaran, P. (2002). Current perspective in nickel bioremediation strategies. *Indian Journal of Microbiology* 42: 1–9.

152 Blazquez, G., Hernainz, F., Calero, M. et al. (2008). The effect of pH on the biosorption of Cr(III) and Cr(VI) with olive stone. *Chemical Engineering Journal* 148: 473–479.

153 Puzon, G.J., Roberts, A.G., Kramer, D.M. et al. (2005). Formation of soluble organo-chromium (III) complexes after chromate reduction in the presence of cellular organics. *Environmental Science and Technology* 39: 2811–2817.

154 Beard, S.J., Hughes, M.N., and Poole, R.K. (1995). Inhibition of the cytochrome bd-terminated NADH oxidase system in Escherichia coli K-12 by divalent metal cations. *FEMS Microbiology Letters* 131: 205–210.

155 Lovely, D.R. (2003). Cleaning up with genomics applying molecular biology to bioremediation. *Nature Reviews Microbiology* 1: 35–44.

156 Leahy, J.G., Tracy, K.D., and Eley, M.H. (2003). Degradation of mixtures of aromatic and aliphatic hydrocarbons by aromatic hydrocarbondegrading bacteria. *FEMS Microbiology and Ecology* 43 (2): 271–276.

157 Gibson, D., Koch, J., and Kallio, R. (1968). Oxidative degradation of aromatic hydrocarbons by microorganisms. I. Enzymatic formation of catechol from benzene. *Biochemistry* 7: 2653–2662.

158 Bewley, R.J.F. and Webb, G. (2001). In situ bioremediation of groundwater contaminated with phenols, BTEX and PAHs using nitrate as electron acceptor. *Land Contamination and Reclamation* 9 (4): 335–348.

159 Coates, J.D., Michaelidou, U., Bruce, R.A. et al. (1999). Ubiquity and diversity of dissimilatory (per)chlorate-reducing bacteria. *Applied and Environmental Microbiology* 65 (12): 5234–5241.

160 Rockne, K.J. and Strand, S.E. (1998). Biodegradation of bicyclic and polycyclic aromatic hydrocarbons in anaerobic enrichments. *Environmental Science and Technology* 32 (24): 3962–3967.

161 Riser-Roberts, E. (1992). *Bioremediation of Petroleum Contaminated Sites*. Boca Raton, FL: CRC Press Inc.

162 Atagana, H.I., Haynes, R.J., and Wallis, F.M. (2003). Optimization of soil physical and chemical conditions for the bioremediation of creosote-contaminated soil. *Biodegradation* 14: 297–307.

163 Wilson, S.C. and Jones, K.C. (1993). Bioremediation of soil contaminated with polynuclear aromatic hydrocarbons (PAHs) a review. *Environmental Pollution* 81: 229–249.

164 Bogan, B.W., Schoenike, B., Lamar, R.T. et al. (1996). Expression of lip genes during growth in soil and oxidation of anthracene by Phanerochaete chrysosporium. *Applied and Environmental Microbiology* 62: 3697–3703.

165 Semple, K.T., Morriss, W.J., and Paton, G.I. (2003). Bioavailability of hydrophobic organic contaminants in soils: fundamental concepts and techniques for analysis. *European Journal of Soil Science* 54: 809–818.

166 Reid, B.J., Jones, K.C., and Semple, K.T. (2000). Bioavailability of persistant organic pollutants in soils and sediments—a perspective on mechanisms, consequences and assessment. *Environmental Pollution* 108: 103–112.

167 Zhu, L., Chen, Y., and Zhou, R. (2008). Distribution of poly aromatic hydrocarbons(PAHs) in water, sediments and soil in drinking water resources of Zhejiang province, China. *Journal of Hazardous Materials* 150 (2): 308–316.

168 Hatzinger, P.B. and Alexander, M. (1995). Effect of aging on chemicals in soil on their biodegradability and extractability. *Environmental Science and Technology* 29: 537–545.

169 Alexander, M. (2000). Aging bioavailability, and overestimation of risk from environmental pollutants. *Environmental Science and Technology* 34 (20): 4259–4265.

170 Balba, M.T., Al-Awadhi, N., and Al-Daher, R. (1998). Bioremediation of oil-contaminated soil: microbiological methods for feasibility assessment and field evaluation. *Journal of Microbiological Methods* 32 (2): 155–164.

171 Loeher, R.C., Mc Millen, S.J., and Webster, M.T. (2001). Predictions of biotreatability and actual results: soils with petroleum hydrocarbons. *Practice Periodical of Hazardous, Toxic and Radioactive Waste Management* 5 (2): 78–87.

172 Delarco, J.P. and De Francea, F.P. (2001). Influence of oil contamination levels on hydrocarbon biodegradation in sandy sediment. *Environmental Pollution* 110: 515–519.

173 El-Tarabily, K.A. (2002). Total microbial activity and microbial composition of a mangrove sediment are reduced by oil pollution at a site in the arabian gulf. *Canadian Journal of Microbiology* 48 (2): 176–182.

174 Ingham, C.J., Sprenkels, A., Bomer, J. et al. (2007). The micro Petri dish, a million-well growth chip for the culture and high- throughput screening of microorganisms. *Proceedings of the National Academy of Science, USA* 104: 18217–18222.

175 Potin, O., Veignie, E., and Rafin, C. (2004). Biodegradation of polycyclic aromatic hydrocarbon (PAHs) contaminated soil. *FEMS Microbiology and Ecology* 51 (1): 71–78.

176 Maphosa, F., Smidt, H., De Vos, W.M. et al. (2010). Microbial community and metabolite dynamics of an anoxic dechlorinating bioreactor. *Environmental Science and Technology* 44: 4884–4890.

177 Toyama, T., Furukawa, T., Maeda, N. et al. (2011). Accelerated biodegradation of pyrene and benzo(a)pyrene in the Phragmites australis rhizosphere by bacteria-root exudates interactions. *Water Research* 45 (4): 1629–1638.

178 Mars, A.E., Kasberg, T., Kaschabek, S.R. et al. (1997). Microbial degradation of chloroaromatics: use of the meta-cleavage pathway for mineralization of chlorobenzene. *Journal of Bacteriology* 179: 4530–4537.

179 Nweke, C.O., Alisi, C.S., Okolo, J.C. et al. (2007). Toxicity of zinc heterotrophic bacteria from a tropical river sediment. *Applied Ecology and Environmental Research* 5 (1): 123–132.

180 Chakraborty, S., Pangga, I.B., and Roper, A.M. (2012). Climate change and multitrophic interactions in soil: the primacy of plants and functional domains. *Global Change Biology* 18: 2111–2125.

181 Vidali, M. (2001). Bioremediation an overview. *Pure and Applied Chemistry* 73 (7): 1163–1172.

182 Zeyaullah, M., Atif, M., Islam, B. et al. (2009). Bioremediation: A tool for environmental cleaning. *African Journal of Microbiology Research* 3 (6): 310–314.

13

Trends in Heavy Metal Remediation

An Environmental Perspective

Baba Uqab[1], Gousia Jeelani[2], Sabeehah Rehman[4], B.A. Ganai[2], Ruqeya Nazir[2], and Javid A. Parray[3]

[1] *Department of Environmental Science, University of Kashmir, Srinagar, Jammu and Kashmir, India*
[2] *Centre of Research for Development, University of Kashmir, Srinagar, Jammu and Kashmir, India*
[3] *Department of Environmental Science, Govt Degree College Eidgah, Srinagar, Jammu and Kashmir, India*
[4] *Centre of Reserach for Development, University of Kashmir, Srinagar, Jammu and Kashmir, India*

13.1 Introduction

Environmental pollution is currently a cause for concern throughout the world. Pollution of every type has serious impacts on all spheres of life [1]. Unwanted pollutants are causing serious impacts on humans and the environment and the result is loss, either in the form of diversity, or the degradation of biosphere. Word pollution became much more prevalent with industrial revolution [2]. Various pollutants have direct impacts on human life, which is a serious concern and needs to be addressed immediately [3]. Development of scientific and industrial technology has benefited the society in numerous ways but has also produced many undesirable and toxic pollutants [4]. A number of pollutants like heavy metals, pesticides, aerosols, sulfur oxides, and nitrogen oxides (SO_X and NO_X) now present in the environment in abundance have direct correlation with industries. Pollution by heavy metals is one of the worldwide problems that is causing serious impacts on the environment [5]. Heavy metals have been defined as metals or metalloids with density greater than 5 g/cm^3. Occurrence of heavy metals in an ecosystem can create modifications in the microbial community and also affect their metabolic activity [6]. Among the 90 known naturally occurring heavy metals only 17 are biologically significant. A precondition for the biological significance of these heavy metals is their

Soil Bioremediation: An Approach Towards Sustainable Technology, First Edition.
Edited by Javid A. Parray, Abeer Hashem Abd Elkhalek Mahmoud, and Riyaz Sayyed.

solubility and physiological condition [7]. The high concentration of heavy metals in an ecosystem can be currently attributed to the excessive use of pesticides [8]. Soil containing heavy metal pollution is one of a several concerns owing to its potential undesirable effects. Even being present in soil in very low concentration, heavy metals are regarded as one of the chief soil contaminants. Heavy metals act as major soil contaminant due to their wide spread occurrence, their toxicity (acute and chronic), and their nature (persistence, nonbiodegradability, and nonthermo-degradability). The reason for accumulation of heavy metals to toxic levels in soil is mostly due to their nonbiodegradable and persistent nature [4, 9, 10].

13.2 Sources of Heavy Metals

With social and economic development, heavy metal pollution has become progressively more widespread in the world. Thus, heavy metal contamination is a major threat to the entire biosphere. Among the world's most important top 10 events related to environment, two events, weathering of rocks and atmospheric inputs of metals, are correlated to heavy metal contamination [11]. Heavy metal contamination unlike other contaminations/pollutions cannot be noticed by change in soil physical properties, hence it needs to be analyzed chemically. Moreover, the damage caused by a heavy metal in the environment is not immediate and can only be noticed once it exceeds or reaches the environmental tolerance level. So heavy metals result in serious ecological damage once the environmental tolerance is reached and this is the reason they are called as chemical time bombs (CTBS) [12].

In terrestrial ecosystems there are two main sources of heavy metals

I) Natural
II) Anthropogenic

Natural sources are further categorized into two main types

1) The parent material
2) Atmosphere

Among various other contributors weathering of rocks and inputs from atmospheric metals are main sources of accumulation of heavy metals in soil [13]. Apart from these natural sources anthropogenic activities like manufacturing of batteries, mining, paints, combustion of fossil fuel, ore refining, metal working industries, cement factories, gas and dust produced by energy, transport, and insecticides and fertilizers has led to the release of heavy metals and accumulation of heavy metals in different ecosystems [7, 14–16]. Besides these there are certain important pathways that load heavy metals in high concentration to terrestrial ecosystems, hence degrading the soil quality [17]. They are as follows.

13.2.1 Atmosphere to Soil

Atmosphere and soil are in direct contact with each other; hence the atmosphere acts as one of the main sources of the heavy metal pollution in the soil. The major sources of heavy metals in the atmosphere are gas and dust, produced from anthropogenic and natural activities like transport, metallurgy, and the production of construction materials like cement and bricks. Metals from industries are released into the atmosphere in the form of aerosols. These aerosols are added to the soil or terrestrial ecosystem by the natural property of sedimentation and precipitation [3].

13.2.2 Sewage to Soil

Wastewater has different categories like sanitary sewage and chemical waste sewage that contain heavy metals, which are further fixed in the soil in different ways [9]. Moreover, a shortage of water in dry regions forces people there to use sewage irrigation, which ultimately results in addition of heavy metals to soils.

13.2.3 Solid Wastes to Soils

Solid wastes contain certain materials that, on processing, produce heavy metals like Pb and Hg. When these wastes pile up, several materials including heavy metals precipitate more easily into surrounding water and soil [9].

13.2.4 Agriculture to Soils

Agricultural practices in the world are considered incomplete without the use of fertilizers, pesticides, and insecticides, etc., because they are directly related to improved production [16]. These chemicals are known heavy metal sources and their excessive use has resulted in heavy metal contaminated soils. The majority of the pesticides are mostly organic compounds and a few are inorganic compounds that contain metals like As, Hg, Cu, Zn [18], therefore they have contributed greatly to heavy metal contamination of the soil.

13.3 Heavy Metal Impacts

Heavy metals are not only hazardous to the environment but also to living organisms. Heavy metals in soil for the most part are found in the following forms: dissolved, exchangeable (inorganic and organic), as structural components of lattices, as insoluble precipitate. The heavy metal toxicity has serious impacts on all biological systems from the microbes to plants and animals, with microorganisms

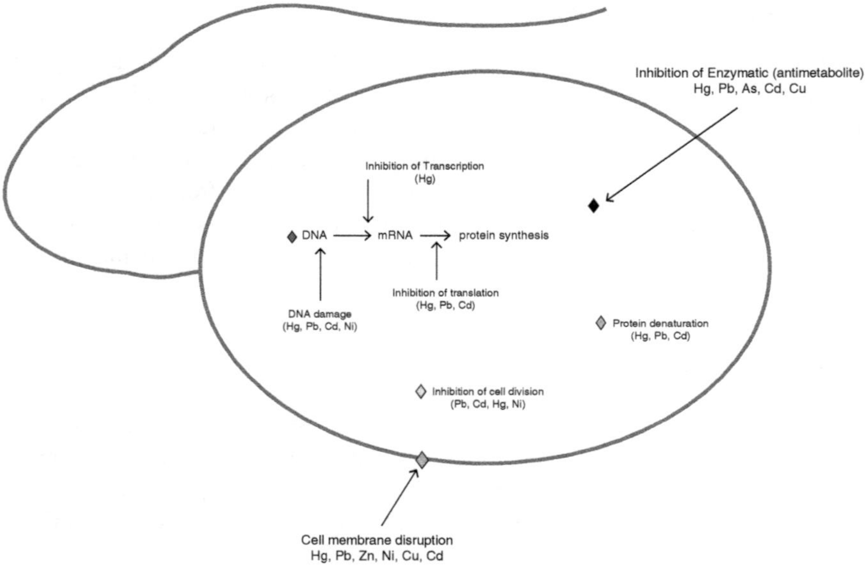

Figure 13.1 Various metabolic processes that are inhibited due to metal toxicity (*Source:* [22] with some modifications).

affected the most due to their minute size and direct interaction with the environment [17, 19–21]. In multicellular organisms, intracellular concentration of heavy metal needs to be highly controlled as metals like Zn^{2+} or Ni^{2+} are highly active in forming toxic complexes. Metals like Hg^{2+}, Cd^{2+}, and Ag^{2+} follow a similar trend. Based on the physical properties, chemical properties, and the molecular mechanism behind heavy metal induced toxicity, there are three eminent molecular mechanisms of heavy metal induced toxicity:

I) Production of reactive oxygen species (ROS) by auto-oxidation and Fenton reaction (for transition metals).
II) Blocking of essential functional groups in biomolecules.
III) Displacement of essential metal ions from biomolecules.

The three mentioned mechanisms directly inhibit various cellular metabolic processes. Figure 13.1 depicts the various processes, inhibited due to metal toxicity.

13.3.1 Impact of Heavy Metals on Humans

Although certain physiological life processes require some heavy metals (e.g., compounds of metallo-enzymes), their accumulation beyond the levels of tolerance is always detrimental in living organisms. Metal toxicity leads to enzyme inactivation, damaged cells as they act as anti-metabolites or form chelates or precipitates with essential metabolites [23]. Many diseases are caused by consumption of few heavy metals such as, Cd, Pb, present in food [24]. Lead toxicity can induce seizures, mental retardation, and behavioral disorders [3].

13.3.2 Risk of Emergence of Developing Antibiotic Resistance Strains Due to Heavy Metals

Contamination by heavy metal and antibiotic pollution serve as the main challenges for microbiologists as these two processes enable microorganisms to develop and maintain resistance in inclusive ranges of emerging bacterial strains, either through cross selection or co-selection. Bacteria rapidly acclimatize to new environmental conditions, enabling them to resist antimicrobials and contemporary exposure to metal pollutants. Heavy metal exposure enhances microbial adaptability by using an array of resistance mechanisms against these antibiotics. The genetic linkage development among constitutive resistance genes in bacteria particularly genetic linkage in metal resistance genes and antibiotic resistance genes is an evolving paradigm in bacteria against myriad pollutants. These linked genes in bacteria may be either present on the plasmid or on bacterial chromosomes. The Scientific Committee on Emerging and Newly Identified Health Risks (SCENIHR) in two successive reports of 2009 and 2010 have prophesized that intermittent exposure to metals and biocides could augment the development of highly antibiotic resistant strains. It may enhance the resistance of hitherto known pathogenic and multiple drug resistant pathogenic strains of bacteria. Metal resistance genes as well as genes for antibiotic resistance are recurrently found on plasmids and metals may therefore be a consistent factor that indirectly selects for antibiotic resistance. Multidrug resistant bacteria *Klebsiella pneumoniae* and *Escherichia coli* have multidrug resistance plasmids that confer heavy metal resistance to Cu, Ag, and Ar besides β-lactams, aminoglycosides, tetracycline, trimethoprim, sulfonamides, and erythromycin. The combined effect of antibiotics and metals present in large quantities in the environment is a further complicating factor with regards to selection thus, it is of great concern. Therefore, several compounds even in a small amount could confer selection when present in combination thereby resulting in selection of resistance. The combined effects of the compounds tested vary from synergistic to less than additive [25].

13.3.3 Impact of Heavy Metals on Soil and Soil Microbes

Contaminations of heavy metals in the soil is primarily accredited to the industrial revolution. Due to change in various soil parameters like pH, clay content, and organic matter, by heavy metal contamination biological as well as biochemical properties of soil are drastically affected [26]. Soil enzymatic activity (dehydrogenase, urease, and phosphatase) is affected indirectly, as heavy metal presence shifts the microbial community, which synthesizes enzymes [27]. Toxic effects are induced by heavy metals on soil biota including inhibition of major microbial processes thereby decreasing the number as well as the activities of microbes in the soil [28]. Hence, heavy metals decrease both diversity and biomass of bacterial communities thereby decreasing bacterial species richness, hence affecting the soil environment [29].

13.3.4 Impacts of Heavy Metals on Plants

Heavy metals such as Cu, Co, and Zn are important for plant growth and their metabolism, but once the heavy metal concentration increases beyond an optimal value they induce toxicity to plants [30–32].

Both macro as well as micronutrients plays a significant role in all physiological processes: synthesis of DNA, biosynthesis of chlorophyll, protein modifications, photosynthesis, redox reactions, nitrogen fixation and sugar metabolism, etc. [33]. However, sometimes plants function as carriers of heavy metals to human, which is a serious cause for concern [34]. The transfer of heavy metals primarily from soils to plants and then from plants to humans provides a path for the entry of these heavy metals into the entire food chain posing a threat to all living organisms including higher plants plus animals [35]. Studies have already shown that certain phytotoxicities are produced by heavy metals, which lead to weak growth in plants, chlorosis, and drastic depression in yield [36]. Presence of lead even in slightest concentrations in the plant can inhibit photosynthesis, mitosis in cells, and absorption of water [33] and can also delay seed germination in plants [37].

13.4 Current Scenario of Heavy Metals

Pollution by heavy metal is a worldwide problem with serious impacts on life [38]. Crop systems contaminated with heavy metals are one of the main human health hazards [39]. So far, many studies have revealed that urban soils are mostly contaminated by sources like transport, industrial waste, and household garbage. However, fertilizers and pesticides are the key sources of heavy metal pollution in most agricultural soils [40, 41]. Environment has a tolerance range for the presence of heavy metal contamination; therefore, there are certain permissible limits for heavy metals.

13.5 Microorganisms and Remediation of Heavy Metals

Heavy metals are dangerous to the environment and to living organisms [42]. Toxicity of heavy metal leads to serious impacts on soil, plants, and humans. A possible way out to this problem lies in microbial remediation. Microbial remediation refers to the utilization of microorganisms for the remediation of heavy metal pollution. Microbes are mainly utilized to perform functions like precipitation, absorption, and oxidation–reduction of these metals in the soils [43]. Currently the attention of researchers is focused on development of alternative but innovative techniques for the treatment of heavy metal pollution in the environment via the use of biological materials mostly bacteria, algae, fungi, and yeast. For this reason, the removal of heavy metals and the recovery of the environment have thus gained importance in the past few decades due to their better performance with lower cost [44–47]. Some of the added advantages of biological clean-up techniques are: the ability to handle the range of physico-chemical parameters in effluents, selectivity to remove only the desired metals, and the cost effectiveness [48]. Many microorganisms have been intensively examined (Tables 13.1 and 13.2) for their abilities to be applied in the bioremediation of heavy metals.

13.5.1 Microbial Processes Concerned with Bioremediation

Biological or microbial processes for elimination and remediation of toxic metals is considered to be economical due to certain features like selective metal binding, high capacity to bind metal, and effective absorption methods. Microbial transformation of heavy metals mostly affects solubility, bioavailability, and mobility [57]. The mechanisms associated with these processes include: chelation of these elements by metabolites; oxidation-reduction in metals, which affects the solubility; change in pH; biosorption; biotransformation; immobilization; biomethylation; and biodegradation. A variety of microorganisms have shown promising results that interact with metals and transform them [58]. The inherent property of metal resistance is present in most microorganisms but many have evolved this property due to exposure to heavy metal polluted environments [59]. Various microbial mediated transformations of metals have been identified [60] many of which may immobilize metals in the environment. The metal resistance mechanisms include complexation, precipitation, and solubilization, which result in microbial transformation of metals.

Table 13.1 Heavy metal accumulating microorganisms and mechanisms of resistance.

Heavy metal	Mechanism of resistance	Microorganism	Reference
Cadmium	Intracellular and extracellular sequestration, efflux, reduction	*Gloeothece magna, Staphylococcus aureus, Fusarium oxysporum, Pseudomonas Azotoformans*	Ahmad et al. [49], Nair et al. [4, 50]
Copper	Intracellular and extracellular sequestration	*Saccharomyces cerevisiae*	Culotta et al. [51]
Chromium	Extracellular sequestration and reduction	*Streptomyces* spp.	Amoroso et al. [52]; Lloyd [53]
Zinc	Efflux	*Ralstonia metallidurans*	Nies [54]
Mercury	Volatilization	*Clostridium glycolicum*	Meyer et al. [55]
Lead	Biosurfactant, extracellular sequestration	*Pseudomonas aeruginosa*	Nair et al. [50]
Arsenic	Reduction, efflux, intracellular sequestration	*Lactobacillus* sp.	Nair and Pradeep [56]

13.5.2 Metal Microbe Interaction

The methodology adopted for bioremediation of contaminated soils entirely depends on the interaction between metals and microbes [61]. The metals in ionic form bind to the cell surface through various interactions for example, electrostatic, Van Dar Waals force, redox reactions, covalent bonding, and extracellular precipitation. Moreover, the interaction can also be a combination of all these processes. Bacterial cell walls help in the retention of the metal cations by mineral nucleation after adsorption by groups that are negatively charged like carboxyl, hydroxyl, and phosphoryl [62]. Biosorption of metals like U, Zn, Pb, Cd, Ni, Cu, Hg, Cs, Th, Au, Ag, Sn, and Mn varies markedly with both the metal and the microorganisms. In nature, microbes immobilize heavy metals via the process of cellular accumulation or via extra-cellular precipitation [22]. *Pseudomonas aerugenosa* shows high specificity for some metals like Pb and Cd by producing

Table 13.2 List of genetically engineered bacteria that have been used for remediation.

Heavy metal	Genetically Engineered Bacteria	Expressed Gene	Reference
Arsenic	*E. coli* strain	Metalloregulatory protein ArsR	Nair and Pradeep [56]
Cadmium	*E.coli* strain, *Methylococcus capsulatus*	SpPCS, CrR	Ahmad et al. [49], Nair et al. [50],
Chromium	*P. putida* strain	Chromate reductase (ChrR)	Amoroso et al. [52]; Lloyd [53]
Mercury	*Deinococcus radiodurans*	merA	Meyer et al. [55]
Nickel	*P. fluorescens*	Phytochelatin synthase (PCS)	Nair et al. [50]
Mercury	*E. coli* strain	Organomurcurial lyase	Meyer et al. [55]
Mercury	*Achromobacter* sp	Mer	Meyer et al. [55]

surfactants such as rhamnolipids. Similarly, certain bacteria such as *Thiobacillus ferroxidans* and *Leptospirillum ferroxidans* are capable of oxidizing iron and sulfur [63]. Reduction of metals like Mn, Fe, Co, AsO_2, and Se by eubacteria and archaea also conserve energy. A strain of *Alcaligenes faecalis* oxidizes AsO_2^- to AsO_4^{3-} and *Pseudomonas fluorescens* LB 300 and *Enterobacter cloacae* reduce CrO_4^{2-} to $Cr(OH)_3$ enzymatically. The nonenzymatic detoxification involves secretion of inorganic metabolic products like carbonate, sulfide, and phosphate ions that precipitate toxic metal.

13.6 Mechanism of Metal Tolerance by Resistant Species

The anthropogenic release of metals into the environment has increased the need for research relating to microbial metal resistance and remediation. Microorganisms provide a key to decreasing the existing contamination by directly influencing the fate of these metals in the environment. In some microorganisms' metal tolerance may have evolved due to exposure to toxic metals shortly after life started [64]. Plasmid-coded mechanisms tend to be resistant to specific metal while other mechanisms confer resistance to variety of metals. In this scenario there are two mechanisms of metal resistance:

- general mechanism, and
- metal dependent mechanism.

13.6.1 General Mechanism of Metal Resistance

The metals are immobilized by binding to the extra cellular molecules and prevent the entry of metals into the cell. Cationic metals like Cd, Zn, Pb, and Fe bind to the anionic cell surface containing carboxylic, amino, thio, hydroxo, hydroxylcarboxylic, and phosphoryl groups and phospholipids in algal and bacterial membrane respectively [22]. Binding of the metals with microbial cells is very important both practically and ecologically. Ecologically, binding of metals to cell surface plays an important role with regards to the distribution of heavy metals, particularly in the aquatic environment. Bioremediation offers practical importance for the removal of metal contaminants from nature. There are five natural, general, mechanism methods of metal resistance in microorganisms.

13.6.2 Exopolymer Binding

Exopolymer binding is one of the common ways of reducing metal bioavailability. The exopolymeric substances or extracellular polymeric substances (EPSs) like polysaccharides, carbohydrates, nucleic acids, and fatty acids play a very important function in binding heavy metals such as Cd, Pb, and U. Metal binding is greatly influenced by pH of the environment because the metal binding functional groups possess a negative charge. Species such as *Staphylococcus aureus*, *Micrococcus luteus*, and *Azotobacter spp.* immobilize Pb through exopolymer binding [22, 65].

13.6.3 Siderophore Complexation

Siderophores are iron chelating organic molecules with low molecular weight. Their biological function is concentration of iron from the environment where their concentration is very low and to transport it from the environment into the cell. Siderophores may also interact with other metals that are chemically analogous to iron, viz. aluminum, gallium, and chromium thus reducing their bioavailability and their toxicity. For instance, siderophores reduce the toxicity of copper in cyanobacteria [60].

13.6.4 Biosurfactants Complexation

Biosurfactants are compounds produce certain bacteria that in some are excreted outside the cell. Biosurfactants have recently been investigated for their capability to bind metals like lead, cadmium, and zinc by making complexes with the metals that then become nontoxic to the cell [66]. Sorption sites for the binding of metals and immobilization are created on the cell wall, these sites are made up of proteins, lipo polysaccharides, and carbohydrates as their functional groups and also exopolysaccharides (EPS). A study by Morillo Pâere et al. [67] showed that EPS produced by *Paenibacillus jamilae*, a gram-negative bacterium, was able to complexate up to 230 mg of Pb (Lead), per gram of EPS produced. Similarly, *Nostoc spongiaeforme* (cyanobacterium) produced EPS that proved to be an effective sorbent of Zn [68]. The main mechanisms via which metal binds to cell surfaces include, Van der Waals forces, electrostatic interactions, redox interactions, or combinations of all these [69].

13.6.5 Precipitation by Metal Reduction

In precipitation by metal reduction the more soluble metals are reduced to the less soluble metal salts. For example, *Citrobacter* spp. under aerobic condition can produce phosphates enzymatically which in turn result in precipitation of metals like lead and copper. *Desulfovibrio* spp., a sulfate reducing bacteria, under anaerobic condition, can cause metal precipitation [60]. In 1970, for the first time, it was reported that some bacteria can reduce the highly toxic waste soluble hexavalent Cr^{6+} to less toxic Cr^{3+}. A study was latter carried by Amoroso et al. [52] that showed that actinomycete and *Streptomyces* sp. are highly resistant to chromate and were found to have 94.7% chromium associated with cell wall, this was due to the reduction of CrO_4^{2-} to Cr^{3+}. Since then many bacterial species have been identified that are known to be capable of reducing chromate, these bacteria mainly include, *Aureobacterium esteroaromaticum*, *Achromobacter* sp., *Acenatobacter calcoaceticus, Aeromonas* sp., *Bacillus licheniformis, Klebsiella oxytoca, Pseudomonas putida, Enterobacter* sp., *Escherichia*, and *Micrococcus* sp. [70].

13.6.6 Efflux System

Plasmid encoded metal efflux systems that are energy dependent are used by a few microorganisms for removal of metals from cells. This system involves certain proteins like ATPases and proton pumps. In both cases the same mechanisms are utilized to pump toxic ions out from inside the cell actively, using cellular energy. Metals like As, Cr, and Cd are mostly associated with efflux resistance microorganisms that use these mechanisms include *S. aureus, Bacillus subtilis, Listeria* spp. *E. coli, A. eutrophus*, and *P. putida* [60].

13.6.7 Metal Dependent Mechanisms of Metal Resistance

Intracellular metal resistance mechanisms are not yet clearly understood in bacteria. To date, metallothioneins (MTs) are best known for the binding or sequestration of metals. MTs play an important role in the management of various forms of stress and also in the heavy metal metabolism. Metallothioneins are responsible for heavy metal detoxification. It has been found that MTs like phytochelatins have protein that contains a high number of cycteine residues. All metallothioneins possess three characteristics:

- low molecular weight proteins,
- high cysteine residue in proteins, and
- high metal content with coordination of metal ions in metal thiolate clusters.

The metallothioneins can be classified into two categories. Class I – MTs that include metallothioneins from animal sources and Class II – includes those metallothioneins that are found in microorganisms and plants [71]. Between the two, Class-I MTs are characterized well and represent the mammalian archetype. The invariant alignment of cys within the protein provides the thiol for mercaptide bonds in arrangements typical of metal-thiolate clusters. For Cd and Zn the alpha-domain in the carboxy terminal region is a metal cys cluster, while the amino terminal beta-domain is a metal $_3cys_9$ cluster. The two domains are linked by 2–4 amino acids. At least 62 such metallothioneins have been described for a variety of mammals, certain fishes, crabs, oysters, and mussels [72]. Metallothionein-like proteins are isolated from a few bacteria like the cyanobacterium *Synechcoccus* spp., *E. coli* and *P. putida* [73].

13.7 Conclusion

Increasing social and economic development is exacerbating the serious problem of heavy metal pollution throughout the modern world and has already affected animals, plants, and ecosystems. Heavy metals affect several human cellular processes and cause several issues, including those of irritation of the throat, skin allergy, chronic anemia, and gastrointestinal disorders. Major heavy metals such as, As, Cd, Cr, Hg, Ni, Hg, and Zn contamination sources are mainly transport, industrial waste, household garbage, fertilizers, and pesticides. Contemporary with this increase in heavy metal pollution, microbes are proving to be a potential tool for bioremediation by involving certain mechanisms. The mechanisms mostly associated with these processes include chelation of certain elements by metabolites, oxidation–reduction of metals that affect the solubility, change in pH, biosorption, biotransformation, immobilization, biomethylation, and biodegradation. However, there is an increasing need for research regarding microbial metal resistance and remediation to understand metal resistance mechanisms.

References

1 Hasin, A.A., Gurman, S.J., Murphy, L.M. et al. (2010). Remediation of chromium (VI). By a methane-oxidizing bacterium. *Environmental Science & Technology* 44: 400–405.

2 Wei, B. and Yang, L. (2010). A review of heavy metal contaminations in urban soils, urban road dusts and agricultural soils from China. *Microchemical Journal* 94 (2): 99–107.

3 Wu, X., Chen, S., Guo, J. et al. (2018). Effect of air pollution on the stock yield of heavy pollution enterprises in China's key control cities. *Journal of Cleaner Production* 170: 399–406.

4 Uqab, B., Mudasir, S., Sheikh, A.Q. et al. (2016). Bioremediation: a management tool. *Journal of Bioremediation & Biodegradation* 7 (331): 2.

5 Alloway, B.J. (1995). *Heavy Metals in Soils*, 2e, 96–101. London: Blackie Academic and Professional.

6 Doelman, P., Jansen, E., Michels, M. et al. (1994). *Biological Fertilized Soil* 17: 177–184.

7 Weast, R.C. (1984). *CRC Handbook of Chemistry and Physics*, 64e. Boca Raton, FL: CRC Press, Inc.

8 Zhang, J., Li, H., Zhou, Y. et al. (2018). Bioavailability and soil-to-crop transfer of heavy metals in farmland soils: a case study in the Pearl River Delta, South China. *Environmental Pollution* 235: 710–719.

9 Rezaei, M.N. et al. (2014). Harvesting yeast (*Saccharomyces cerevisiae*). at different physiological phases significantly affects its functionality in bread dough fermentation. *Food Microbiology* 39: 108–115.

10 Zojaji, F., Hassani, A.H., and Sayadi, M.H. (2014). Bioaccumulation of chromium by *Zea mays* in wastewater-irrigated soil: an experimental study. *Proceedings of the International Academy of Ecology and Environmental Sciences* 4 (2): 62–67.

11 Yang, Y.B. and Sun, L.B. (2009). Status and control counter measures of heavy metal pollution in urban soil. *Environmental Protection Science* 35 (4): 79–81.

12 Wood, J.M. (1974). Biological cycles for toxic elements in the environment. *Science* 183: 1049–1052.

13 Ahamed, M. (2012). Implications of bacterial resistance against heavy metals in bioremediation: a review. *IIOABJ* 3 (3): 39–46.

14 Nies, D.H. (1999). Microbial heavy metal resistance. *Applied Microbiology and Biotechnology* 51: 730–750.

15 Qin, Y.S., Zhao, J., Liu, Z.Q. et al. (2008). Study on the influences of combined pollution of heavy metals Cu and Pb on soil respiration. *Journal of Anhui Agricultural Sciences* 36 (3): 1117–1128.

16 Zhang, W.J., Jiang, F.B., and Ou, J.F. (2011). Global pesticide consumption and pollution: with China as a focus. *Proceedings of the International Academy of Ecology and Environmental Sciences* 1 (2): 125–144.

17 Su, C. (2014). A review on heavy metal contamination in the soil worldwide: situation, impact and remediation techniques. *Environmental Skeptics and Critics* 3 (2): 24.

18 Arao, T., Ishikawa, S., Murakam, I.M. et al. (2010). Heavy metal contamination of agricultural soil and counter measures in Japan. *Paddy and Water Environment* 8 (3): 247–257.

19 Giller, K.E., Wittwer, E., and McGrath, S.P. (1999). Assessing risks of heavy metal toxicity in agricultural soils. *Human and Ecological Risk Assessment* 5: 683–689.

20 Patel, P.C., Goulhen, F., Boothman, C. et al. (2007). Arsenate detoxification in a *Pseudomanoad hypertolerant* to arsenic. *Archives of Microbiology* 187: 171–183.

21 Sarret, G., Laure, A., Marie, C. et al. (2005). Chemical forms of selenium in the metal-resistant bacterium *Ralstonia metallidurans* CH34 exposed to selenite and selenate. *Applied and Environmental Microbiology* 71 (5): 2331–2337.

22 Maier, R.M., Pepper, I.L., and Gerba, C.P. (2009). *Environmental Microbiology*. San Diego, CA: Academic Press.

23 Sobolev, D. and Begonia, M.F.T. (2008). Effects of heavy metal contamination upon soil microbes: lead-induced changes in general and denitrifying microbial communities as evidenced by molecular markers. *International Journal of Environmental Research and Public Health* 5 (5): 450–456.

24 Nogawa, K., Honda, R., Kido, T. et al. (1987). Limits to protect people eating cadmium in rice, based on epidemiological studies. *Trace Substances in Environmental Health* 21: 431–439.

25 Stanković, J.D., Sabovljević, A.D., and Sabovljević, M.S. Bryophytes and heavy metals: a review. *Acta Botanica Croatica*, 77 (2): 109–119. (2018).

26 Speira, T.W., Kettlesb, H.A., Percivalc, H.J. et al. (1999). *Soil Biology and Biochemistry* 31: 1953–1961.

27 Shun-hong, H., Bing, P., Zhi-hui, Y. et al. (2009). Chromium accumulation microorganism population and enzyme activities in soils around chromium-containing slag heap of steel alloy factory. *Transactions of Nonferrous Metals Society of China* 19: 241–248.

28 Mora, A.P., Calvo, J.J.O., Cabrera, F. et al. (2005). Changes in enzyme activities and microbial biomass after "in situ" remediation of a heavy metal-contaminated soil. *Applied Soil Ecology* 28: 125–137.

29 Chen, G.Q., Chen, Y., Zeng, G.M. et al. (2010). Bacterial communities in red soil following application of cadmium-polluted compost. *Environmental Engineering Science* 27 (12): 1019–1026.

30 Garrido, S., Campo, G.M.D., Esteller, M.V. et al. (2002). Heavy metals in soil treated with sewage sludge composting, their effect on yield and uptake of broad bean seeds (*Vicia faba* L.). *Water, Air, and Soil Pollution* 166: 303–319.

31 Kamran, M.A., Bibi, S., Xu, R.K. et al. (2017). Phyto-extraction of chromium and influence of plant growth promoting bacteria to enhance plant growth. *Journal of Geochemical Exploration* 182: 269–274.

32 Rascio, N. and Izzo, F.N. (2011). Heavy metal hyperaccumulating plants: how and why do they do it? And what makes them so interesting? *Plant Science* 180: 169–181.

33 Singh, J. and Kalamdhad, A.S. (2011). Effects of heavy metals on soil, plants, human health and aquatic life. *International Journal of Research in Chemistry and Environment* 1 (2): 15–21.

34 Jordao, C.P., Nascentes, C.C., Cecon, P.R. et al. (2006). Heavy metal availability in soil amended with composted urban solid wastes. *Environmental Monitoring and Assessment* 112: 309–326.

35 Sprynskyy, M., Kosobucki, P., Kowalkowski, T. et al. (2007). Influence of clinoptilolite rock on chemical speciation of selected heavy metals in sewage sludge. *Journal of Hazardous Materials* 149: 310–316.

36 Guala, S.D., Vega, F.A., and Covelo, E.F. (2010). The dynamics of heavy metals in plant–soil interactions. *Ecological Modelling* 221: 1148–1152.

37 Gjorgieva, A.D. (2018). Heavy metals and their general toxicity on plants. *Plant Science Today* 5 (1): 15–19.

38 Kamunda, C., Mathuthu, M., and Madhuku, M. (2016). Assessment of radiological hazards from gold mine tailings in Gauteng province, South Africa. *International Journal of Environmental Research and Public Health* 13: 138.

39 Aeliona, C.M., Davisa, H.T., and McDermottb, S. (2008). Metal concentrations in rural topsoil in South Carolina: potential for human health impact. *Science of the Total Environment* 402: 149–156.

40 He, Z., Hu, Y., Yin, Z. et al. (2016). Microbial diversity of chromium-contaminated soils and characterization of six chromium-removing bacteria. *Environmental Management* 57 (6): 1319–1328.

41 Montagne, D., Cornu, S., and Bourennane, H. (2007). Effect agricultural practices on trace-element distribution in soil. *Communications in Soil Science and Plant Analysis* 38: 473–491.

42 Kabata-Pendias, A. (2011). *Trace Elements in Soil and Plants*, 4e. Boca Raton, FL, USA: Taylor & Francis.

43 Siegel, S.M., Galun, M., Keller, P. et al. (1986). *Proceedings of the International Symposium on Metal Speciation*. Chicago, IL: Kluwer.

44 Iqbal, M. and Edyvean, R.G.J. (2004). Biosorption of lead, copper and zinc on loofa sponge immobilized biomass of *Phanerochaete chrysosporium*. *Minerals Engineering* 17: 217–223.

45 Kratochvil, D. and Volesky, B. (1998). Advances in the biosorption of heavy metals. *Trends in Biotechnology* 16: 291–300.

46 Veglio, F., Beolchini, F., and Gasbarro, A. (1997). Biosorption of toxic heavy metals: an equilibrium study using free cells of Arthrobacter spp. *Process Biochemistry* 32: 99–105.

47 Volesky, B. (2001). Detoxification of metal-bearing effluents biosorption for the next century. *Hydrometallurgy* 44: 203–216.

48 Malik, A. (2004). Metal bioremediation through growing cells. *Environment International* 30: 261–278.

49 Ahmad, A., Mukherjee, P., Mandal, D. et al. (2002). *Journal of the American Chemical Society* 124: 12108–12109.

50 Nair, A., Juwarkar, A.A., and Singh, S.K. (2007). Production and characterization of siderophores and its application in arsenic removal from contaminated soil. *Water, Air, and Soil Pollution* 180: 199–212.

51 Culotta, V.C., Howard, W.R., and Liu, X.F. (1994). CRS5 encodes a metallothionein-like protein in *Saccharomyces cerevisiae*. *The Journal of Biological Chemistry* 269: 25295–25302.

52 Amoroso, M.J., Castro, G.R., Durain, A. et al. (2001). Chromium accumulation by two Streptomyces spp. isolated from riverine sediments. *Journal of Industrial Microbiology & Biotechnology* 26: 210–215.

53 Lloyd, J.R. (2003). Microbial reduction of metals and radionuclides. *FEMS Microbiology Reviews* 27: 411–425.

54 Nies, D.H. (2003). Efflux mediated heavy metal resistance in prokaryotes. *FEMS Microbiology Reviews* 27: 313–339.

55 Meyer, J., Schmidt, A., Michalke, K. et al. (2007). Volatilization of metals and metalloids by the microbial population of an alluvial soil. *Systematic and Applied Microbiology* 31: 81–87.

56 Nair, B. and Pradeep, T. (2002). Coalescence of nanoclusters and formation of submicron crystallites assisted by Lacto strains. *Crystal Growth & Design* 2: 293–298.

57 Francis, J.A. (1997). Biotechnology of radioactive wastes: a general overview. In: *Biotechnology for Waste Management and Site Restoration*, NATO ASI Series, 2/34 (eds. C. Ronneau and O. Bitchaeva), 18–28. Dordrecht–Boston–London: Klüwer.

58 Poole, R.K. and Gadd, G.M. (1989). *Metal-Microbe Interaction*. Oxford: IRL Press.

59 Banjerdkij, P., Vattanaviboon, P., and Mongkolsuk, S. (2003). Cadmium-induced adaptive resistance and cross-resistance to zinc in *Xanthomonas campestris*. *Current Microbiology* 47: 260–262.

60 Roane, T.M. and Pepper, I.L. (2000). Microorganisms and metal pollution. In: *Environmental Microbiology*, vol. 55 (eds. R.M. Maier, I.L. Pepper and C.B. Gerba), 403–423. London: Academic Press.

61 Tabak, H.H., Lens, P., van Hullebusch, E.D. et al. (2005). Developments in bioremediation of soils and sediments polluted with metals and radio nuclides – 1. Microbial processes and mechanisms affecting bioremediation of metal contamination and influencing metal toxicity and transport. *Reviews in Environmental Science and Biotechnology* 4: 115–156.

62 Wase, J. and Forster, C.F. (1997). *Biosorbents for Metal Ions*. London: Taylor & Francis.

63 Sand, W., Rohde, K., Sobotke, B. et al. (1992). Evaluation of Leptospirillum ferrooxidans for leaching. *Applied and Environmental Microbiology* 58 (1): 85–92.

64 Endo, G., Narita, M., Huang, C. et al. (2002). Microbial heavy metal resistance transposons and plasmids: potential use for environmental biotechnology. *Journal of Environmental Biotechnology* 2 (2): 71–82.

65 Hesse, E., O'Brien, S., Tromas, N. et al. (2018). Ecological selection of siderophore producing microbial taxa in response to heavy metal contamination. *Ecology Letters* 21 (1): 117–127.

66 Miller, R.M. (1995). Biosurfactant-facilitated remediation of metal-contaminated soils. *Environmental Health Perspectives* 103: 59–62.

67 Morillo Pâere, A.J., Garcia-Ribera, R., Quesada, T. et al. (2006). Biosorption of heavy metals by the exopolysaccharide produced by *Paenibacillus jamilae*. *World Journal of Microbiology and Biotechnology* 24 (11): 2699–2704.

68 Singh, N., Asthana, R.K., and Singh, S.P. (2003). Characterisation of an exopolysaccharide mutant of Nostoc spongiaeforme: Zn^{2+} sorption and uptake. *World Journal of Microbiology and Biotechnology* 19: 851–857.

69 Blanco, A. (2000). *Immobilization of Nonviable Cyanobacteria and their Use for Heavy Metal Adsorption from Water in Environmental Biotechnology and Cleaner Bioprocesses*, 135–151. Philadelphia: Taylor & Francis.

70 Batool, R., Yrjala, K., and Hasnain, S. (2012). Hexavalent chromium reduction by bacteria from tannery effluent. *Journal of Microbiology and Biotechnology* 22 (4): 547–554.

71 Wilfried, E.R. (2000). The role of thiols in plants under metal stress, sulfur nutrition and sulfur assimilation in higher plants. In: *Sulfur Nutrition and Sulfur Assimilation in Higher Plants: Molecular Biochemical, Physiological Aspects* (eds. C. Burnold, H. Rennenberg, L.J. De Kok, et al.), 169–183. Bern: Paul Haupt.

72 Fowler, B.A., Hildebrand, C.E. et al. (1987). Nomenclature of metallothionein. In: *Metallothionein II* (eds. J.H.R. Kagi and Y. Kojima), 19–22. Boston: Birkhäuser.

73 Gupta, A., Morby, A.P., Turner, J.S. et al. (1993). Deletion within the metallothionein locus of cadmium-tolerant Synechococcus PCC-6301 involving a highly iterated palindrome (hip1). *Molecular Microbiology* 7: 189–195.

Index

Soil Bioremediation: An Approach Towards Sustainable Technology, First Edition.
Edited by Javid A. Parray, Abeer Hashem Abd Elkhalek Mahmoud, and Riyaz Sayyed.

www.ingramcontent.com/pod-product-compliance
Lightning Source LLC
LaVergne TN
LVHW080031301125
826312LV00003B/38

* 9 7 8 1 1 1 9 5 4 7 9 5 2 *